FUNGAL NUTRITION
AND PHYSIOLOGY

FUNGAL NUTRITION
AND PHYSIOLOGY

MICHAEL O. GARRAWAY
Department of Plant Pathology
The Ohio State University
Columbus, Ohio

ROBERT C. EVANS
Biology Department
Rutgers University
Camden, New Jersey

A Wiley-Interscience Publication

JOHN WILEY & SONS

New York / Chichester / Brisbane / Toronto / Singapore

Copyright © 1984 by John Wiley & Sons, Inc.

All rights reserved. Published simultaneously in Canada.

Reproduction or translation of any part of this work
beyond that permitted by Section 107 or 108 of the
1976 United States Copyright Act without the permission
of the copyright owner is unlawful. Requests for
permission or further information should be addressed to
the Permissions Department, John Wiley & Sons, Inc.

Library of Congress Cataloging in Publication Data:

Garraway, Michael O. (Michael Oliver)
 Fungal nutrition and physiology.

 "A Wiley-Interscience publication."
 Includes index.
 1. Fungi—Physiology. 2. Plants—Nutrition. I. Evans,
Robert C. (Robert Church) II. Title.
QK601.G27 1984   589.2'0413   83-23450
ISBN 0-471-05844-0

Printed in the United States of America

10 9 8 7 6 5 4 3 2 1

PREFACE

The motivation for writing this book came with the insights gained from 10 years of teaching a one-quarter graduate level course in fungal physiology to students in botany, mycology, microbiology, and plant pathology at The Ohio State University. During that period many excellent books were published on all facets of fungal physiology; they included monographs, symposium volumes, and long treatises. But none of these books presented the subject in a manner adequate for use in a fungal physiology course. Books on fungal physiology either provide more detailed coverage of topics than students need or emphasize molecular and biochemical aspects without commensurate coverage of nutritional aspects. Our experience has shown that students of fungal physiology are primarily interested in fungal nutrition and its implications for growth, development, and metabolite production. Our book is written primarily for seniors and undergraduate students who seek this kind of information and knowledge, but we believe that our approach to the subject will also make the book appealing to academic and nonacademic professionals in biological, agricultural, and medical sciences.

The aim of this book is to present in one volume the principles and concepts relating to nutrition and physiology of fungi with supporting evidence from the current literature. In pursuing this goal we have tried to impart a level of coherence and uniformity of style often lacking in similar texts. We think this approach will help make the book useful and interesting to individuals with varying levels of competence in the fundamentals of physiology, biochemistry, and molecular biology. Also, we have tried to present the material with sufficient rigor and detail for it to be useful in explaining and interpreting fungal nutrition and physiology even after the literature sources are no longer current. These features should make the book appealing to a diverse group of individuals with an interest in nutrition and physiology of fungi.

We have also attempted to select chapter themes that are of current interest on the one hand and of lasting importance on the other. The first chapter emphasizes the nutritional theme of the book and points to its evolutionary, ecological, and morphological implications for fungi. The chapter concludes with a discussion of the historical approaches to the study of fungal physiology.

Chapter 2 describes the principal components of fungal cells and some of the ways in which they function. Such knowledge is indispensable to a study of fungal nutrition and physiology.

v

Chapters 3, 4, 5, and 6 relate to the central theme of the book, fungal nutrition. Here we use numerous examples from the recent literature to illustrate the diverse roles of carbon, nitrogen, inorganic, and vitamin nutrition not only in growth, development, and reproduction but also in metabolism and metabolite production.

The chapters on fungal development include physiological aspects of spore dormancy and germination (Chapter 7), growth (Chapter 8), and reproduction (Chapter 9). The writing of these chapters presented a special challenge to us. We had to reconcile the voluminous literature dealing with the many facets of each topic with our self-imposed mandate to be concise. Moreover, we sought to present the judiciously selected topics in sufficient depth to stimulate the curiosity of interested readers.

The concise reviews of intermediary metabolism (Chapter 10) and of nucleic acid and protein synthesis (Chapter 11) serve at least two purposes. They provide basic information helpful in interpreting and explaining aspects of fungal nutrition and development, and they point to the potential usefulness of fungi in basic research in biochemistry and molecular biology.

As with other themes of this book, metabolite production in fungi is an enormously complex subject about which there is a voluminous literature. Therefore, the presentation on secondary metabolism (Chapter 12), even when read in conjunction with Chapter 10, is merely an introduction to the subject. Nevertheless, these two chapters should provide the reader with a sense of the strategies available to fungi for the synthesis of a wide array of compounds that are potentially harmful or beneficial, either to the fungi which produce them or to other organisms including humans.

Many thanks to our wives, Marie Garraway and Sue Evans, for their encouragement and support over the years. We are also grateful to our colleagues and friends in our respective departments for their interest and their support, tangible and intangible, during the period of preparation of the manuscript. We are especially indebted to Dr. Henry Stempen, Biology Department, Rutgers University, Camden, whose artistic skills have greatly improved this book. To friends and colleagues who have assisted us by contributing original photographs of their published or unpublished works we say many thanks.

MICHAEL O. GARRAWAY
ROBERT C. EVANS

Columbus, Ohio
Camden, New Jersey
May 1984

CONTENTS

FUNGAL NUTRITION
AND PHYSIOLOGY

1
NUTRITION AS A BASIS
FOR THE STUDY OF FUNGI

The manner in which fungi obtain nutrients from the environment distinguishes them from plants and animals and is largely responsible for their growth habit. Fungi lack chlorophyll and thus are not capable of utilizing the sun's radiant energy to manufacture organic molecules as plants do. In terms of their mode of nutrition fungi are similar to animals in being heterotrophic: they must obtain organic substances ("food") preformed from the environment. Animals, however, make use of a mouth to ingest food particles, which are then digested internally; fungi digest their food externally by releasing hydrolytic enzymes into their immediate surroundings. These enzymes break down the substrate into smaller subunits, which are then absorbed by the fungus. This absorptive mode of nutrition is one of the bases on which Whittaker (1969) places the fungi in a kingdom all their own (Figure 1.1). In this kingdom, the Kingdom Fungi, are grouped all those eukaryotic multicellular organisms that obtain nutrients by absorption rather than by photosynthesis (Kingdom Plantae) or ingestion (Kingdom Animalia).

There is enormous variety in the types of organic substances which fungi as a group can utilize as energy sources. Individual species, however, are often quite selective in their nutritional requirements and will grow only on certain substrates. In order for a substrate to be utilized as a nutrient three criteria must first be met: (1) the fungus must be able to synthesize and secrete the enzymes necessary to hydrolyze the substrate into molecules of relatively small size, (2) the fungus must possess the uptake mechanisms necessary to transport these small molecules into the cell, and (3) the fungus must possess the metabolic machinery necessary to convert these molecules into cellular energy as well as into building blocks for growth and development. Thus a fungus may well starve in the midst of plenty if it lacks an enzyme required for the hydrolysis, uptake, or metabolism of a particular substrate. Utilization may also be prevented if environmental conditions cause any of these systems to be operating at a critically low rate.

These three criteria, which underlie the absorptive mode of nutrition, determine

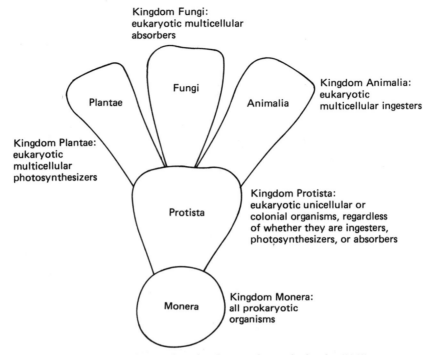

Figure 1.1   The five-kingdom classification scheme of Whittaker (1969).

to a large extent the habitats and ecological niches that a fungus can exploit. For example, many types of fungi are saprophytes; they obtain their primary sources of energy from nonliving organic materials of plant or animal origin. As fungi degrade these materials the nutrients are recycled into the biosphere for use by other organisms. In terms of human needs this decay may be deleterious, as in those cases where fungi cause the spoilage of food, the "rotting" of wooden houses and boats, the deterioration of paint, and the mildewing of clothing and other fabrics.

Other types of fungi possess the enzymes necessary to obtain nutrients directly from living organisms and are then called parasites. There are varying degrees of parasitism. In certain instances the absorption of nutrients by the fungus from the host organism is not detrimental and may even by beneficial to the host. This type of symbiotic relationship is found in associations between fungi and algae in a lichen and between fungi and the root of a vascular plant in mycorrhizae. Other fungal parasites cause significant damage to the host and are then called pathogens. The diseases of plants and animals, including humans, caused by fungal pathogens range from debilitating to deadly and are the subject of worldwide concern.

An understanding of fungal nutrition and of the ways nutrients function at the biochemical level offers a means of controlling the activity of these organisms, thereby maximizing their beneficial effects and minimizing their harmful ones. In addition, fungi serve as ideal experimental organisms for studying the biochemical bases for certain types of developmental and physiological processes because the metabolism of fungi is not additionally complicated by the presence of photosynthesis. Such information may be important not only to fungal physiologists but also to mycologists, plant pathologists, microbiologists, and others.

MORPHOLOGICAL ADAPTATIONS TO THE ABSORPTIVE LIFE— THE EXTERNAL FEATURES OF FUNGI

Over many millions of years of evolutionary time fungi have become well adapted to the absorptive mode of life. Fungi grow within their food supply, and structures have evolved that facilitate the exploitation of the substrate. Parasitic fungi, in addition, have developed a wide variety of structural features that aid in the penetration of the host and the subsequent ramification through its tissues. The fungal physiologist is interested in understanding the manner in which these various structures are formed, the mechanisms by which they function, and the ways in which changes in the fungal environment trigger their development. In preparation for examining these aspects of fungal physiology let us briefly review some of the principal morphological features found in fungi.

The Fungal Thallus

The majority of fungi are composed of long, branching filaments called **hyphae** or hyphal strands. A mass of hyphae make up a **mycelium.** A hyphal strand is actually an elongated tubular cell consisting of a cell wall and a membrane-bound cytoplasm in which are embedded numerous organelles and one or more nuclei. In some species the hyphae are partitioned by cross walls called **septa.** Because most septa are interrupted by one or more pores, these cross walls are not an appreciable barrier to the passage of cellular materials and are thought instead to aid in mechanically strengthening the tubular filament. Apparently, the only solid septa found in fungi are those that delimit reproductive or resistant structures. The delicate, filamentous nature of the hyphae provides an extensive surface area over which enzymes can be released and substrates absorbed. The growth of hyphal strands takes place at the tip, and branches usually arise at some distance behind the apex. By a combination of tip growth and subsequent proliferation of branches, a mycelial fungus can quickly ramify throughout a substrate.

A number of fungi do not form hyphae, and the thallus consists of only a few cells. A yeast, for example, is a nonmycelial fungus that is unicellular. Such an organism is placed in Whittaker's Kingdom Protista, but we shall include such eukarotic, unicellular absorbers with the fungi because mycologists have traditionally done so. Certain fungi have the capability of existing in two alternate forms, depending on the environment, and can switch back and forth between a mycelial and a nonmycelial phase. Figure 1.2 illustrates the growth habit of *Mucor rouxii*. In air this species grows as a filament, but it changes to a yeastlike form under elevated

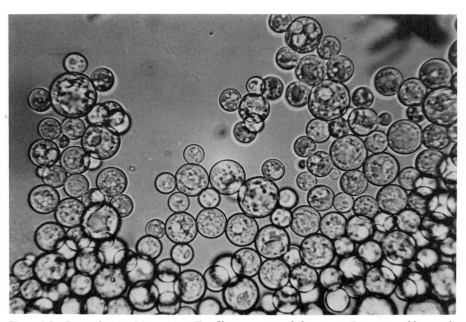

Figure 1.2 Dimorphism in *Mucor rouxii*. Top: filamentous growth form in air. Bottom: yeastlike growth form under CO_2. ×625. (Bartnicki-Garcia and Nickerson, 1962)

CO_2 levels. Such dimorphic fungi are important pathogens of animals, including man.

Still other fungi, the plasmodial slime molds, are unicellular and motile for part of their life cycle but fuse to form a multinucleate, wall-less mass of protoplasm called a plasmodium. The plasmodium creeps along the surface of the substrate and eventually develops fruiting bodies. The stages in the development of the fruiting

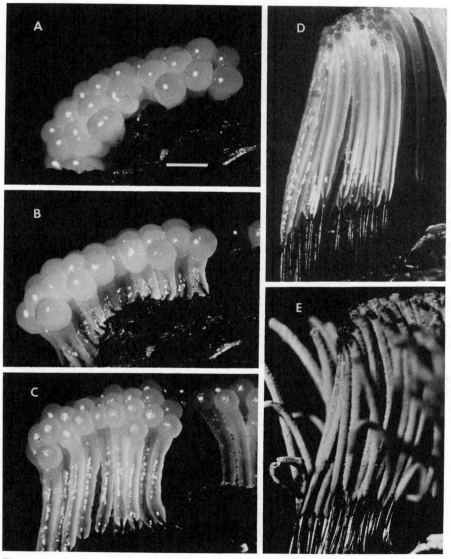

Figure 1.3 Development of fruiting bodies (sporangia) of *Stemonitis fusca*. Time sequence is as follows: (A) 0 time, (B) 2.25 h, (C) 3.25 h, (D) 6 h, (E) 48 h. Bar equals 1 mm. (Courtesy of H. Stempen.)

Figure 1.4 Cluster of sporangia is a single fruiting of *Stemonitis fusca* on a rotting log. × 5.5. (Courtesy of H. Stempen.)

bodies of *Stemonitis fusca* is illustrated in Figure 1.3. A cluster of mature fruiting bodies of this species found on a rotting log is shown in Figure 1.4. The cells of another group, the cellular slime molds typified by *Dictyostelium discoideum*, aggregate and behave as a unit but do not fuse.

MODIFICATIONS OF MYCELIA AND HYPHAE

Modifications Associated with Nutrient Procurement.   Fungal parasites of plants often have modified hyphal branches called **haustoria**. A pathogen growing on the surface of a plant will at intervals produce hyphae which penetrate the host cell wall and ramify intracellularly as haustoria (Figure 1.5). As the haustorium grows it distends the plasma membrane of the host cell but does not enter the cytoplasm. In response to this invasion the host cell secretes a polysaccharide sheath which lies between the haustorial wall and the host cytoplasm. Haustoria are thus structural adaptations for bringing the fungus into an intimate association with the host cell and for increasing the surface area available for nutrient absorption. Because the haustorium does not penetrate the plasma membrane and thus does not immediately kill the cell, the pathogen can "tap" nutrients from the host for an extended period of time.

The hyphae of certain other fungi are modified into structures used for trapping invertebrate animals, particularly nematodes. Some of these **hyphal traps** are networks of loops and constricting rings that snare a nematode as it attempts to pass through. Figure 1.6 shows the steps in the formation of a constricting ring trap, and Figure 1.7 illustrates the results of the trapping. In this type of trap the inner surface

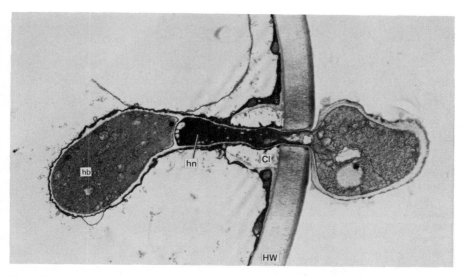

Figure 1.5 Haustoria of *Erysiphe graminis* infecting a wheat leaf epidermal cell. In this transverse section the fungus has penetrated the host cell wall (*HW*) and once inside has formed a haustorial neck (*hn*) and a haustorial body (*hb*) cell. A ring of electron-transparent material associated with the inside of the host cell wall surrounds the proximal end of the neck to form the collar (*Cl*). (Bracker and Littlefield, 1973.)

of the ring is sensitive to touch; the presence of the nematode causes the ring cells to become inflated and prevent the nematode from escaping. Other fungi secrete a sticky fluid from portions of the hyphal strands; when nematodes come into contact with the fluid, they become stuck to the hyphae. Once a nematode is trapped, infection hyphae penetrate the cuticle of the animal and grow through the body, absorbing the contents.

Modifications Associated with Nutrient Transport. **Strands** and **rhizomorphs** are aggregations of parallel hyphal strands that aid in the transport of nutrients over relatively long distances. In a strand the hyphae are loosely associated but may fuse at intervals in an anastomosing pattern. Such strands grow in diameter by the addition of hyphae at the base. Rhizomorphs, though similar in appearance to strands, are much more highly organized and grow from an apical meristem in a manner resembling the growth of a plant root. The rhizomorph is thus an autonomous structure that involves a greater degree of coordination among component hyphae than that found in strands. Strands and rhizomorphs are usually subterranean structures that infect roots of higher plants. The rapid rate at which nutrients are transported via these structures enables a pathogen to infect a host at some distance from its original food supply.

Modifications Associated with Survival of Environmental Extremes. At the onset of harsh environmental conditions such as dessication, extremes of temperature, or

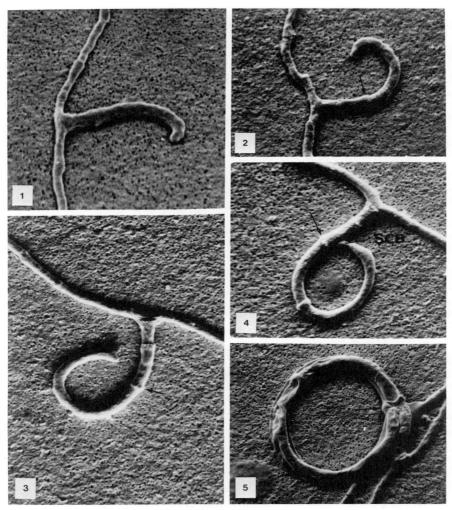

Figure 1.6   SEM micrographs depicting the formation of a constricting ring trap in *Dactylaria brocho-paga*. (By permission of Dowsett et al., 1977.)

nutrient deprivation certain fungi form resistant structures such as **chlamydospores** and **sclerotia.** Both structures arise from either terminal or intercalary cells of the hyphal strand. In the formation of chlamydospores, an individual cell enlarges, rounds up, and develops a thick wall. When sclerotia form, one or more cells begin to branch profusely and develop into a tightly interwoven mass of hyphae surrounded by a thick rind. Both chlamydospores and sclerotia contain stored nutrients and may survive for years under unfavorable conditions. Once the environment becomes suitable for growth, germination occurs and new hyphae emerge.

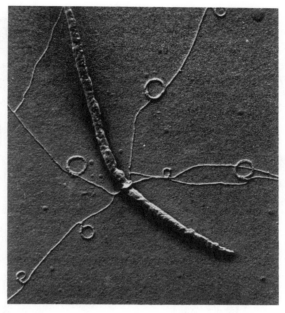

Figure 1.7 SEM micrograph of a nematode trapped by *Dactylaria brochopaga*. Notice the regular arrangement of rings along the vegetative hyphae. × 825. (By permission of Dowsett et al., 1977.)

Modifications Associated with Entry. Frequently when a spore from a plant parasite lands on a host and germinates, a localized swelling called an appressorium develops at the end of the germ tube in response to contact with the host cell. **Appressoria** secrete a mucilaginous substance that firmly attaches the fungus to the plant. Once this attachment is made, one or more **infection hyphae** (or infection pegs) arise from the lower surface of the appressorium and penetrate the host cell. The formation of appressoria and infection hyphae are often crucial steps in the establishment of the disease, and factors influencing the development of these structures are of particular interest to plant pathologists.

Modifications Associated with Reproduction. Sexual and asexual spores are frequently produced on an aggregation of mycelial tissue that has a different organization from that of the vegetative mycelium yet does not have a direct role in the reproductive process. The component hyphae of these structures range from loosely woven and poorly differentiated to closely packed and highly differentiated. The simplest of these structures is the **stroma** (pl. stromata), which is a compact mat of interwoven hyphae. In some species, spores are borne on the surface of the stroma; in others, the spores are embedded within. Figure 1.8 illustrates a stroma of the latter type. Stroma are not only important in reproduction; they may also function as resistant structures for the survival of the fungus during adverse conditions.

A more highly differentiated hyphal aggregation that is analogous to stromatic

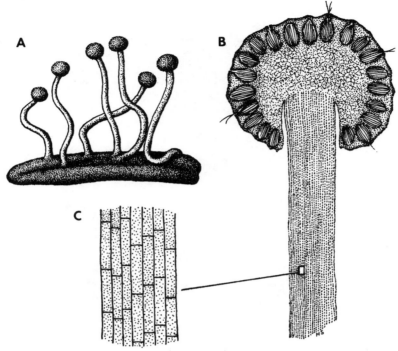

Figure 1.8 Stroma of *Claviceps purpurea*. (A) Several stromata developing from a sclerotium. (B) Longitudinal section through the head and upper portion of the stalk of a stroma. Perithecia lie just beneath the surface of the head. (C) Enlargement of stalk section to show hyphae. (Drawing by H. Stempen.)

tissue is found in the **fruiting bodies** or **fructifications** of certain Ascomycetes and Basidiomycetes. These structures, such as those found in mushrooms, puffballs, and morels, can be quite large and diverse in morphology. As with stroma the vast majority of the tissue in a fruiting body serves to support the actual spore-forming apparatus.

The Fungal Spore

Spores are minute propagating units of various shapes, sizes, and colors, produced by almost all fungi in at least one phase in their life cycle. Spores are usually dispersed by wind, water, or insects or other animals and provide a mechanism for establishing the fungus in new habitats. Although a considerable terminology has built up around the multitude of spore types, in general there are two basic kinds of spores depending on whether they are produced by sexual or asexual means. Sexual spores are formed following the fusion of compatible nuclei from two individuals; asexual spores are formed in the absence of such fusions.

During the fungal life cycle, asexual spores are typically produced in greatest numbers when environmental conditions are favorable and nutrients freely available. Under such conditions several generations of asexual spores may be produced; these spores are usually capable of germinating immediately after being released. For saprophytic fungi these repeating cycles of spore production, dispersal, and germination increase the chances of colonizing a wide variety of substrates. Similar patterns of asexual spore production in plant-parasitic fungi result in repeated cycles of infection that increase the severity of the disease in a particular area and accelerate the spread of the pathogen to other locations.

The production of sexual spores is usually associated with the onset of adverse environmental conditions such as extremes of temperature or the depletion of water or nutrients. Accordingly, sexual spores frequently exhibit adaptations such as thick, resistant cell walls, reserves of lipid and carbohydrate, and dormancy mechanisms which enable them to remain viable until conditions are favorable.

The following is a brief summary of the principal types of fungal spores.

ASEXUAL SPORES

Some spores are formed within a modified fungal cell called a **sporangium** and are thus called **sporangiospores.** Motile sporangiospores are termed **zoospores;** motility is provided by one or two flagella, depending on the species. When zoospores are formed, the multinucleate contents of the sporangium are cleaved into uninucleate portions that develop a surrounding membrane, a cell wall, and finally flagella. Eventually one or more pores develop in the sporangial wall and the zoospores are released. Nonmotile sporangiospores are called **aplanospores** and are also formed by cleavage of the sporangial cytoplasm. These spores may be released through pores or by disintegration of the sporangium. However, some fungi have evolved rather elaborate mechanisms for the dispersal of aplanospores. For example, *Pilobolus* (the "hat-thrower") forcibly ejects the entire sporangium from the underlying stalk or sporangiophore. This process is powered by a squirt of vacuolar sap and can propel the sporangium as far as two meters.

Asexual spores that are not produced in a sporangium are called **conidia.** Conidia are thought to have evolved from a sporangium containing only one spore whose wall is fused to the sporangial wall. Conidia may be formed directly at the surface of a hyphal strand or at the tips of stalks called **conidiophores.** Figure 1.9 illustrates some of the many different types of conidia produced by fungi. In some species the conidiophores are clustered on a stroma forming a **sporodochium;** in others the conidiophores are cemented together resulting in a **synnema** (Figure 1.10). Lastly, conidiophores may occur within discrete fruiting bodies such as a flask-shaped **pycnidium** or a saucer-shaped **acervulus.** The morphology and arrangement of conidia and conidiophores are important taxonomic criteria.

SEXUAL SPORES

In general, sexual spores are those associated with the meiotic divisions that follow the sexual fusion of nuclei. The differences in the morphology and manner of development of sexual spores are the principal criteria for distinguishing the major

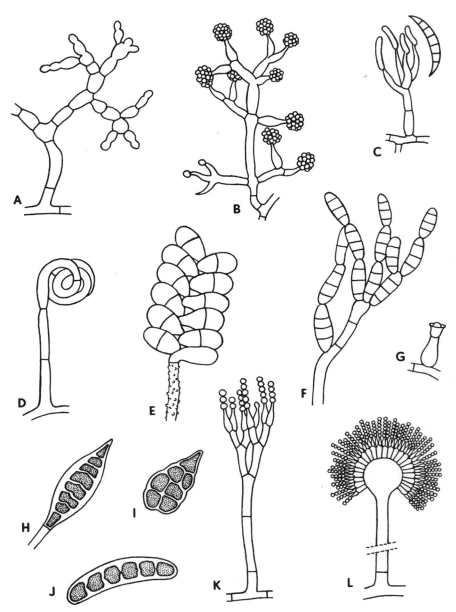

Figure 1.9 Some examples of conidia. Several types are those produced by members of genera mentioned in the text. Not drawn to scale. (A) *Neurospora sitophila*. (B) *Trichoderma* sp. (C) *Fusarium* sp. (Redrawn by H. Stempen from Conant et al., 1954.) (D) *Helicodendron tubulosum*. (E) *Trichothecium roseum*. (F) *Dendryphiella vinosa*. (Redrawn by H. Stempen from Cole and Samson, 1979) (G) *Phialophora* sp. (H) *Microsporum* sp. (I) *Alternaria* sp. (J) *Bipolaris maydis*. (K) *Penicillium* sp. (L) *Aspergillus niger*. (Drawings by H. Stempen.)

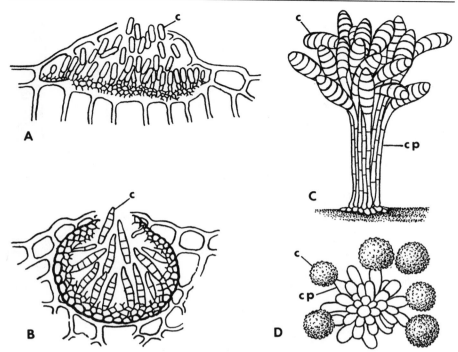

Figure 1.10 Examples of asexual fruiting bodies. (A) Acervulus (*Discogloeum*). (B) Pycnidium (*Scolecosporiella*). (Redrawn by H. Stempen from Morgan-Jones et al., 1972) (C) Synnema (*Phragmocephala*) (Redrawn by H. Stempen from Barnett, 1960) (D) Sporodochium (*Epicoccum nigrum*). c, conidium; cp, conidiophore. (Redrawn by H. Stempen from Bessey, 1950)

groups of fungi. In fact, the classification of one taxonomic group, the Deuteromycetes or Fungi Imperfecti, is based solely on the absence of sexual spores.

Oospores and **zygospores** are thick-walled zygotes that divide by meiosis after a period of dormancy. An oospore results from the fertilization of a nonmotile egg that is enclosed within a swollen cell called an oogonium. A zygospore results from the fusion of two specialized hyphal branches called gametangia.

Ascospores and **basidiospores,** on the other hand, are the products of meiotic divisions. Ascospores are usually produced in groups of 4 to 8 within a saclike cell called an **ascus** (Figure 1.11). Asci may be naked or borne within fruiting bodies called **ascocarps**, some of which may grow to appreciable size. Some asci are completely enclosed within a spherical ascocarp (a cleistothecium), others within an ascocarp that is flask-shaped (a perithecium), and still others are borne on the surface of a disk-shaped ascocarp (an apothecium). Basidiospores are formed as a result of meiosis within a swollen cell called a **basidium** followed by migration of the nuclei through specialized branches (sterigmata) to the outside (Figure 1.12). The basidium may be septate and naked as in the rusts and smuts, or it may be nonseptate and

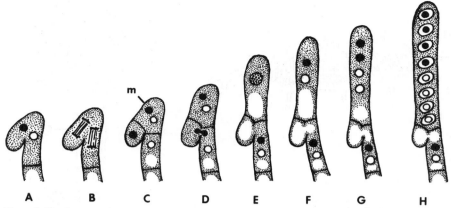

Figure 1.11 Diagram showing stages in the development of an ascus with ascospores. (A) Crozier with two haploid nuclei (dikaryon). (B) Mitotic division. (C) Formation of ascus mother cell (*m*). (D) Nuclear migration. (E) Nuclear fusion. (F) Ascus after meiosis I. (G) Ascus after meiosis II. (H) Stage G is followed by mitotic divisions to produce eight nuclei which are incorporated into an equal number of ascospores. (Drawing by H. Stempen modified from Smith, 1955.)

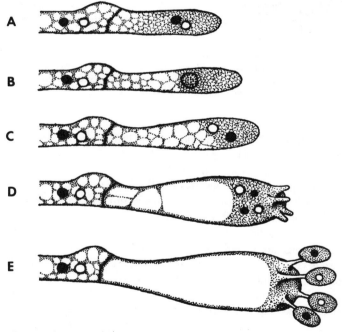

Figure 1.12 Diagram showing development of a basidium with basidiospores. (A) Young basidium with a pair of haploid nuclei (dikaryon). (B) Nuclear fusion. (C) After meiosis I. (D) After meiosis II. Sterigmata are forming on distil end of basidium. (E) Mature basidium with basidiospores. Each of the four nuclei has migrated through a sterigma and is now located in a basidiospore formed on the end of the sterigma. (Drawing by H. Stempen modified from Smith, 1955.)

borne within a fruiting body called a **basidiocarp,** as in bracket fungi, puffballs, and mushrooms. The large size and striking morphological differences among many basidiocarps make them quite noticeable in the field.

FUNGAL CLASSIFICATION—AN OVERVIEW FOR PHYSIOLOGISTS

In theory, a classification scheme for any group of organisms should indicate the degree of evolutionary relatedness among the members of that group. For many organisms, the fossil record provides the basis for understanding their phylogeny. For fungi, the fossil record is unfortunately so meager that the taxonomists are forced to base their classification schemes on criteria whose phylogenetic importance is difficult to verify. Historically, morphological characteristics have been the principal basis for fungal classification schemes. More recently, schemes using cell wall chemistry (Bartnicki-Garcia, 1968), nutrition (Cantino, 1966), and cytochrome c amino acid sequence (Nolan and Margoliash, 1968) as taxonomic criteria have been published. Ainsworth (1976) predicts that morphological characteristics will continue to serve as the basis for classifying fungi but that nontraditional approaches are helpful in verifying and refining these relationships.

In sum, there is much speculation among fungal taxonomists and thus much variation among published classification schemes. This is evident by comparing those in the recent works of Ainsworth (1973), Alexopoulos and Mims (1979), Burnett (1977), Bold (1973), and Whittaker (1969). Obviously, a discussion of the relative merits of these schemes is beyond the scope of this book. However, just as Whittaker's five-kingdom hypothesis illustrates the relationship of fungi to other organisms, the fungal classification scheme of Ainsworth et al. (as a recent representative example) presented in Table 1.1 illustrates the relationship of fungi to other fungi. This scheme should help not only to categorize the many kinds of fungi but also to put a particular group in perspective when its name is mentioned in later chapters.

FUNGAL PHYSIOLOGY—A HISTORICAL PERSPECTIVE

Fungal physiologists are interested in understanding the cellular and biochemical mechanisms responsible for the many aspects of the growth and development of fungi. These mechanisms are largely unknown, and those that have been studied are extremely complex. Therefore, investigations of fungal growth and development usually proceed in two stages. First, a descriptive study is made of the phenomenon of interest, be it spore formation, spore germination, hyphal growth, development of sexual and asexual structures, or the production of metabolites. Such studies are of particular interest to fungal physiologists when they include information on how the phenomenon is affected by stimuli such as nutrients, light, temperature, or other environmental factors. These descriptive studies provide the basis for examining the underlying fundamental mechanisms by which these phenomena occur and the ways in which they are regulated. Such fundamental processes include the function-

Table 1.1   A Simplified Fungal Classification Scheme

Division Myxomycota—has a plasmodium or pseudoplasmodium at some stage in the life cycle
    Class Acrasiomycetes—has a pseudoplasmodium
    Class Myxomycetes—has a plasmodium that is saprophytic
    Class Plasmodiophoromycetes—has a plasmodium that is parasitic
Division Eumycota—lacks a plasmodium or pseudoplasmodium; has an assimilative phase that is typically filamentous
  Subdivision Mastigomycotina—forms zoospores; sexual spores are typically oospores
    Class Chytridiomycetes—zoospores posteriorly uniflagellate (flagella whiplash)
    Class Hypochytridiomycetes—zoospores anteriorly uniflagellate (flagella tinsel)
    Class Oomycetes—zoospores biflagellate (posterior flagellum is whiplash; anterior is tinsel)
  Subdivision Zygomycotina—motile cells are absent; sexual spores are zygospores
    Class Zygomycetes—saprophytic, or if parasitic or predacious having mycelium immersed in the host tissue
    Class Trichomycetes—associated with arthropods and attached to the cuticle or digestive tract by a holdfast, and not immersed in the host tissue
  Subdivision Ascomycotina—lacks motile cells and zygospores; sexual spores are ascospores
    Class Hemiascomycetes—lacks ascocarps and ascogenous hyphae; thallus mycelial or yeastlike
    Class Loculoascomycetes—asci bitunicate; ascocarp an ascostroma
    Class Laboulbeniomycetes—asci regularly arranged within the ascocarp as a basal or peripheral layer; exoparasites of arthropods
    Class Plectomycetes—asci evanescent; ascocarp typically a cleistothecium
    Class Pyrenomycetes—ascocarp typically a perithecium
    Class Discomycetes—ascocarp an apothecium or modified apothecium
  Subdivision Basidiomycotina—lacks motile cells and zygospores; sexual spores are basidiospores
    Class Teliomycetes—lacks basidiocarp
    Class Hymenomycetes—basidiocarp with basidia exposed before the spores are mature
    Class Gasteromycetes—basidiocarp remains closed at least until the spores have been released from the basidia
  Subdivision Deuteromycotina—sexual stage is absent
    Class Blastomycetes—contains yeast or yeastlike cells; true mycelium lacking or not well-developed
    Class Hyphomycetes—mycelium well-developed; spores not borne in pycnidia or acervuli
    Class Coelomycetes—spores borne in pycnidia or acervuli

Source: Ainsworth (1973).

ing of metabolic pathways, gene activation and regulation, the production and utilization of energy, and the uptake of nutrients.

Fungal physiology can thus be defined as the study of (1) phenomena in fungi in relation to the effects of various stimuli and (2) the fundamental processes that determine or control these phenomena. Historically, these components developed sequentially; the phenomenological aspect is rooted in the early history of mycology and the fundamental aspect is quite a recent development. To illustrate these two components we shall emphasize nutritional aspects, partly because nutrition was one of the first areas of fungal physiology to be explored and partly because of its importance as the foundation on which later work is based. This material is part of a broader history of mycology recently compiled by Ainsworth (1976).

Phenomenological Aspects of Fungal Development: Nutrition

Among the first to investigate phenomena in fungi were the gastronomically oriented mycologists who were interested in the cultivation of edible mushrooms. They noticed that certain fungi grew more readily on particular substrates and under particular environmental conditions than others. For example, the following "technique" for obtaining "toad-stools" was published in 1589 by Giambattista della Porta (Ainsworth, 1976):

> . . . *if an upland or hilly field that hath in it much stubble and many stalks of corn, be set on fire at such a time as there is rain brewing in the clouds, then the rain falling, will cause many toad-stools there to spring up of their own accord: but if, after the field is thus set on fire, happily the rain which the clouds before threatened does not fall; then, if you take a thin linnen cloth, and let the water drop through by little and little like rain, upon some part of the field where the fire hath been, there will grow up toad-stools, but not so good as otherwise they would be, if they had been nourished with a shower of rain.*

By the early 1700s, the French had progressed to the stage where mushrooms were cultivated in specially prepared beds, much like they are grown today. These early studies indicate that increasing attention was being paid to nutritional and environmental factors that affect the biological activities of fungi.

The first significant experimental work done with fungi was by Antonio Micheli (1679–1737), a botanist in charge of the public gardens in Florence. Although Micheli is known primarily for his taxonomic descriptions of *Aspergillus, Mucor, Botrytis*, and others (for which he used the newly invented microscope), he also pioneered some elegant field and laboratory experiments. In one study Micheli collected various fungi from the field, brought them into the laboratory, and inoculated their spores onto the leaves of different plant species. He then returned the leaves to the field, placing replicate sets of leaves in each of two locations. By periodically making observations at each site, Micheli demonstrated not only that fungi consistently reproduce their own kind by means of spores but also that the growth of different species varies both with the type of substrate and the particular environmental conditions (Ainsworth, 1976).

Micheli was able to draw some of the same conclusions from a more controlled laboratory experiment. Using melon as a substrate, he placed spores of *Mucor, Aspergillus*, and *Botrytis* separately on different sides of the fruit and recorded the development of each species. As in the field, individuals of each species gave rise to more of the same species. When different individuals appeared on the melon,

Micheli concluded that this was due not to spontaneous generation (a theory in vogue at that time) but rather to contamination from the air. Micheli was one of the earliest mycologists to recognize the need for control of nutrition and environment in order to study phenomena such as growth and reproduction in fungi. His work gave added focus to the persistent concerns regarding the true nature of the causes of diseases of plants and animals and to the growing controversy surrounding spontaneous generation.

Progress in understanding the roles of microorganisms in disease and decay (or putrefaction, as it was then called) was slowed by the problems of contamination and the lack of chemically defined media. Although by the mid-1700s it was known that media could be sterilized by boiling, for decades the only way to keep these solutions sterile was to hermetically seal them. Only after the introduction of the swan-necked "Pasteur flask" in the early 1800s and particularly the cotton plug in 1874 was it possible to easily grow pure cultures of fungi.

With the development of techniques to grow fungi in sterile media, the way was clear for examining the components of those media and the roles these components play in fungal growth and development. The goal was to avoid utilizing, if possible, parts of plants (such as leaves, stems, or fruit) or animals (such as hair or nails) as components of the growth media and instead develop media whose chemical composition was more precisely known. The first critical study of fungal nutrition in such a defined medium was achieved by Jules Raulin, a student of Pasteur, in the 1860s. Raulin set out to examine the mineral nutrition of *Aspergillus niger* and in so doing established a protocol for controlling the physical as well as the chemical environment.

Raulin grew his *Aspergillus* in a specially designed incubator that maintained a constant temperature and humidity. Beginning with the fungus growing in an undefined medium containing sugar, ammonium bitartrate, and "ashes of yeast," he

Table 1.2   **Raulin's Formula for his *Aspergillus* Growth Medium**

'Eau	1500 parts
Sucri candi	70
Acide tartrique	4
Nitrate d'ammoniaque	4
Phosphate d'ammoniaque	0.6
Carbonate de potasse	0.6
Carbonate de magnesie	0.6
Sulfate d'ammoniaque	0.25
Sulfate de zinc	0.07
Sulfate de fer	0.07
Silicate de potasse	0.07

Source: Ainsworth (1976).

painstakingly developed a defined medium (outlined in Table 1.2) that resulted in optimal growth. By omitting individual nutrients or varying the chemical form in which a particular element was added, he amassed considerable data on the nutritional effects of certain compounds. He also became aware of the effects of pH on growth and was able to identify those components of the medium whose influence on growth was primarily one of pH rather than nourishment.

One major outcome from Raulin's work was the discovery of a trace element requirement for fungi. As his experiments progressed he noticed that the higher the purity of the chemicals in his medium, the poorer was the resulting fungal growth. This trend could be reversed by the addition of small amounts of powdered baked earth, powdered porcelain, or wood ashes but not be increasing the concentration of any of the known constituents in his medium. Eventually Raulin determined that the lack of growth was owing to an absence of trace elements of copper and zinc in the medium. This study provided the foundation for the identification of other trace elements required for fungal growth and reproduction (e.g., the classical work of Steinberg beginning in the 1920s) and for experiments using vitamins in the 1930s (Ainsworth, 1976).

Fundamental Aspects of Fungal Physiology

By the mid-1800s the development of improved microscopes and the enunciation of the cell theory set the stage for disproving the theory of spontaneous generation, on the one hand, and for the experimental verification of the germ theory of disease on the other. These events not only paved the way for but were part of the rapid advances in microbiology and cell physiology that characterized this era. In terms of fungal physiology these developments facilitated experiments that led to an improved understanding of the effects of nutrition and environment on growth, reproduction, and metabolite production in fungi. Moreover, they led to new insights into mechanisms underlying these phenomena. Two developments resulting from studies on the process of alcoholic fermentation in yeast during this period illustrate the rate of progress made in understanding fundamental aspects of fungal physiology.

Although the use of fungi for alcoholic fermentation had been known for thousands of years, the nature of the process and even the role played by yeast cells was unknown even as late as the 1830s. Yeast cells had been observed by van Leeuwenhoek in 1680, and in 1826 Desmazieres described five different species growing in beer. No one, however, associated yeast with fermentation until 1836–37, when the idea was proposed independently by three workers: Cagniard-Latour in France, and Schwann and Kutzing in Germany.

Despite the increasing amounts of evidence, the idea that fermentation was a biological process met considerable resistance in the scientific community, particularly among chemists. One of the most prestigious spokesmen for this group was Justis Liebig, who based his arguments mostly on "theoretical" rather than experimental evidence. Liebig felt that the yeast cells found in fermenting mixtures were not living organisms (Brock, 1961).

It is only a chemical product which precipitates in the fermentation and which takes the ordinary form of a non-crystalline precipitate, even inorganic, of small balls which group themselves one after the other and form chains . . . . It is thus quite natural that they imitate the forms of the simplest organisms of plant life. Nevertheless form alone does not constitute life.

In the late 1850s, Pasteur conducted a series of elegant experiments that ended the controversy concerning the role of yeast in fermentation. Pasteur demonstrated that yeast could be grown in a defined medium containing sugar, a nitrogen source, and minerals without the addition of proteinaceous "ferments" from aged cultures. Second, he showed that no alcoholic fermentation occurs in the absence of yeast cells, nor when yeast cells are placed in a medium lacking a carbon or nitrogen source. Third, by careful quantitative analysis he demonstrated that fermentation is accompanied by a loss of carbon and nitrogen from the medium with a subsequent increase in the carbon and nitrogen concentrations of the yeast population (Brock, 1961).

In 1897 the role of yeast in fermentation was further clarified when Buchner reported that it was not the yeast cells themselves but rather a proteinaceous substance that could be extracted from them that was most directly responsible for fermentation. Buchner's discovery gave impetus to a yet more fundamental approach to biology: that the structure and function of organisms and their component cells are understandable in terms of enzymes and the metabolic pathways and processes in which these enzymes are involved. These observations, marking the rise of modern biochemistry, received a decisive thrust forward when Sumner and his co-workers in 1926 isolated the enzyme urease from bean and made crystalline preparations (Brock, 1961; Gabriel and Fogel, 1955). The advances in knowledge in biochemistry and molecular biology that followed have had a profound impact on the study and understanding of fungal physiology.

SUMMARY

Fungi are heterotrophic organisms that obtain nutrients by extracellular digestion and absorption. The extent to which a fungus can hydrolyze and take up and metabolize nutrients from a particular chemical environment determines whether or not it can survive in that environment and whether it functions as a saprophyte or parasite, or even both.

In response to this absorptive mode of nutrition, fungi have evolved a characteristic growth habit and adaptive morphological structures. Hyphae, strands, rhizomorphs, haustoria, appressoria, and traps all aid the fungus in obtaining and absorbing nutrients. Resistant structures such as chlamydospores and sclerotia provide mechanisms for enduring periods when nutrients are in short supply or when the physical environment becomes limiting. Except for aquatic fungi that produce zoospores, the most common means of reproduction in fungi is by the production of large numbers of nonmotile spores at the surface of the substrate where they can be carried to new environments by wind, splashing raindrops, or animals. Many fungi have evolved structures such as stroma and fruiting bodies that aid in this dispersal.

These morphological features of fungi, particularly the types of sexual and asexual spores, have traditionally been used as criteria in fungal taxonomy and classification, although these features are now supplemented by physiological and biochemical criteria.

Fungal physiology relates to the study of phenomena in fungi and the underlying fundamental processes that control these phenomena. The phenomenological component of fungal physiology has its origins somewhere in antiquity; the fundamental component did not arise until the second half of the 19th century. Advances in microbiology, biochemistry, and molecular biology over the past hundred years have given direction and impetus to both the phenomenological and fundamental aspects of fungal physiology. The evidence of this progress is presented in the following chapters.

REFERENCES

Ainsworth, G. C. 1973. Introduction and keys to higher taxa. In C. C. Ainsworth, F. K. Sparrow, and A. S. Sussman (Eds.), *The Fungi*, Vol. IVA, pp. 1–7. New York: Academic.

Ainsworth, G. C. 1976. *Introduction to the History of Mycology*. Cambridge University Press.

Alexopoulos, C. J., and C. W. Mims. 1979. *Introductory Mycology*. New York: Wiley.

Barnett, H. L. 1960. *Illustrated Genera of Imperfect Fungi*, 2nd. ed. Minneapolis: Burgess.

Bartnicki-Garcia, S. 1968. Cell wall chemistry, morphogenesis and taxonomy of fungi. *Ann. Rev. Bacteriol.* 22: 87–108.

Bartnicki-Garcia, S., and W. J. Nickerson. 1962. Induction of yeast-like development in *Mucor* by carbon dioxide. *J. Bacteriol.* 84: 829–840.

Bessey, E. A. 1950. *Morphology and Taxonomy of Fungi*. Philadelphia: Blakiston.

Bold, H. C. 1973. *Morphology of Plants*, 3rd ed. New York: Harper & Row.

Bracker, C. E., and L. J. Littlefield. 1973. Structural concepts of host-pathogen interfaces. In R. J. W. Byrde C. V. Cutting (Eds.), *Pathogenicity and the Plant's Response*, pp.159–317. New York: Academic.

Brock, T. D. 1961. *Milestones in Microbiology*. Englewood Cliffs: Prentice-Hall.

Burnett, J. H. 1977. *Fundamentals of Mycology*, 2nd ed. New York: St. Martin's.

Cantino, E. C. 1966. Morphogenesis in aquatic fungi. In G. C. Ainsworth and A. S. Sussman (Eds.), *The Fungi*, Vol. 2, pp. 283–237. New York: Academic.

Cole, G. T., and R. A. Samson. 1979. *Patterns of Development in Conidial Fungi*. London: Pitman.

Conant, N. F., D. T. Smith, R. D. Baker, J. L. Callaway, and D. S. Martin. 1954. *Manual of Clinical Mycology*. Philadelphia: Saunders.

Dowsett, J. A., J. Reid, and L. van Caeseele. 1977. Transmission and scanning electron microscope observations on the trapping of nematodes by *Dactylaria brochopaga*. *Can. J. Bot.* 55: 2945–2955.

Gabriel, M. L., and S. Fogel. 1955. *Great Experiments in Biology*. Englewood Cliffs: Prentice-Hall.

Morgan-Jones, G., T. R. Nag Raj, and B. Kendrick. 1972. *Icones Genera Coelomycetarum. Fascicle II*. University of Waterloo Biology Series No. 3, 4, Waterloo.

Nolan, C., and E. Margoliash. 1968. Comparative aspects of primary structures of proteins. *Ann. Rev. Biochem.* 37: 727–790.

Smith, G. M. 1955. *Crytogamic Botany*, Vol. I. New York: McGraw-Hill.

Whittaker, R. H. 1969. New concepts of kingdoms of organisms. *Science* 163: 150–160.

2
THE FUNGAL CELL

Fungal cells are similar to cells of other eukaryotic organisms in being composed of a membrane-bound cytoplasm containing a variety of organelles and at least one nucleus. However, most fungal cells are distinctive in that they possess a cell wall of unique composition, certain peculiar organelles, and characteristic cross walls, or septa, between adjacent cells. Fungal groups differ in the extent to which these characteristics are present, and these differences have been useful in fungal taxonomy. In this chapter we briefly review the principal components of fungal cells and some of the ways in which they function.

FUNGAL CELL WALLS

Fungal cell walls are composed principally of polysaccharides with relatively small amounts of proteins, lipids, and inorganic ions. The polysaccharides occur in two major types of structures: cablelike **microfibrils** and a less highly organized **matrix**. Microfibrils are the principal structural components of the wall and are composed of separate polysaccharide chains wound around each other forming a strong, coarse strand (Figure 2.1). Networks of these strands are embedded in the matrix, an aggregation of smaller polysaccharides as well as proteins and lipids, which appears amorphous and granular. Portions of the fungal cell wall thus resemble reinforced concrete with the microfibrils functioning as the steel rods and the matrix the surrounding cement. This organization may explain the mechanical rigidity and strength of the wall.

In fungi microfibrils are composed of chitin, cellulose, or noncellulosic glucan (Figure 2.2). Chitin is an unbranched polymer of β-1,4-linked N-acetylglucosamine units, and its presence in cell walls is one feature distinguishing fungi from higher plants. Chitin is found in several of the major fungal groups including the Ascomycetes, Basidiomycetes, Deuteromycetes, and Chytridiomycetes. Cellulose is an unbranched polymer of β-1,4-linked glucose units and is present in the Acrasiales, the Oomycetes, the Hyphochytridiomycetes, and a few species of the Ascomycetes. The

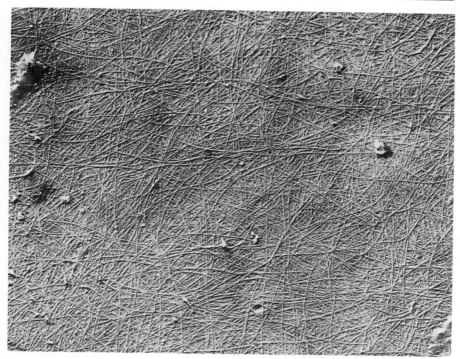

Figure 2.1 Freeze-fracture electron micrograph of microfibrils from the wall of *Phytophthora cactorum*. (Hegnauer and Hohl, 1978)

cellulose present in these groups is slightly different from that found in the walls of green plants; it is less crystalline when examined by X-ray diffraction. Noncellulosic glucans occur in the microfibrils and matrix of many fungal groups. The term glucan is used to describe a wide variety of polysaccharides composed entirely of glucose units, regardless of the type and extent of branching. Because of their complexity the structure of most glucans is not completely known. In addition, there is evidence that some fungi possess microfibrils composed in part of polysaccharide linked to protein.

The carbohydrates present in the matrix vary considerably from one taxonomic group to another. The major matrix polysaccharides are chitin, cellulose, polymers of galactosamine, and noncellulosic glucans of various types including glycogen (a polymer of α-1,4-linked glucose), mannans (polymers of mannose), chitosan (polymers of glucosamine), and galactans (polymers of galactose) (Figure 2.2). In addition, minor amounts of xylose, rhamnose, fucose, and uronic acids may be present. The arrangement of these monosaccharides in polymers and their roles in the architecture of the cell wall is not known.

The occurrence of microfibrils and matrix polysaccharides in fungal groups serves as one basis for fungal classification. As is evident from Table 2.1, there is a fairly

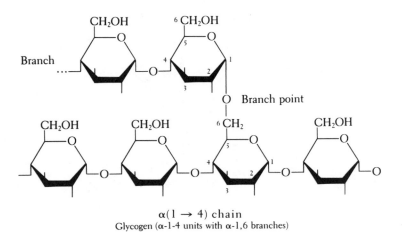

α-D-Glucose (showing numbering system)

β-D-Glucose

α(1 → 4) chain
Glycogen (α-1-4 units with α-1,6 branches)

Cellulose (β-1,4-glucan)

Chitin (β-1,4-N-acetylglucosamine)

Galactan

Figure 2.2 Some basic cell wall components of fungi.

Table 2.1   Relationships Between Wall Composition and Taxonomic Grouping in Fungi

Principal Polymers in the Wall	Taxonomic Class	Examples of Organisms Analyzed
Cellulose; glycogen	Acrasiales	*Dictyostelium discoideum*
Cellulose; glucan	Oomycetes	*Pythium debaryanum*
Cellulose; chitin	Hyphochytridiomycetes	*Rhizidiomyces sp.*
Chitin; chitosan	Zygomycetes	*Mucor rouxii*
Chitin; glucan	Chytridiomycetes	*Allomyces macrogynus*
	Ascomycetes (mycelial forms)	*Neurospora crassa*
	Basidiomycetes (mycelial forms)	*Schizophyllum commune*
	Deuteromycetes	*Aspergillus niger*
Glucan; mannan	Ascomycetes (yeast forms)	*Saccharomyces cerevisiae*
	Deuteromycetes (yeast forms)	*Candida utilis*
Chitin; mannan	Basidiomycetes (yeast forms)	*Sporobolomyces rosei*
Galactan; polygalactosamine	Trichomycetes	*Amoebidium parasiticum*

Source: Bartnicki-Garcia (1968).

close relationship between classical fungal taxonomy based on morphology and a taxonomy based on the major wall polysaccharides. Thus, seemingly minor changes in the composition of the wall are correlated with rather striking morphological alterations and support the hypothesis that cell morphogenesis in fungi is controlled by patterns in cell wall differentiation (Bartnicki-Garcia, 1968).

Proteins typically make up less than 10% of the cell wall matrix. Cell wall proteins function both as structural components and as enzymes. In the walls of *Saccharomyces cerevisiae* and *S. carlsbergenesis*, mannans are linked to protein via molecules of N-acetylglucosamine (Figure 2.3). Other glycoproteins present in dermatophytes such as *Trichophyton*, *Aspergillus*, and *Microsporum* species occur on the outside of the wall and are responsible for the hypersensitive skin reactions of allergic humans and other animals (Gander, 1974). Numerous enzymes, particularly hydrolases such as acid phosphatase, α-amylase, and protease have been located in the cell wall. These enzymes may be important in hydrolyzing substrates in the environment to subunits that can be transported into the cell.

Lipids usually constitute less than 8% of the matrix, although *Blakeslea trispora* is reported to have a total lipid content of 30% (Taylor and Cameron, 1973). Wall lipids characteristically are composed of saturated rather than unsaturated fatty acids,

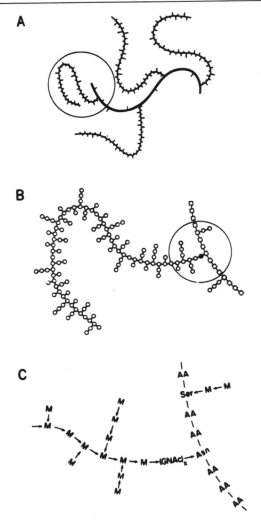

Figure 2.3 Diagrammatic representation of a tentative structure for yeast mannan. In (A), the heavy line represents a peptide chain to which long and short mannose chains are attached. (B) is an enlargement of the circled portion of (A). The empty circles represent mannosyl residues, the dark circle an unspecified number of N-acetylglucosamine residues (possibly two), and the squares amino acid residues. The circled portion of (B) is enlarged in (C) to show that the short chains are linked to serine (*Ser*), whereas the large polysaccharide is attached to asparagine (*Asn*) through an N-acetylglucosamine (GNAc) residue. (Cabib, 1975)

and phospholipids are common. Small amounts of glycolipids and sphingosines have been identified in the walls of certain yeasts although the presence of glycolipids may be due to their being trapped in the wall en route from the cytoplasm into the medium (Phaff, 1971). In *Candida tropicalis*, Kappeli et al. (1978) identified a mannan that is covalently linked to fatty acids. This mannan–fatty acid complex is found at the outer surface of the wall when the yeast is incubated on a medium

Table 2.2   Chemical differentiation of the cell wall in the life cycle of *Mucor rouxii.*[a]

Wall Component	Yeasts	Hyphae	Sporangiophores	Spores
Chitin	8.4	9.4	18.0	2.1[b]
Chitosan	27.9	32.7	20.6	9.5[b]
Mannose	8.9	1.6	0.9	4.8
Fucose	3.2	3.8	2.1	0.0
Galactose	1.1	1.6	0.8	0.0
Glucuronic acid	12.2	11.8	25.0	1.9
Glucose	0.0	0.0	0.1	42.6
Protein	10.3	6.3	9.2	16.1
Lipid	5.7	7.8	4.8	9.8
Phosphate	22.1	23.3	0.8	2.6
Melanin	0.0	0.0	0.0	10.3

Source: Bartnicki-Garcia (1968).

[a]Values are percent dry wt of the cell wall.

[b]Not confirmed by X ray. Value of spore chitin represents N-acetylated glucosamine; chitosan is nonacetylated glucosamine.

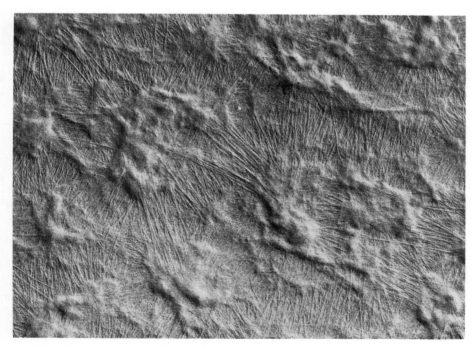

Figure 2.4   Mature sporangial wall from *Phytophthora palmivora*. (Hegnauer and Hohl, 1978)

containing *n*-alkanes as the carbon source but is absent when alkanes are replaced with glucose. The hydrophobic portion of the mannan–fatty acid complex apparently is induced by the presence of hydrocarbons and serves as a medium in which the alkanes can dissolve, thus permitting their entry into the cell.

Cell walls also contain variable amounts of inorganic ions, which are usually listed as "ash" following complete wall combustion. Usually the most abundant element in ash is phosphorus, a component of phosphorylated sugars and phospholipids. Calcium and magnesium are often present in smaller amounts.

Not only does the wall composition vary among fungal taxonomic groups but differences are also seen in the walls of a single individual as it passes through its life cycle. We have seen that the dimorphic fungus *Mucor rouxii* can occur either as a filament or a unicell; the hyphal form produces spores. Glucans are the main structural components of spore walls, but chitosan predominates in the yeast and hyphal forms (Table 2.2). Protein and melanin pigments are also higher in spore walls compared with walls of vegetative cells, but the phosphate content of spores is much lower. Considering the striking morphological differences between unicell and

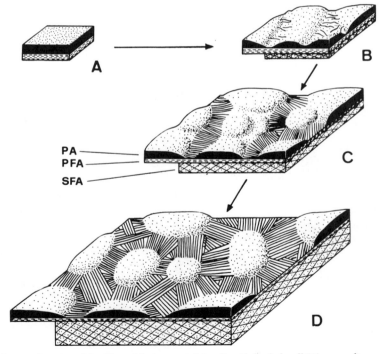

Figure 2.5 Development of the *Phytophthora* sporangial wall, with hyphal wall (A), expanding sporangial wall (B) and (C), and mature sporangial wall (D) with characteristic stellate pattern. The diagram mainly illustrates the proposed mechanical stretching of the primary wall during sporangial expansion. PA, amorphous layer of primary wall; PFA, fibrillar layer of primary wall; SFA, fibrillar layer of secondary wall. (Hegnauer and Hohl, 1978)

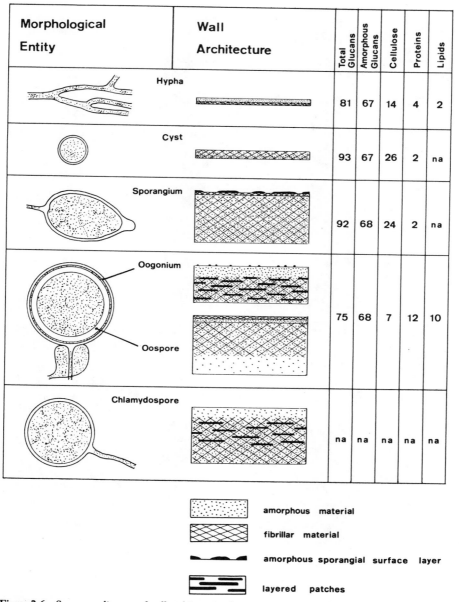

Morphological Entity	Wall Architecture	Total Glucans	Amorphous Glucans	Cellulose	Proteins	Lipids
Hypha		81	67	14	4	2
Cyst		93	67	26	2	na
Sporangium		92	68	24	2	na
Oogonium / Oospore		75	68	7	12	10
Chlamydospore		na	na	na	na	na

amorphous material

fibrillar material

amorphous sporangial surface layer

layered patches

Figure 2.6 Summary diagram of wall architecture of various morphological structures in *Phytophthora*. (Hegnauer and Hohl, 1978)

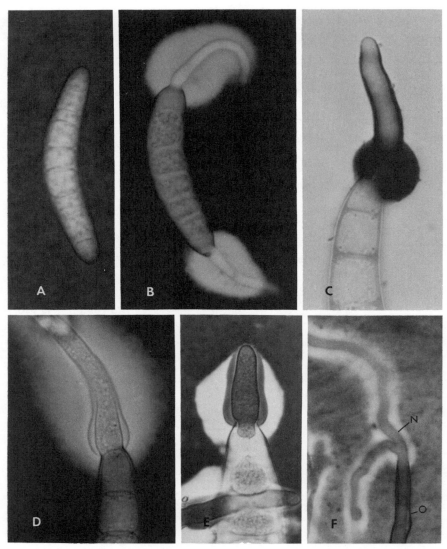

Figure 2.7 Hyphal sheaths of *Bipolaris maydis*. (A) Ungerminated conidium mounted in India ink. (B) Conidium incubated 120 min and mounted in India ink. Note the broad sheath on the germ tube. (C) Germinated conidium mounted in nigrosin showing the collarlike sheath at the base of the germ tube. (D and E) The two-layered structure of the sheath is made visible by mounting the conidium in Black Magic India ink. (F) A sheath is present on new growth, but absent on the older portions of the hypha. (Photographs courtesy of H. Stempen.)

hypha it is surprising that their wall compositions are so similar; the major differences are due to protein and mannose concentrations (Bartnicki-Garcia, 1968).

When the walls of unicells and filaments from *M. rouxii* are examined by electron microscopy, it is evident that although the two types of wall have a similar composition the arrangement of these components are quite different. The walls of unicells are thick and two-layered; those of hyphae are one-tenth as wide and are unilayered. These architectural differences may be the key to understanding how the shape of the cell is determined.

Thus, the cell wall architecture of a species is not static but changes with time concurrent with the morphological changes of the life cycle. The hyphal walls of *Phytophthora* species, for example, are composed of a layer of amorphous glucan overlying a layer of cellulose microfibrils embedded in protein. Hegnauer and Hohl (1978) have shown that sporangia arise from swellings of the hyphal wall. As the sporangium expands the outer glucan layer appears to stretch and tear, revealing the underlying network of microfibrils (Figure 2.4) . Once expansion is complete a thick secondary wall is laid down underneath the stretched existing wall (Figure 2.5). Cysts, chlamydospores, oogonia, and oospores also have distinct wall structures that differ in the number, thickness, and order of appearance of the various components (Figure 2.6). Thus, unique morphological structures have unique cell walls.

In some species the wall of a particular cell may be differentiated into two or more regions that appear more physically distinct than mere wall layers. For example, germ tubes of *Bipolaris maydis*, the southern leaf blight pathogen, develop a thick, fibrillar sheath that extends several hyphal diameters away from the "normal" wall (Figure 2.7A,B). This sheath is visualized only by negative staining, such as

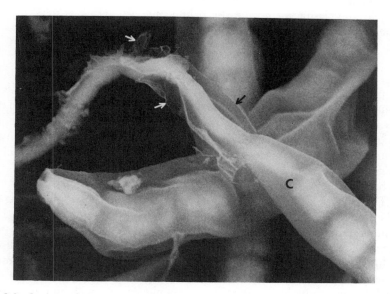

Figure 2.8 Scanning electron micrograph of a germinated conidium (C) of *Bipolaris maydis* showing a ragged sheath (*arrows*). (Evans et al., 1982)

with India ink. A variety of other stains react with a smaller, collar-like sheath at the base of the germ tube (Figure 2.7C), and thus the sheath appears to be two-layered (Figure 2.7D,E). The sheath is present only on hyphae that are less than three days old, but older hyphal strands that are rejuvenated by transfer to fresh medium develop sheaths on the growing regions (Figure 2.7F). Although the sheath can be seen in scanning electron micrographs (Figure 2.8), it apparently is a relatively fragile structure that is easily damaged during preparation of the specimen (Evans et al., 1982). It has been suggested that the sheath might enable the fungus to attach to the leaf surface and thus facilitate infection (Hau and Rush, 1982).

For the fungal physiologist the ultimate purpose of characterizing cell walls is to understand how wall structure is related to aspects of growth and development such as hyphal branching, formation of reproductive structures, cell enlargement, and differentiation. We shall deal with these topics in later chapters.

PLASMA MEMBRANE

The plasma membrane regulates the passage of materials into and out of the cell. Because the membrane is selectively permeable it maintains the cell at a chemical composition much different from that of the external environment. Organelles and most macromolecules are effectively "trapped" within the membrane and are thus concentrated to the extent that the multitude of cellular reactions can take place at rates compatible with life. Indeed, life could not have evolved without this specific type of compartmentation. In those fungi that lack cell walls during a part of their life cycle, the plasma membrane also delimits the cell and gives some measure of physical as well as chemical protection from the environment.

Membrane Structure

The plasma membrane is composed mostly of lipid and protein in roughly equal amounts; in addition, small quantities of carbohydrates and occasionally even nucleic acids have been reported as being present. The basic composition of the plasma membrane from several fungal species is given in Table 2.3. The difficulty in isolating and assaying membrane components is evident from the varied results obtained by different workers for *Saccharomyces cerevisiae*.

The major lipids present in fungal membranes are phospholipids and sphingolipids. These are polar molecules containing a hydrophilic "head" and a long hydrophobic "tail." Phosphatidylcholine and phosphatidylethanolamine (Figure 2.9) are the most common phospholipids with phosphatidylserine and phosphatidylinositol present in smaller amounts. The fatty acid components present in phospholipids are roughly correlated with evolutionary relationships. In higher fungi the hydrocarbon tails tend to have an even number of carbons and are either saturated or monounsaturated; in lower fungi there are more odd-numbered fatty acids and most are polyunsaturated (Rose, 1976).

Table 2.3 Composition of Isolated Plasma Membranes from Yeasts and Filamentous Fungi

Organism	Reference	Content (% dry wt)			
		Protein	Lipid	Carbohy-drate	Nucleic Acid
Yeasts					
Candida albicans (yeast form)	Marriott (1975)	52.0	43.0	9.0	0.3
C. utilis	Garcia-Mendoza & Villanueva (1967)	38.5	40.4	5.2	1.1
Saccharomyces cerevisiae	Boulton (1965)	46.5	41.5	3.2	7.5
	Christensen & Cirillo (1972)	65.0	29.0	3.0	4.0
	Longley, Rose, & Knights (1968)	49.3	39.1	5.0	7.0
	Matile, Moor, & Muhlethaler (1967)	26.6	45.5	30.8	—
	Schibeci, Rattray, & Kidby (1973)	49.3	2.3	5.0	7.0
	Suomalainen, Nurminen, & Oura (1967)	35.0	35.0	25.0	—
Filamentous Fungi					
C. albicans (mycelial form)	Marriott (1975)	45.0	31.0	25.0	0.5
Fusarium culmorum	Nombela, Uruburu, & Villanueva (1974)	25.0	40.0	30.0	—

Source: Rose (1976).

33

A. Phospholipids

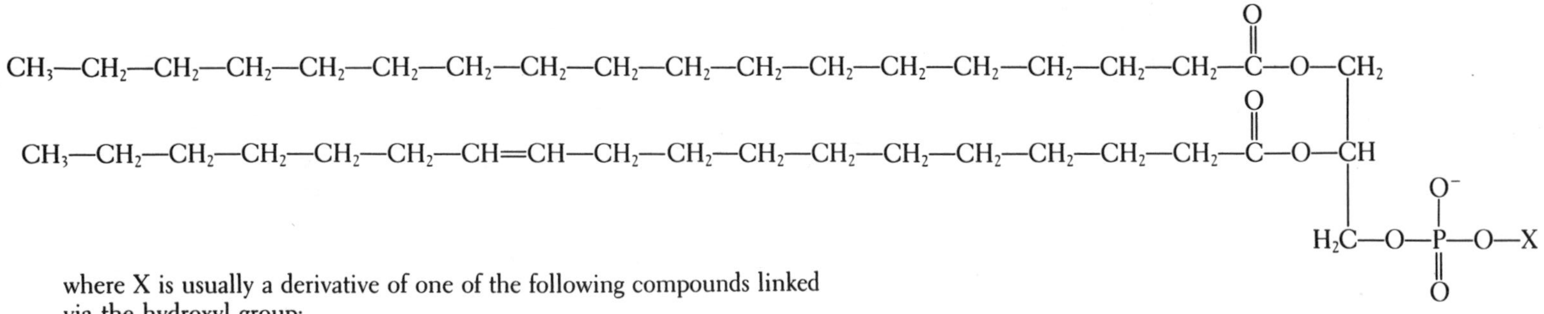

$$CH_3-CH_2-CH_2-CH_2-CH_2-CH_2-CH_2-CH_2-CH_2-CH_2-CH_2-CH_2-CH_2-CH_2-CH_2-CH_2-\overset{\overset{\displaystyle O}{\|}}{C}-O-CH_2$$

$$CH_3-CH_2-CH_2-CH_2-CH_2-CH=CH-CH_2-CH_2-CH_2-CH_2-CH_2-CH_2-CH_2-\overset{\overset{\displaystyle O}{\|}}{C}-O-CH$$

$$H_2C-O-\overset{\overset{\displaystyle O^-}{|}}{\underset{\underset{\displaystyle O}{\|}}{P}}-O-X$$

where X is usually a derivative of one of the following compounds linked via the hydroxyl group:

Ethanolamine $HO-CH_2-CH_2-{}^+NH_3$

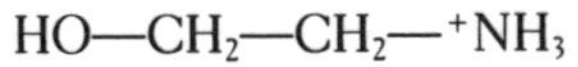

Choline $HO-CH_2-CH_2-{}^+N(CH_3)_3$

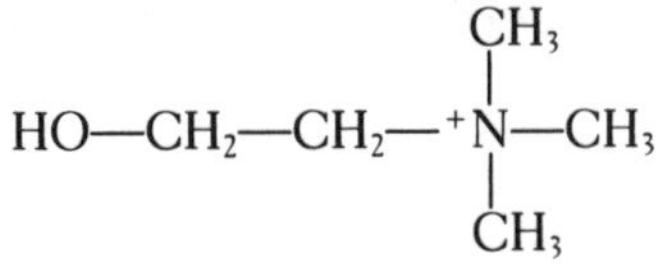

Serine $HO-CH_2-\underset{\underset{\displaystyle {}^+NH_3}{|}}{CH}-COOH$

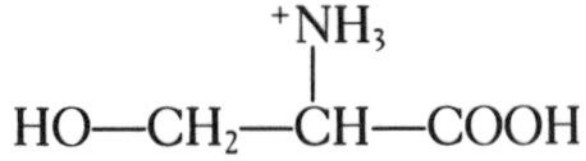

Inositol

34

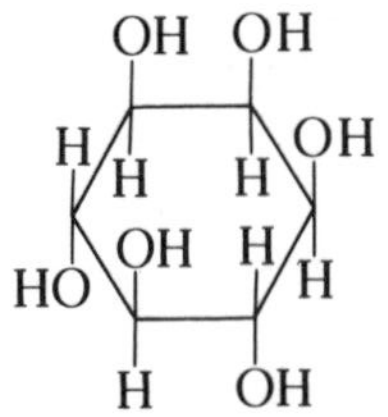

B. Sphingolipids

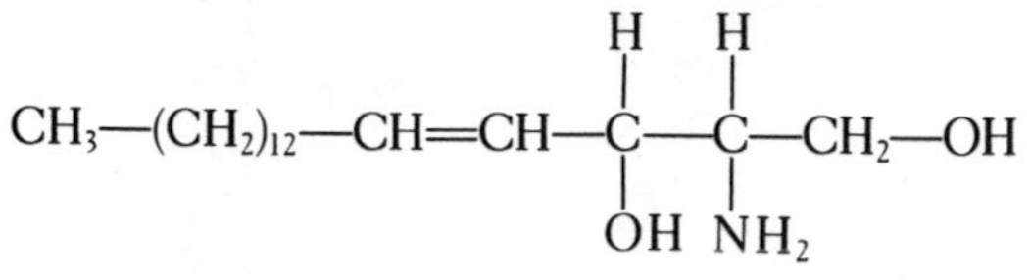

$CH_3-(CH_2)_{12}-CH=CH-\overset{\overset{\displaystyle H}{|}}{\underset{\underset{\displaystyle OH}{|}}{C}}-\overset{\overset{\displaystyle H}{|}}{\underset{\underset{\displaystyle NH_2}{|}}{C}}-CH_2-OH$

Sphingosine

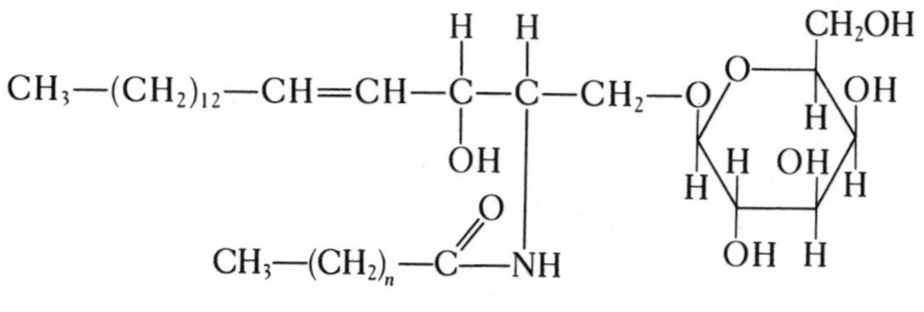

A Cerebroside

Figure 2.9 Representative phospholipids and sphingolipids.

Sphingolipids are composed of one fatty acid, one polar head group, and the long-chain amino alcohol sphingosine or its derivative (Figure 2.9). This type of sphingolipid is also called a ceramide; if the polar group is a sugar the sphingolipid is a cerebroside. Although the structure of most fungal sphingolipids present in membranes has not been determined, ceramides and cerebrosides have been isolated from a wide variety of fungi. The term glycolipid is also used to denote lipid classes such as sphingolipids or phospholipids that contain a sugar moiety. Glycolipids composed of a carbohydrate linked to a fatty acid have been reported for several species, although it is uncertain whether they are membrane components or merely contaminants from the cytoplasm. Similarly, the reports of nonpolar lipids such as sterols and triglycerides in fungal membranes may be misleading due to the difficulty in isolating uncontaminated membrane fractions (Rose, 1976).

Experimental evidence suggests that phospholipids are arranged "back-to-back" in a bilayer (Figure 2.10). On either surface of the bilayer are the hydrophilic heads; the hydrophobic tails are buried in the interior of the membrane away from the aqueous environment. The proteins are interspersed within the lipid bilayer. Some proteins are weakly bound and are easily removed by treatment with salts or chelating agents; these are termed peripheral proteins. Most, however, are integral proteins and are tightly bound. Membranes cleaved by freeze-fracturing in a plane parallel to the surface show these proteins forming a rough, pebblelike array in the interior (Figure 2.11). Labeling studies indicate that surface proteins, and lipids as well, tend to remain on the same surface but are able to move laterally. This observation has given rise to the "fluid mosaic" model of membrane structure in which proteins "float upon a sea of lipids" (Rothman and Lenard, 1977).

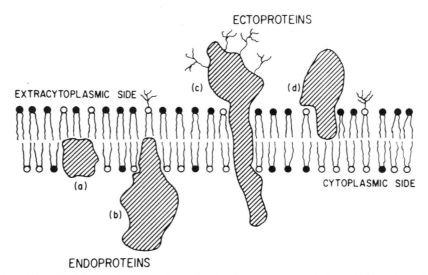

Figure 2.10   Model of a plasma membrane showing the arrangement of phospholipids and proteins. (a,b) Internal peripheral proteins; (c) integral protein; (d) external peripheral protein. (Rothman and Lenard, 1977)

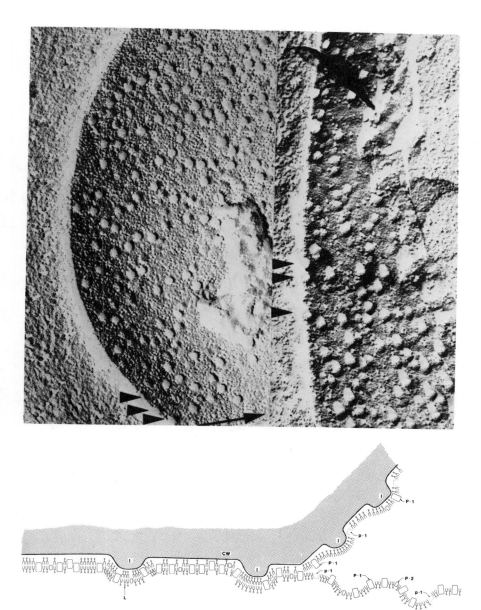

Figure 2.11 (Top) Portions of complementary fractured faces of a plasma membrane of a mature spore of *Phycomyces blakesleeanus*. The left-hand figure shows indentations on the plasmic fracture face being rather uniform in size. The right-hand figure shows corresponding protrusions on the exoplasmic fracture face. These indentations and protrusions result from invaginations of the plasma membrane (*arrows*). ×69,000. (Bottom) Diagrammatic representation of the plasma membrane of the mature spore of *Phycomyces* with its two fracture faces. CW, cell wall; *L*, lipid bilayer; *I*, indentations. P-1 and P-2 are intramembranous particles, and the complementary depressions of P-1 are indicated by p-1. Depressions of P-2 are not discernible in replicas. Not to scale. (Tu and Malhotra, 1977)

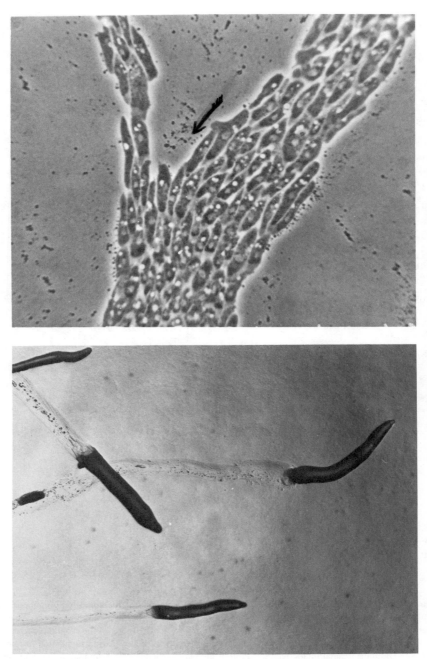

Figure 2.12 Aggregation of *Dictyostelium discoideum*. (Top) Highly magnified section showing aggregation of individual amoebae. (Bottom) A lower magnification showing a pseudoplasmodium resulting from such an aggregation. (Photographs by K. B. Raper.)

Membrane proteins have a wide variety of functions. Santos and co-workers (1978) demonstrated that plasma membranes of S. *cerevisiae* are able to bind labeled colchicine. Colchicine is fairly specific for tubulin, the proteinaceous subunits of microtubules. Rothman and Lenard (1977) suggest that microtubules present in the membrane may form a cytoskeleton that regulates both the shape of the membrane and the various processes occurring there. In addition, proteins function in the transport of nutrients and as enzymes involved in the synthesis of cell wall components.

Carbohydrates also occur in the plasma membrane but appear to be localized on the exterior surface. One role for these carbohydrates is in cellular recognition. For example, the cellular slime molds *Dictyostelium discoideum* and *Polysphondylium pallidum* exist for a portion of their life cycle as single-celled amoebae lacking cell walls. When nutrients become limiting, these amoebae become attracted to each other and aggregate into a multicellular pseudoplasmodium (Figure 2.12). The ability of these cells to recognize each other and thus properly aggregate is due to lectins and complementary receptor sites on the membrane surface (Rosen et al., 1977). Lectins are proteins that have affinities for certain carbohydrates and are produced by many organisms. The carbohydrate portion of glycoproteins on the exterior membrane surface apparently functions as the receptor site for complementary lectin proteins on the membranes of other cells and enables the amoebae to form stable associations.

The Movement of Materials Across Membranes

The lipoprotein structure of the plasma membrane makes it an effective barrier to the passage of most molecules. Those that can penetrate this barrier possess certain special attributes of size, shape, or solubility. These molecules enter the cell via one of three basic processes: nonmediated diffusion, facilitated diffusion, and active transport. These processes play a crucial role in fungal nutrition, and they have been studied extensively. In order to distinguish them unambiguously, several criteria must be used; the most basic one concerns energy.

ENERGETICS OF TRANSPORT

All molecules at temperatures above absolute zero ($-273.16°C$) possess kinetic energy and thus are in constant motion. It is difficult to measure energy, however, and most often one measures the changes in energy that occur during a particular event such as a chemical reaction or the movement of molecules from one region to another. The type of energy changes of most interest to biologists are called free-energy changes, abbreviated ΔG. The change in free energy during a particular process is equal to the change in total energy (ΔE) minus the change in wasted energy, or entropy (ΔS) times the absolute temperature (T):

$$\Delta G = \Delta E - T\Delta S$$

The free energy change that occurs when one mole of an uncharged molecule moves from a location where it is present at concentration C_1 to a second location where it is present at concentration C_2 is given by another equation:

$$\Delta G = RT \ln(C_2/C_1)$$

where R is the gas constant (1.98 cal mole^{-1} deg^{-1}), T is the absolute temperature in degrees Kelvin, and ln is the natural logarithm (equal to 2.303 log$_{10}$). Thus, the free energy change involved in moving one mole of galactose from a medium containing galactose at 100 mM to a cell containing 10 mM galactose at 27°C (300 K) is as follows:

$$
\begin{aligned}
\Delta G &= RT \ln(10/100) \\
&= (1.98 \text{ cal mole}^{-1} \text{ deg}^{-1})(300 \text{ deg})(2.303 \log 0.1) \\
&= -1368 \text{ cal mole}^{-1}
\end{aligned}
$$

Because ΔG is negative the reaction is said to be exergonic, meaning that energy in the form of 1368 calories is released into the environment. The molecule will move spontaneously from the region of higher concentration to the region of lower concentration. This is also described as moving down a concentration gradient. To move galactose in the opposite direction, that is, from the cell into the environment, would yield a ΔG of +1368 cal. Because ΔG is positive the reaction is endergonic, meaning that energy in the form of 1368 calories must be added to the molecules before they will move against the concentration gradient. Endergonic reactions will occur only after the input of energy; they are not spontaneous.

If the molecule to be transported is an ion, then another gradient comes into play. This is a gradient of electrical charge, also called a potential. Membranes of all cells have a potential difference of usually 60–90 millivolts (mV) between the inside and the outside, with the inside being negative. The ease or difficulty with which a charged molecule enters the cell is expressed quantitatively by the following equation:

$$\Delta G = zF\Delta\psi$$

where z is the number of charges on the molecule, F is the faraday (23,062 cal volt-equiv^{-1}), and $\Delta\psi$ is the membrane potential in volts. We can calculate the free energy change involved in moving one mole of chloride ions (Cl$^-$) across a membrane having a potential of -90 mV. (By convention the negative sign indicates that the inside of the membrane is negative relative to the outside).

$$
\begin{aligned}
\Delta G &= (-1)(23,062 \text{ cal volt}^{-1})(-0.09 \text{ volts}) \\
&= +2076 \text{ cal}
\end{aligned}
$$

It appears that even without taking the concentration gradient into consideration the transport of Cl$^-$ into the cell is an endergonic process requiring the input of 2076 cal. The transport of the same amount of sodium ions (Na$^+$), however, is an exergonic process with $\Delta G = -2076$ cal.

The total free energy change accompanying the movement of a molecule across a membrane is made of the algebraic sum of the two components, the free energy change due to differences in concentration and the free energy change due to differences in electrical charge:

$$\Delta G(total) = \Delta G(conc) + \Delta G(elect)$$

Or more specifically,

$$\Delta G(total) = RT\ln(C_2/C_1) + zF\Delta\psi$$

For an uncharged molecule the second term ($zF\Delta\psi$) is zero, but for ions both gradients (collectively termed the electrochemical gradient) must be considered.

With this introduction in mind, let us consider the three basic transport processes in detail.

NONMEDIATED DIFFUSION

Certain molecules enter a cell by moving down an electrochemical gradient unassisted by other molecules. This is termed nonmediated diffusion and is a passive process, not involving the expenditure of metabolic energy. This process is characterized by having a transport rate directly proportional to the magnitude of the gradient and having a Q_{10} of about 1.4, equal to that of other purely physical processes. Once sufficient molecules have entered the cell so that the gradient no longer exists, diffusion ceases and the molecules across the membrane are said to be in equilibrium.

It has been known for some time that lipids and lipid-soluble molecules diffuse readily across membranes. This is not surprising when we recall the high lipid content of membranes. Molecules that are lipid soluble can dissolve directly in the membrane lipids and move from one side to the other "in solution." Both CO_2 and O_2 are relatively lipid soluble and move across the membrane by diffusion.

FACILITATED DIFFUSION

Most fungal nutrients such as sugars, amino acids, and various ions penetrate the plasma membrane much faster than would be expected on the basis of their lipid solubility—they are simply too polar. Measurements indicate that transport is still passive, that is, transport occurs down an electrochemical gradient, but it is too rapid to be explained by nonmediated diffusion through the membrane lipids. In this case, polar molecules are "helped" across the membrane by specialized carrier proteins. This process is called facilitated diffusion; it is passive transport facilitated by a carrier.

Three principal criteria are used to determine if the movement of a particular nutrient is carrier-mediated. First, the rate of transport exhibits Michaelis–Menton saturation kinetics. This means that as the concentration of the molecule to be transported (called the substrate) increases, the rate of its uptake first increases linearly and then levels off at a certain maximum velocity, abbreviated V_{max} (Figure

2.13a). Once the V_{max} plateau is reached, further addition of substrate does not change the uptake rate, and the system is said to be saturated with substrate. In nonmediated diffusion a plateau is not reached. The presence of saturation kinetics suggests that the substrate must first bind to another molecule (the carrier) before transport can take place. Saturation occurs when all the carriers are occupied by substrate; excess substrate molecules must then "wait their turn" until a carrier becomes available. A parameter that gives an indication of the affinity of a substrate for a particular carrier is the Michaelis constant, K_m, which can be obtained from a graph like Figure 2.13a. The K_m of a carrier-mediated process is that substrate concentration which yields a rate of uptake that is one-half the maximum velocity. A high K_m value means that a relatively large amount of substrate is required before half the carriers are saturated; hence the carrier has a low affinity for the substrate. A small K_m value indicates a high affinity of the carrier for the substrate because half the carriers are saturated by a relatively low substrate concentration. An alternative and often more accurate method of obtaining V_{max} and K_m is to plot the reciprocal of the reaction velocity versus the reciprocal of the substrate concentration. This double-reciprocal plot (Figure 2.13b) gives a straight line that intercepts the $1/v$ axis at $1/V_{max}$ and intercepts the $1/S$ axis at $-1/K_m$.

Carrier-mediated transport is also highly specific. Each class of nutrients (e.g., sugars, amino acids, vitamins, and metal ions) has its own array of carrier proteins, and within each class there may be specific carriers for different stereochemical groups. For example, in S. cerevisiae there are two principal carrier systems for the facilitated diffusion of monosaccharides. The first carrier is specific for monosaccharides having an equatorial hydroxyl group on carbons 1 and 4 when the sugar is in the chair conformation. Sugars transported by this carrier include D-glucose, D-mannose, D-xylose, D-arabinose, D-lyxose, and L-glucose. The second carrier is specific for monosaccharides with an equatorial hydroxyl group on carbon 2 and an equatorial $-CH_2OH$ on carbon 5 in the chair conformation. This carrier transports

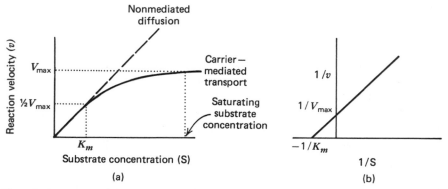

Figure 2.13 Kinetics of transport across membranes. (a) Transport via a carrier (——) becomes saturated at high substrate concentrations, but transport via nonmediated diffusion (– – –) does not. (b) Data for carrier-mediated transport graphed in a double-reciprocal plot.

D-glucose, D-galactose, D-fucose, L-rhamnose, D-ribose, L-arabinose, and L-xylose (Cirillo, 1968). These data imply that a highly specific, structure-related binding occurs between the carrier and its substrate similar to that occurring at the active site of an enzyme. Because all carrier molecules appear to be proteins, this is an appropriate analogy for transport processes as well.

The third principal characteristic of carrier-mediated transport is that the uptake of a certain substrate can be inhibited by the addition of a similarly shaped molecule. For example, Rand and Tatum (1980) measured the rate of uptake by *Neurospora crassa* of different concentrations of ^{14}C-fructose with and without the presence of 20 mM sorbose. When these data are graphed in a double-reciprocal plot (Figure 2.14a), two straight lines are obtained which intersect at $1/V_{max}$ but have different intercepts on the 1/S axis. Because the V_{max} for inhibited transport is the same as for uninhibited transport, the inhibition of fructose uptake by sorbose can be overcome by adding sufficient fructose. Fructose and sorbose thus compete for the same site on the carrier protein. This competitive inhibition is often observed in carrier-mediated uptake when two substrates of similar structure are present.

In addition, uptake can be noncompetitively inhibited by compounds that alter the structure of the carrier by acting at sites other than the binding site. For example, fructose uptake by *N. crassa* is inhibited by the presence of glucose. A double-reciprocal plot of fructose uptake in the presence of different glucose concentrations gives straight lines that intersect at $-1/K_m$ (Figure 2.14b). Because the V_{max} for inhibited uptake is different from the V_{max} for uninhibited uptake, this type of inhibition cannot be overcome by adding fructose; competition is said to be noncompetitive. Noncompetitive inhibition may be due to the inhibitor repressing the genes for the synthesis of the carrier protein or by interfering with the shape of the protein itself. For example, N-ethylmaleimide, which blocks protein sulfhydryl groups, and 2,4-dinitrofluorobenzene, which binds free amino groups, are noncompetitive inhibitors of certain uptake systems.

Many different models have been proposed to explain how facilitated diffusion might take place. The most satisfactory model at present proposes that the molecule to be taken up passes through a pore or channel in the center of certain transmembrane proteins. Consider the uptake of K^+ by facilitated diffusion (Figure 2.15). The ion first binds to a portion of the protein that has the complementary shape (Figure 21.5a,b). This binding causes a conformational change in the protein that results in the K^+ being carried through the protein interior to the inside (Figure 2.15c). During this process, the shape of the binding site also changes so that the affinity of the protein for K^+ is reduced (Figure 2.15d). The K^+ is thus released into the cell, and the protein changes back to its original shape (Figure 2.15e). One of the attractive aspects of this model is that only a small portion of the protein actually moves. The hydrophobic parts of the protein remain in contact with the hydrophobic membrane lipids, and the hydrophilic ends of the protein at the inner and outer membrane surfaces remain stationary as well. However, the interior of the protein forms a hydrophilic channel to which the polar K^+ can easily bind and through which the ion can be transported without contacting the nonpolar hydrophobic groups. Transport by this mechanism is thus energetically quite feasible.

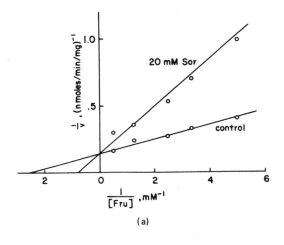

(a)

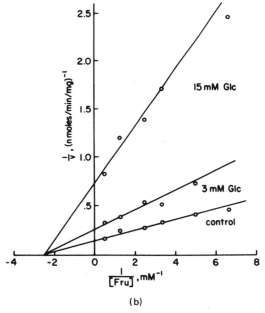

(b)

Figure 2.14 Competitive and noncompetitive inhibition of fructose uptake in *Neurospora crassa*. (a) Double reciprocal plot of fructose uptake with and without (*control*) the presence of 20 mM sorbose (*Sor*). (b) Double-reciprocal plot of fructose uptake with and without the presence of 3 or 15 mM glucose (*Glc*). (Rand and Tatum, 1980)

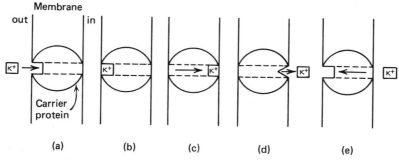

Figure 2.15 The pore model for the facilitated diffusion of K^+.

ACTIVE TRANSPORT

In many cases, however, fungi are able to transport nutrients against an electrochemical gradient. This uphill transport requires the use of metabolic energy, usually in the form of ATP, to actively "pump" molecules across the membrane. Because fungi possess these active transport systems they are able to scavenge nutrients present in small amounts in the environment by continuing to transport them into the cell even when their intracellular concentration exceeds their concentration in the environment.

Active transport, like facilitated diffusion, is a carrier-mediated process and thus exhibits saturation kinetics, high specificity, and inhibition phenomena. However, three principal characteristics make active transport unique. First, as we mentioned above, molecules are transported up an electrochemical gradient. But because both charge and concentration must be taken into consideration, it is often difficult to determine if such a gradient actually exists. Second, active transport is dependent on metabolic energy; it is not a passive process. Thus, transport can be inhibited by molecules or environmental conditions that affect the production of energy in the cell. For example, active transport is inhibited by anaerobicity, a condition that also reduces the cellular ATP supply. Fluoride, an inhibitor of glycolysis, also inhibits active transport. Third, active transport processes are unidirectional; a particular carrier can move its substrate in only one direction across the membrane. This is due to a decrease in affinity of the carrier for the substrate once the substrate has been transported.

In general, ATP hydrolysis drives active transport by powering a conformational change in the carrier protein. However, in the uptake of certain sugars, amino acids, and metal ions it has been found that ATP hydrolysis is linked only indirectly to substrate transport but is directly linked to the transport of hydrogen ions. A model illustrating this type of active transport is shown in Figure 2.16. ATP is thought to be hydrolyzed at the inner membrane surface via an ATPase that spans the membrane. The energy released in this hydrolysis is used to actively transport hydrogen ions from the cytoplasm into the environment. The pH of the medium thus drops

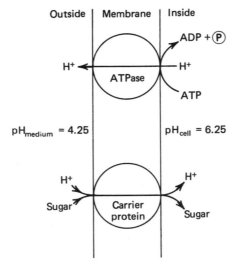

Figure 2.16   Active transport via ATP-induced hydrogen ion symport.

due to the accumulation of hydrogen ions. Another protein in the membrane has binding sites for both a hydrogen ion and the substrate. When one of the accumulated hydrogen ions in the environment binds to this carrier it induces a conformational change that increases the affinity of the carrier for the substrate. Both the hydrogen ion and the substrate are then co-transported across the membrane down a hydrogen ion gradient. On the inside the hydrogen ion leaves the carrier, thus lowering its affinity for the substrate, and the substrate is released. The carrier then assumes its original conformation and is ready to bind an additional hydrogen ion and substrate molecule on the exterior membrane surface (Eddy, 1978).

By this two-step mechanism sugars and other substrates can be transported agasinst a gradient in substrate concentration because the H^+-substrate-carrier complex is transported down a H^+ gradient. Hydrogen symport, as this type of active transport is called, can be inhibited by molecules such as 2,4-dinitrophenol and azide, which increase the permeability of the membrane to H^+ and thus destroy the hydrogen ion gradient.

Still another variety of active transport involves the exchange of ions between the cytoplasm, the cell wall, and the incubation medium. For example, when potassium is taken up by *N. crassa*, K^+ first binds to negatively charged groups (perhaps proteins or phosphates) in the cell wall, thus displacing other cations such as Na^+ that were previously bound there. This process does not depend on metabolic energy, but the extent of the ion exchange process depends on the pH of the medium. There is negligible exchange below pH 5.8, and exchange increases with increasing pH up to pH 9. This suggests that the cell wall constituents that bind cations are protonated at low pHs and thus have no affinity for cations. As the pH increases, these constituents become ionized and are thus effective ion exchangers. Following this exchange in the wall, K^+ binds to a carrier protein and is actively transported into the cell. To avoid a buildup of positive charge in the cell, the influx of K^+ is accompanied by the efflux of other cations such as Na^+, H^+, or even K^+. Although electrical neutrality can also be accomplished by the co-transport of an anion with K^+, this does not occur in *N. crassa* (Slayman and Slayman, 1970).

CYTOPLASM

The cytoplasm, found immediately inside the plasma membrane, is a watery mass of dissolved solutes in which are embedded highly organized, membranous organelles such as mitochondria, Golgi apparatus, and microbodies as well as nonmembranous structures such as ribosomes, microtubules, and microfilaments. Each organelle contains specialized enzymes which are compartmentalized within the organelle membrane and thus are separated from solutes in the amorphous portion of the cytoplasm. Compartmentation increases the efficiency of cellular metabolism by grouping together in one location the enzymes involved in a particular metabolic pathway. Also, the organelle membrane often serves as a surface area on which many of these reactions occur.

There are many enzymes which are not compartmentalized in organelles, and these along with a wide assortment of both large and small molecules are found in the amorphous portion of the cytoplasm. The presence of these solutes, particularly proteins, give the cytoplasm the capability of sol-gel transformations. Under certain conditions portions of the cytoplasm can be a semisolid (a gel) and under other conditions a liquid (a sol). Often the boundary between these two physical states within a single cell can be very sharp. For example, in the plasmodial stage of the slime mold *Physarum polycephalum*, cytoplasm in the gel state forms channels through which rivers of sol-state cytoplasm flow. After a few minutes these cytoplasmic states change so that channels become rivers and the rivers solidify to channels. In the past several years electron microscopy has also revealed tiny proteinaceous microfilaments and larger microtubules traversing the cytoplasm. Although their cytoplasmic functions are largely unknown, they may play a role in changing the physical state of the cytoplasm and the associated cytoplasmic streaming. It appears that as techniques improve the "amorphous" cytoplasm may be seen to have more structure than once imagined.

NUCLEUS

The fungal nucleus is a relatively large, membrane-bound structure that contains the cell's genetic material. Nuclei vary widely in shape and size. They have been observed squeezing through septal pores and are thus quite elastic. At a particular point in their life cycle, certain Ascomycetes and Basidiomycetes contain two genetically distinct nuclei and are said to be dikaryotic. The dikaryotic condition arises after two hyphae undergo cytoplasmic fusion (plasmogamy) without nuclear fusion (karyogamy). The nuclei in this binucleate cell then divide by mitosis, and the resulting dikaryotic hyphal strand can ramify throughout the medium, sometimes for extended periods of time, before the nuclei fuse.

The nucleus is surrounded by a double membrane called the nuclear envelope. The two membranes fuse at intervals, forming pores through which material can pass between the nucleus and cytoplasm. In yeast cells there may be as many as 200 pores per nucleus, with the pores occupying almost 7% of the nuclear surface area (Burnett, 1975).

Inside the nuclear envelope is the nucleoplasm, most of the time appearing amorphous and granular. The only clearly visible structure is the nucleolus, a densely staining spherical region containing large quantities of ribosomal RNA and other precursors for the synthesis of ribosomes.

The genetic material itself is largely invisible except when mitosis or meiosis is occurring. The DNA and associated proteins form long thin threads of chromatin which condense when the nucleus begins to divide. Fungi exhibit considerable variation in the behavior of their chromosomes during cell division. Heath (1980) has reviewed the data on mitosis in fungi and has speculated on which characteristics are primitive and which are advanced. For example, in "primitive" groups the chromatin does not condense during prophase nor line up on the equatorial plate in metaphase. In addition, the nuclear membrane and nucleolus tend to remain intact throughout mitosis rather than dispersing in prophase and reforming during telophase as in "advanced" groups. Differences also exist in the number and arrangement of spindle microtubules and kinetochores. Using these criteria, Heath proposed a scheme showing that in terms of their mitotic behavior the Zygomycetes are the most primitive group, the Oomycetes, Ascomycetes, and Basidiomycetes are intermediate, and the Myxogastria and Protostelidia are the most advanced.

THE ENDOMEMBRANE SYSTEM: ENDOPLASMIC RETICULUM, GOLGI, VESICLES, AND VACUOLES

Endoplasmic Reticulum

Fungal cells are criss-crossed by a system of narrow membrane-bound channels called the endoplasmic reticulum, or ER for short. The morphology and abundance of this membrane system varies with the physiological state of the cell and appears to be most abundant in cells that are rapidly growing (Bracker, 1967). The ER undergoes continuous turnover; under certain conditions portions of the ER may break up into spherical microbodies which later coelesce and form new channels. Some of the ER is continuous with the nuclear membrane and may conduct material from the nucleus to the cytoplasm.

The ER may have ribosomes bound to its outer surface; it is then called rough ER, abbreviated RER. Ribosomes, whether attached to the ER or free in the cytoplasm, are the sites for the attachment of mRNA and the subsequent synthesis of proteins. Once synthesized, the proteins may be transported to the lumen of the ER, from which they migrate to different parts of the cell. Smooth ER lacks ribosomes and is the site of lipid synthesis. Vesicles containing the newly synthesized material may then form at the ER surface and be transported through the cell.

Golgi Apparatus

At certain locations in some cells portions of smooth ER are in close association with another membrane system called the Golgi apparatus. This structure is composed of a stack of flattened sacs with tubules and vesicles at the periphery. The flattened sacs are called cisternae, and a stack of cisternae makes up a dictyosome. Strictly speaking a Golgi apparatus is composed of several interconnected dictyosomes, but the term is also used to denote an isolated dictyosome. Macromolecules such as lipids and proteins that have been synthesized elsewhere are transported to the dictyosomes, where they are chemically modified and packaged in a vesicle for transport through the cytoplasm. Precursors for the synthesis of additional plasma membrane and cell wall material are packaged in vesicles at the dictyosome and released at the cell surface. Hydrolytic enzymes are packaged in other vesicles that remain in the cell and function as lysosomes.

There is a functional continuum between the nuclear envelope, ER, dictyosomes, vesicles, and the plasma membrane (Figure 2.17). The ER can be thought of as an extension of the nuclear envelope that ramifies throughout the cell. Near the Golgi apparatus portions of smooth ER form small vesicles that migrate toward a dictyosome and then coelesce to form cisternae on the dictyosome's proximal, or forming, face. These cisternae then migrate from the forming face to the opposite, or maturing, face where they give rise to vesicles. A steady-state is achieved between the

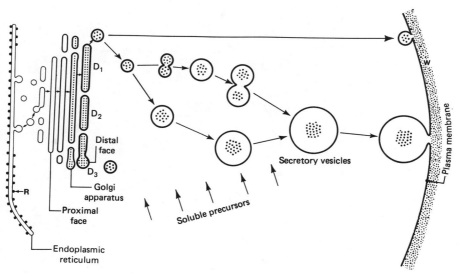

Figure 2.17   Relationship of the Golgi apparatus to other associated structures. (Grove et al., 1970)

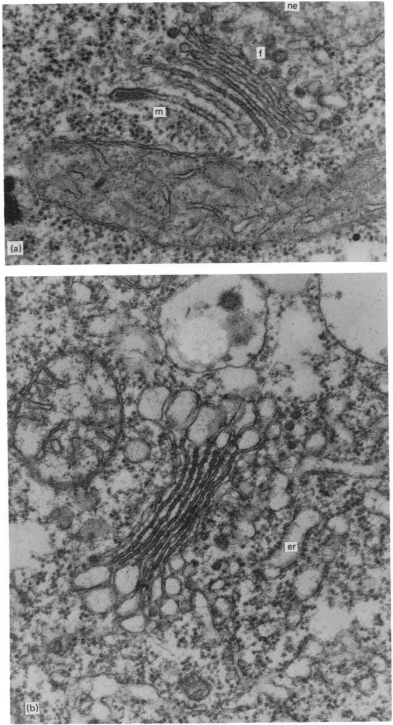

Figure 2.18 The Golgi apparatus of *Chytridium confervae* 1, 3, 4, 5, and 6 h after induction of zoosporogenesis (a–e, respectively). *ne*, nuclear envelope; *f*, forming face; *m*, maturing face; *er*, endoplasmic reticulum; *v*, cleavage vesicles; *cf*, cleavage furrow; *mt*, microtubules; *pcb*, paracrystalline body. (Taylor and Fuller, 1981)

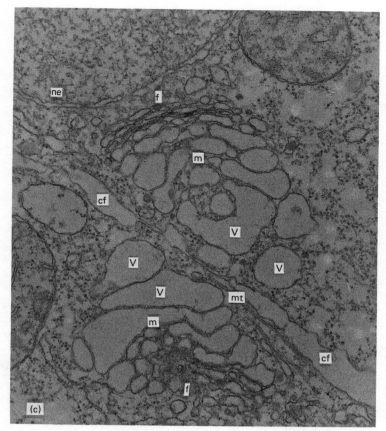

Figure 2.18 The Golgi apparatus of *Chytridium confervae* 4 h after induction of zoosporogenesis.

formation of cisternae on one side of the dictyosome and their breakup on the other, so that the number of cisternae remains relatively constant. The vesicles thus produced may migrate to the plasma membrane, fuse with it, and release their contents to the outside; other vesicles apparently remain within the cell for some time. The Golgi apparatus is thus part of a complex endomembrane system composed of a variety of membrane-bound structures (Morré et al., 1971).

The morphology of the Golgi apparatus may change markedly throughout the life cycle of a fungus. In *Chytridium confervae*, for example, the appearance of the dictyosome is relatively constant throughout the growth period and during the first two hours after the organism is transferred to a minimal medium to induce zoosporogenesis (Figure 2.18a). Three hours after induction (Figure 2.18b) the dictyosome possesses an additional cisternae and has developed large peripheral swellings. Four hours after induction (Figure 2.18c) the dictyosome has atrophied and become wider. However, after further incubation, the dictyosome becomes increasingly smaller

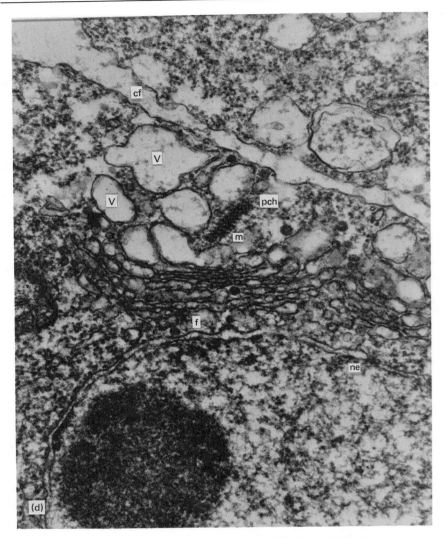

Figure 2.18   The Golgi apparatus of *Chytridium confervae* 5 h after induction of zoosporogenesis.

with smaller vesicles (Figure 2.18d,e). During zoosporogenesis the dictyosomes produce a variety of vesicles in succession: (1) vesicles for cell wall formation, (2) secretory vesicles that form a portion of the discharge apparatus, (3) cleavage vesicles that fuse to form the plasma membrane of the developing zoospore, and (4) vesicles that contain cell coat material for the zoospore (Taylor and Fuller, 1981).

Members of the Zygomycetes, Ascomycetes, Basidiomycetes, and Deuteromycetes lack a discrete Golgi apparatus, but this organelle is common in Oomyetes (Morre et al., 1971). In *Allomyces javanicus*, enzymes characteristically found in the Golgi

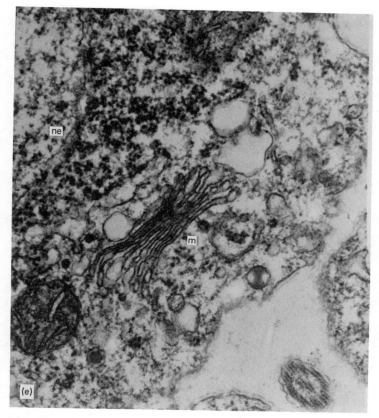

Figure 2.18 The Golgi apparatus of *Chytridium confervae* 6 h after induction of zoosporogenesis.

are found instead in the smooth ER (Figure 2.19). Thus, the structure probably functions in a capacity analogous to the Golgi in this organism (Feeney and Triemer, 1979). *Fusarium acuminatum* possesses an endomembrane system consisting of interconnecting tubules and fenestrated sheets with closely associated vesicles. This system also appears to be the functional equivalent of the Golgi apparatus (Howard, 1981).

Vacuoles

A vacuole is usually thought of as a large vesicle originating from smooth ER or a dictyosome just as other vesicles do. In addition, some vacuoles are formed at the plasma membrane as a result of pinocytosis or phagocytosis. In these processes, collectively called endocytosis, a portion of the plasma membrane surrounds a food particle (in phagocytosis) or some liquid (in pinocytosis). This membrane-enclosed material is then pinched off from the rest of the plasma membrane and enters the

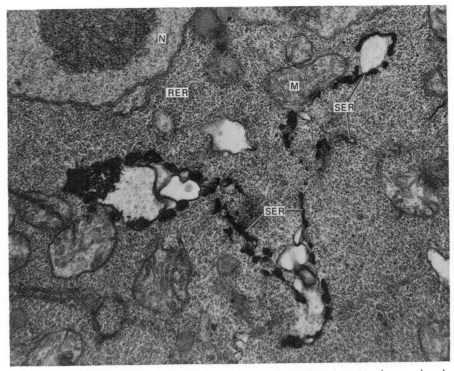

Figure 2.19 Localization of thiamine pyrophosphatase (a Golgi marker enzyme) in the smooth endoplasmic reticulum (SER) of *Allomyces*. N, nucleus; M, mitochondrion; RER, rough endoplasmic reticulum. (Feeney and Triemer, 1979)

cytoplasm as a vacuole. In fungi, phagocytosis is common only in cells lacking cell walls, such as the amoeba stages of the slime molds. Pinocytosis is of more general occurrence.

Once in the cytoplasm, vacuoles may increase in size by fusing with other vacuoles. When phagocytotic vacuoles fuse with lysosomes the food particle is digested and its subunits released into the cytoplasm. Lysosomes can also digest other organelles such as mitochondria and ER and thus recycle their molecular building blocks back into the cell.

Vacuoles also act as storage areas for certain nutrients. For example, in *Saccharomyces cerevisiae* excess phosphate is polymerized into 20–200 unit chains; these polyphosphate molecules are found largely in the cell's vacuole (Urech et al., 1978). Certain amino acids, particularly arginine, are also stored in yeast vacuoles (Matile, 1978). Vacuoles may also serve as a mechanism for storing water and thus maintaining cell turgor. In filamentous fungi, vacuoles accumulate in older sections of the hyphal strand, leaving room for only a thin film of cytoplasm. Because the vacuoles fill such a large portion of the cell the bulk of the cytoplasm remains at the growing tip of the filament. Thus, vacuoles enable the fungus to utilize a relatively small amount of cytoplasm to maintain quite a substantial volume of hyphae. This cellular economy is important to the survival of the organism (Wiebe, 1978).

MITOCHONDRIA

The mitochondria of fungi are basically similar in structure and function to those found in cells of plants and animals. Mitochondria are organelles that contain the enzymes involved in respiration, the degradation of fatty acids, and various other processes. All fungal cells possess at least one mitochondrion, but there is much structural variation not only among species but also during the life cycle of a single individual.

All mitochondria are composed of a cytoplasmic matrix surrounded by a double membrane. The outer membrane is smooth and resembles the plasma membrane, but the inner membrane is normally convoluted into a series of projections called cristae that may be either platelike or tubular. Platelike cristae (Figure 2.20a) are generally found in those fungi with chitinous cell walls (e.g., Chytridiomycetes, Zygomycetes, Ascomycetes, and Basidiomycetes). Tubular cristae (Figure 2.20b,c) occur in Myxomycetes and in fungi with cellulosic walls such as Oomycetes and Hyphocytridiomycetes (Beckett et al., 1974). Cristae provide an increased membrane surface area to which enzymes such as those associated with the respiratory electron transport chain are attached. Other enzyme systems are found in the matrix.

Mitochondria possess their own DNA, ribosomes, and protein-synthesizing machinery. The DNA and ribosomes in mitochondria are different from those found in the nucleus and cytoplasm of the parent cell and instead resemble those of bacteria. These and other observations suggest that mitochondria evolved from bacterial cells that were trapped inside a larger cell and were able to form a symbiotic relationship with this surrounding host (Margulis, 1970). Mitochondria appear to reproduce by a process resembling the binary fission of bacterial cells. For example, Kuroiwa and co-workers (1978) reported a division cycle of mitochondria in *Physarum polycephalum* that parallels the division cycle of the cell.

Mitochondrial structure is closely related to the physiological state of the cell. An actively growing cell of *Saccharomyces cerevisiae* has 10 or fewer irregularly shaped mitochondria that are often elongated, branched, or constricted (Figure 2.21). Once rapid growth ceases, the mitochondria increase to as many as 50 per cell, and they become smaller and ovoid with numerous platelike cristae. When serial sections are made through such a cell, many of the observed small mitochondria are found to be branches of a "giant" mitochondrion which in some cases extends from one end of the cell to the other (Figure 2.22). Each yeast cell contains at least one of these organelles; they may be so highly branched that the branches rejoin, forming a type of reticulum. The size of this mitochondria varies with the phase of growth; it is much larger in rapidly growing cells than in cells in which growth has slowed. One hypothesis is that as the cell enters the phase of rapid growth the numerous small mitochondria fuse and form a larger branched organelle. As growth slows the giant

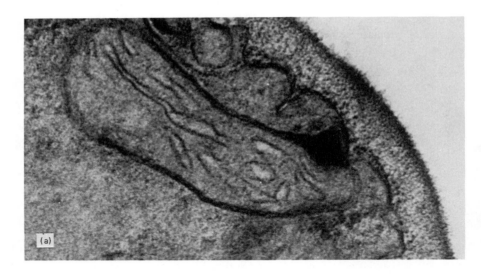

Figure 2.20 Mitochondria with platelike and tubular cristae. (a) Platelike cristae in a mitochondrion from *Bipolaris maydis*. (b and c) Tubular cristae in mitochondria from *Fuligo septica*. In (c) the cell has been treated with diaminobenzidine (DAB) and H_2O_2, which form a reaction product in the cristae and enhance their visualization. (Courtesy of H. Stempen.)

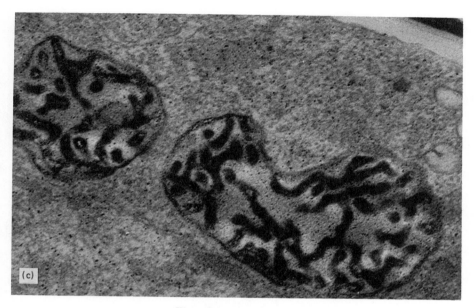

Figure 2.20 (*Continued*)

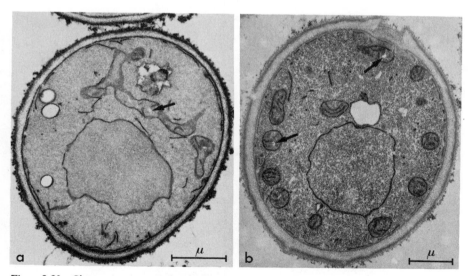

Figure 2.21 Changes in yeast mitochondrial structure with growth conditions. (a) In exponential phase, mitochondria have irregular, branched, and contorted shapes. (b) In stationary phase mitochondria are more numerous and more regular in shape and contain numerous cristae. (By permission of Stevens, 1977.)

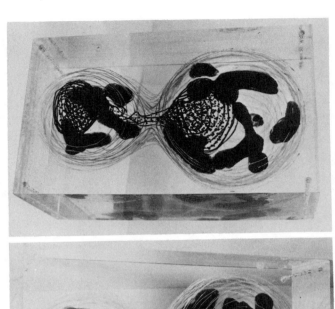

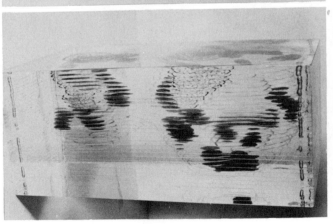

Figure 2.22 Plexiglass model of a budding yeast cell viewed from three different sides. Mitochondria are in black, nucleus in dotted black. (By permission of Stevens, 1977.)

mitochondrion fragments into smaller structures. Thus, the total mitochondrial volume of the cell remains the same throughout the growth period but the number of these organelles changes (Stevens, 1977).

MICROBODIES—PEROXISOMES AND GLYOXYSOMES

Microbodies are electron dense, membrane-bound structures characterized by the presence of catalase and other distinctive enzymes. There are two general classes of microbodies: peroxisomes and glyoxysomes.

Peroxisomes contain oxidase enzymes, particularly those involved in reactions generating hydrogen peroxide as a by-product. Peroxides are very reactive molecules and are toxic to cells because of their ability to alter the structure of nucleic acids and proteins. The peroxisome can thus be thought of as the cell's mechanism for sequestering this by-product in a safe location. The H_2O_2 thus produced can then be degraded in the peroxisome by catalase:

$$H_2O_2 \xrightarrow{\text{Catalase}} \text{Catalase } H_2O + \frac{1}{2} O_2$$

The size and number of peroxisomes in a cell is increased when the cell is grown in the presence of a nutrient that requires an oxidase for its utilization. For example, the development of peroxisomes in *Hansenula polymorpha* is enhanced when methanol is the sole carbon source (van Dijken, 1975). Methanol is metabolized via an alcohol oxidase, and hydrogen peroxide is a product of the reaction. Often peroxisomes have a distinct crystalline interior (Figure 2.23). Assays show the crystals to be composed of octameric alcohol oxidase molecules arranged in a tetragonal lattice

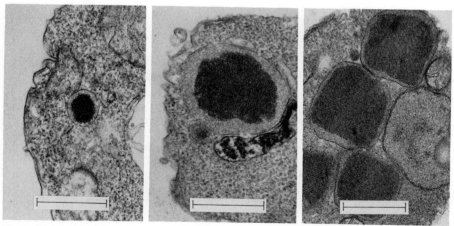

Figure 2.23   The ultrastructure of developing peroxisomes in *Hansenula polymorpha*. Bar equals 0.5 micron. (Veenhuis et al., 1978)

(Figure 2.24). Catalase molecules are also found in the crystal but are freely mobile in the spaces within the lattice (Veenhuis et al., 1978, 1981).

Glyoxysomes are microbodies that contain the enzymes of the glyoxylate cycle, a pathway important in the conversion of lipids to carbohydrates (Chapter 10). In *Botryodiplodia theobromae*, glyoxysomes are often found appressed to lipid bodies and in close proximity with the endoplasmic reticulum (Figure 2.25). The number

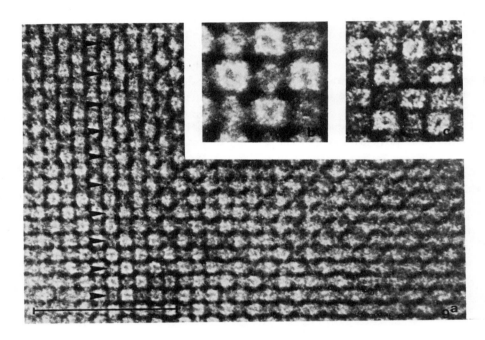

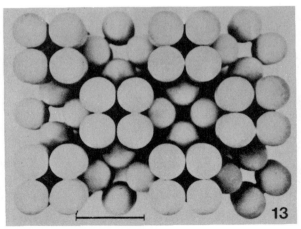

Figure 2.24 (Top) Detail of a horizontal section of a peroxisome from *Hansenula polymorpha* showing crystalline arrangement. (Bottom) A model showing the three-dimensional arrangement of alcohol oxidase molecules. (Veenhuis et al., 1981)

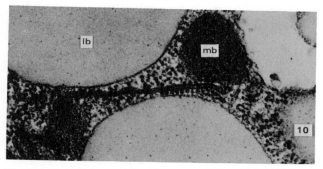

Figure 2.25   Microbody (*mb*) with attached ER from *Botryodiplodia theobromae. lb*, lipid body. (Armentrout and Maxwell, 1981)

of glyoxysomes increases during spore germination, presumably as storage lipids are utilized as sources of energy (Armentrout and Maxwell, 1981).

Microbodies appear to arise from outpocketings of the ER, but there is also evidence that they are formed from existing microbodies by budding (Armentrout and Maxwell, 1981; Veenhuis et al., 1978). Zoospores of *Blastocladiella emersonii* contain one large branched microbody (termed a symphyomicrobody) which fragments during germination; these fragments later coelesce when the new zoospore is being formed (Mills and Cantino, 1979).

WORONIN BODIES

Woronin bodies are small (0.2 μm diameter) spherical organelles associated with the septal pores of certain Ascomycetes and Deuteromycetes (Figure 2.26). They consist of a homogeneous, electron-dense matrix surrounded by a single membrane. This

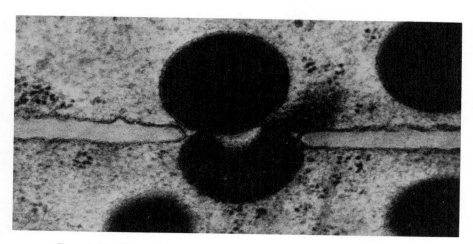

Figure 2.26   Woronin bodies blocking septal pore of *Aleuria aurantia.* (Gull, 1978)

matrix is proteinaceous and frequently has a lattice-like appearance. Woronin bodies develop within microbodies as small electron-dense inclusions, later budding outward and eventually pinching off from the parent organelle. They then become distributed near the septum; frequently several lie on either side of the septal pore. Woronin bodies are believed to act as plugs that can regulate the flow of cytoplasm between adjacent cells and may be an important means of sealing off regions of the hyphal strand that have little remaining cytoplasm (Wergin, 1973; Gull, 1978). Some workers have suggested that Woronin bodies contain hydrolytic enzymes and are thus a type of lysosome, but others have disputed this observation (e.g., McKeen, 1971).

MICROTUBULES, CILIA, AND FLAGELLA

Microtubules are long hollow cylinders, approximately 25 nm in diameter, that appear to be present in all cells and that affect a variety of processes including cell

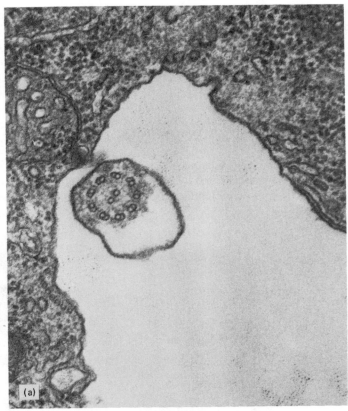

Figure 2.27 (a) Cross section of a flagellum from *Saprolegnia* showing the 9 + 2 arrangement of microtubules. (b) Cross section through the base of a kinetosome (*k*) showing the microtubules arranged in nine triplets. (Holloway and Heath, 1977)

shape and motility. Microtubules are composed of two protein subunits, called "a" and "b" tubulin, which exist as a dimer. Microtubules are assembled from an intracellular pool of these dimers at special locations in the cell called microtubular organizing centers. This assembly is a reversible process and is strongly influenced by cellular conditions. For example, at 37°C microtubule assembly is favored, but dissociation occurs at 0°C. Magnesium, calcium, and the nucleotide GTP are also required for polymerization (Hepler, 1974).

Microtubules are found free in the cytoplasm and as part of larger structures. Cilia and flagella contain a ring of nine pairs of microtubules surrounding two microtubules in the center (Figure 2.27a). The microtubules make up the axoneme, and the "9 + 2" arrangement is similar to that found in the flagella and cilia of other eukaryotes. The microtubules apparently slide past each other causing repeated bending first on one side and then on the other, which powers the cell through the medium. In the cellular interior these microtubules terminate in a kinetosome, or basal body, in which the microtubules are arranged in nine triplets surrounding a

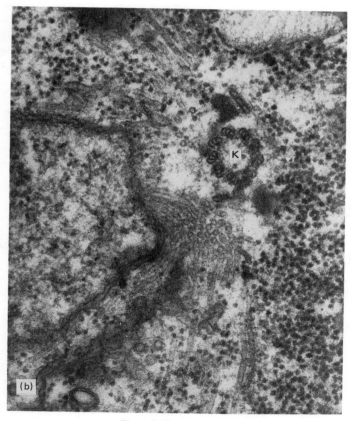

Figure 2.27   (*Continued*)

granular central region (Figure 2.27b). At the base of the kinetosome is a microtubule "rootlet" anchoring the flagellar apparatus to the nucleus. Notice in Figure
2.28 that there is an extensive microtubule system radiating from the nuclear cone
of *Saprolegnia* zoospores. Changes in the number and arrangement of these microtubules are associated with changes in the morphology of the zoospore during its life
cycle and suggest that microtubules act as a type of cytoskeleton and are involved in
morphogenesis (Holloway and Heath, 1977).

In addition to these morphological roles, microtubules are also involved in the
movement of organelles, nuclei, and chromosomes. When spores of the rust fungus
Uromyces phaseoli are germinated on media containing antimicrotubule agents, the
movement of nuclei, mitochondria, vacuoles, and apical vesicles is disrupted (Figure
2.29a,b) (Herr and Heath, 1982). Microtubules mediate the transport of Golgi
vesicles containing cell wall precursors to the actively growing hyphal tip (Howard
and Aist, 1980). In addition, microtubules are the principal components of spindle
fibers and thus play an important role in the movement of chromosomes during cell
division. In fungi with flagellated cells, nuclear microtubules appear to arise from
structures called centrioles, which resemble kinetosomes. Centrioles usually occur
in pairs and lie at right angles to each other just outside the nuclear envelope. At the
beginning of nuclear division the centrioles replicate, and the two pairs migrate to
opposite sides of the nucleus where they direct the formation of the spindle apparatus. The mechanism by which microtubules and new centrioles are formed is not
known, although the assembly of each takes place in a granular, rather fibrous region
50–100 nm away from the centriole itself (Fulton, 1971). In fungi lacking centrioles
the polymerization of spindle fibers occurs at a "nucleus-associated organelle" formerly
called the "spindle pole body" (Heath, 1980). Neither the structure nor the mode of
action of these noncentriolar regions is well understood.

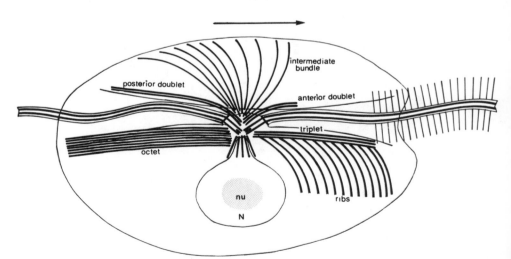

Figure 2.28   Diagrammatic representation of the microtubule root system of the maturing secondary
zoospore of *Saprolegnia*. Arrow indicates the direction of swimming. (Holloway and Heath, 1977)

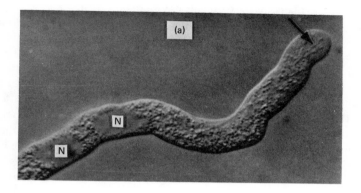

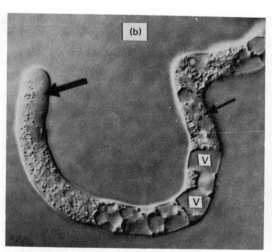

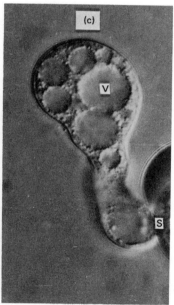

Figure 2.29 Light micrographs of germ tubes of *Uromyces phaseoli*. (a) Control germ tube showing nuclei (N) and filamentous mitochondria (large arrow). (b) Germ tube from a spore incubated in 10^{-4} M vincristine sulfate, an antimicrotubule agent. Note absence of particles at the apex (*arrow*), many vacuoles (*v*), and one nucleus (*small arrow*) considerably distal to the apex. (c) Swollen, vacuolate (V) germ tube growing from a urediospore (S) incubated in 10^{-4} M cytochalasin B. (Herr and Heath, 1982)

MICROFILAMENTS, LOMASOMES, AND CHITOSOMES

Microfilaments

The cytoplasm of most cells contains long filaments that have a much smaller diameter (5–8 nm) than microtubules. Although their exact composition is not known, these microfilaments are apparently composed of actin, the globular protein present in muscle fibers. Microfilaments are thought to play a role in cytoplasmic streaming as well as in the amoeboid movements of cells such as *Physarum* that lack cell walls for a portion of their life cycle (Hepler, 1974).

In order for movement to occur, the actin filaments must slide against something else. In *Physarum* this function is served by myosin molecules, and an actin–myosin complex is formed just as in muscle cells (Nachmias et al., 1970). In other cells the actin filaments may slide against each other. Microfilaments have been found attached to the plasma and vacuolar membranes in *Physarum* cells and may also be bound to other organelles or to microtubules. Such attachments may anchor the microfilaments and permit waves to be set up through the cytoplasm as the filaments slide past each other.

Microfilaments, like microtubules, may be involved in the fungal cytoskeleton. Spores of *Saprolegnia* germinated on the antimicrofilament agent cytochalasin-B have germ tubes that have lost the capacity for apical growth and have ballooned out (Figure 2.29c) (Herr and Heath, 1982). In *Trametes versicolor,* a belt of microfilaments accumulates in the incipient septal region during cytokinesis (Girbardt, 1979).

Lomasomes

The lomasome is a small, membranous structure lying between the plasma membrane and the cell wall in many fungi (Figure 2.30). The internal structure of the

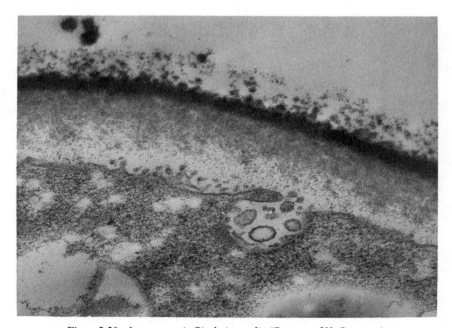

Figure 2.30   Lomosomes in *Bipolaris maydis.* (Courtesy of H. Stempen.)

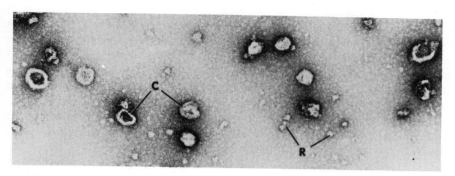

Figure 2.31 Chitosomes from hyphal cells of *Mucor rouxii*. (Bartnicki-Garcia et al., 1978)

lomasome varies and may consist of granules, tubules, or vesicles (Bracker, 1967). In some cases, the outer membrane of the lomasome is continuous with the plasma membrane, suggesting it may arise as a membrane invagination. Often such invaginations appear inside the plasma membrane in the cytoplasm as well as on the outside, and many workers consider these internal structures lomasomes as well. Ghosh (1974) suggests that lomasomes are merely a variation of a general kind of membrane invagination called a mesosome, common in prokaryotic cells and frequently found in eukaryotes. Care must be taken in the interpretation of lomasomes because similar structures are found as artifacts when membrane preparations are incompletely fixed or when cells are undergoing lysis.

Although lomasomes have been implicated in secretion, wall synthesis, pinocytosis, membrane proliferation, and stress response, their role in the cell is largely unknown (Bracker, 1967).

Chitosomes

Chitosomes are small, 40–70 nm particles that are mostly spherical in shape and possess a membranelike shell (Figure 2.31). These structures contain chitin synthetase and are able to synthesize chitin microfibrils *in vitro*. Chitosomes apparently function in transporting chitin synthetase to the cell surface (Bartnicki-Garcia et al., 1978).

SUMMARY

Although there is no such thing as a typical fungal cell, in this chapter we have presented a survey of the cellular components most commonly found in fungi.

The cell can be thought of as a series of small compartments (organelles) residing within a larger compartment bounded by the plasma membrane. By virtue of its encompassing, differentially permeable membrane, each compartment has a composition that is different from the surrounding environment, and thus it is able to carry out certain biochemical reactions at a highly efficient rate. Even though each compartment is morphologically distinct, the functioning of all compartments is integrated; and materials pass from one compartment to another in a highly organized manner. This integration appears to be facilitated by certain nonmembranous structures such as microtubules and microfilaments as well as by the physical nature of the cytoplasm through which materials must pass.

The staggering diversity in the types of cells present in the various fungal groups is intriguing to the physiologist—even fascinating! What, for example, are the physiological consequences of a particular organelle being absent in one species but present in another? What can be learned about fungal development from the changes occurring in cell components as an organism passes through its life cycle? What roles do nutrients play in the structure, function, and occurrence of a particular cellular component? We examine some of these questions in the chapters to follow.

REFERENCES

Armentrout, V. N., and D. P. Maxwell. 1981. A glyoxysomal role for microbodies in germinating conidia of *Botryodiplodia theobromae*. *Exp. Mycol.* **5**: 295–309.

Bartnicki-Garcia, S. 1968. Cell wall chemistry, morphogenesis, and taxomony of fungi. *Ann. Rev. Bacteriol.* **22**: 87–108.

Bartnicki-Garcia, S., C. E. Bracker, E. Reyes, and J. Ruiz-Herrera. 1978. Isolation of chitosomes from taxonomically diverse fungi and synthesis of chitin microfibrils in vitro. *Exp. Mycol.* **2**: 173–192.

Beckett, A., I. B. Heath, and D. J. McLaughlin. 1974. *An Atlas of Fungal Ultrastructure*. London: Longman.

Bonner, J. T. 1967. *Development of Cellular Slime Molds*. Princeton University Press.

Boulton, A. A. 1965. Some observations on the chemistry and morphology of the membranes released from yeast protoplasts by osmotic shock. *Exp. Cell Res.* **37**: 434–459.

Bracker, C. E. 1967. Ultrastructure of fungi. *Ann. Rev. Phytopathol.* **5**: 343–374.

Burnett, J. H. 1975. *Fundamentals of Mycology*. New York: St. Martin's Press.

Cabib, E. 1975. Molecular aspects of yeast morphogenesis. *Ann. Rev. Microbiol.* **29**: 191–214.

Christensen, M. S., and V. P. Cirillo. 1972. Yeast membrane vesicles; isolation and general characteristics. *J. Bacteriol.* **110**: 1190–1205.

Cirillo, V. 1968. Relationship between sugar structure and competition for the sugar transport system in baker's yeast. *J. Bacteriol.* **95**: 603–611.

Eddy, A. A. 1978. Proton-dependent solute transport in microorganisms. *Curr. Top. Membr. Transp.* **10**: 279–360.

Evans, R. C., H. Stempen, and P. Frasca. 1982. Evidence for a two-layered sheath on germ tubes of three species of *Bipolaris*. *Phytopathology* **72**: 804–807.

Feeney, D. M., and R. E. Triemer. 1979. Cytochemical localization of Golgi marker enzymes in *Allomyces*. *Exp. Mycol.* **3**: 157–163.

Fulton, C. 1971. Centrioles. In J. Reinert and H. Ursprung (Eds.), *Origin and Continuity of Cell Organelles*, pp. 170–221. New York: Springer-Verlag.

Gander, J. E. 1974. Fungal cell wall glycoproteins and peptido-polysaccharides. *Ann. Rev. Microbiol.* **28**: 103–119.

Garcia-Mendoza, C., and J. R. Villanueva. 1967. Preparation and composition of the protoplast membrane of *Candida utilis*. *Biochim. Biophys. Acta* **135**: 189–195.

Ghosh, B. K. 1974. The mesosome: A clue to the evolution of the plasma membrane. *Sub-Cell. Biochem.* **3**: 311–367.

Girbardt, M. 1979. A microfilamentous septal belt (FSB) during induction of cytokinesis in *Trametes versicolor* (L. ex Fr.). *Expr. Mycol.* **3**: 215–228.

Grove, S. N., C. E. Bracker, and D. J. Morré. 1970. An ultrastructural basis for hyphal tip growth in *Pythium ultimum. Amer. J. Bot.* **57**:245–266.

Gull, K. 1978. Form and function of septa in filamentous fungi. In J. E. Smith and D. R. Berry (Eds.), *The Filamentous Fungi*, Vol. 3, pp. 78–93. New York: Wiley.

Hau, F. C., and M. C. Rush. 1982. Preinfectional interactions between *Helminthosporium oryzae* and resistant and susceptible rice plants. *Phytopathology* **72**: 285–292.

Heath, I. B. 1980. Fungal mitoses, the significance of variations on a theme. *Mycologia* **72**: 229–443.

Hegnauer, H., and H. Hohl. 1978. Cell wall architecture of sporangia, chlamydospores, oogonia and oospores in *Phytophthora. Exp. Mycol.* **2**: 216–233.

Hepler, P. 1974. Microtubules and microfilaments. *Ann. Rev. Plant Physiol.* **25**: 309–362.

Herr, F. B., and M. C. Heath. 1982. The effects of antimicrotubule agents on organelle positioning in the cowpea rust fungus, *Uromyces phaseoli* var. vigne. *Exp. Mycol.* **6**: 15–24.

Holloway, S. A., and I. B. Heath. 1977. Morphogenesis and the role of microtubules in synchronous populations of *Saprolegnia* zoospores. *Exp. Mycol.* **1**: 9–29.

Howard, R. J. 1981. Ultrastructural analysis of hyphal tip cell growth in fungi: Spitzenkörper, cytoskeleton, and endomembranes after freeze-substitution. *J. Cell Sci.* **48**: 89–103.

Howard, R. J., and J. R. Aist. 1980. Cytoplasmic microtubules and fungal morphogenesis: Ultrastructural effects of methyl benzimidazole-2-ylcarbamate determined by freeze-substitution of hyphal tip cells. *J. Cell Biol.* **87**: 55–64.

Kappeli, O., M. Muller, and A. Fiechter. 1978. Chemical and structural alterations at the cell surface of *Candida tropicalis*, induced by hydrocarbon substrate. *J. Bacteriol.* **133**: 952–958.

Kuroiwa, T., M. Hizume, and S. Kawano. 1978. Studies on mitochondrial structure and function in *Physarum polycephalum*. IV: Mitochondrial division cycle. *Cytologia* **43**: 119–136.

Lehninger, A. L. 1975. *Biochemistry*, 2nd edition. New York: Worth.

Longley, R. P., A. H. Rose, and B. A. Knights. 1968. Composition of the protoplast membrane from *Saccharomyces cerevisiae. Biochem. J.* **108**: 401–412.

Margulis, L. 1970. *Origin of Eukaryotic Cells*. New Haven: Yale University Press.

Marriott, M. S. 1975. Isolation and chemical characterization of plasma membranes from the yeast and mycelial form of *Candida albicans. J. Gen. Microbiol.* **86**: 115–132.

Matile, P. 1978. Biochemistry and function of vacuoles. *Ann. Rev. Plant Physiol.* **29**: 193–213.

Matile, P., H. Moor, and K. Muhlethaler. 1967. Isolation and properties of plasmalemma in yeast. *Arch. Mikrobiol.* **58**: 201–211.

McKeen, W. E. 1971. Woronin bodies in *Erysiphe graminis* DC. *Can. J. Microbiol.* **17**: 1557–1560.

Mills, G. L., and E. C. Cantino. 1979. Trimodal formation of microbodies and associated biochemical and cytochemical changes during development in *Blastocladiella emersonii. Exp. Mycol.* **3**: 53–69.

Morré, D. J., H. H. Mollenhauer, and C. E. Bracker. 1971. Origin and continuity of Golgi apparatus. In J. Reinert and H. Ursprung (Eds.), *Origin and Continuity of Cell Organelles*, pp. 82–126. New York: Springer-Verlag.

Nachmias, V. T., D. Kessler, and H. E. Huxley. 1970. Electron microscope observations on actinomysin and actin preparations from *Physarum polycephalum* and on their interaction with heavy meromyosin subfragment I from muscle myosin. *J. Mol. Biol.* **50**: 83–90.

Nombela, C., F. Uruburu, and J. R. Villanueva. 1974. Studies on membranes isolated from extracts of *Fusarium culmorum. J. Gen. Microbiol.* **81**: 247–254.

Phaff, H. J. 1971. Structure and biosynthesis of the yeast cell envelope. In A. H. Rose and J. S. Harrison (Eds.), *The Yeasts*, Vol. 2, pp. 135–210. New York: Academic.

Rand, J. B., and E. L. Tatum. 1980. Fructose transport in *Neurospora crassa. J. Bacteriol.* **142**: 763–767.

Rose, A. H. 1976. Chemical nature of membrane components. In J. E. Smith and D. R. Berry (Eds.), *The Filamentous Fungi*, Vol. 2, pp. 308–327. New York: Wiley.

Rosen, S. D., C. M. Chang, and S. H. Barondes. 1977. Intracellular adhesion in the cellular slime mold *Polysphondylium pallidum* inhibited by interaction of asialofetuin or specific univalent antibody with endogenous cell surface lectin. *Dev. Biol.* **61**: 202–213.

Rothman, J. E., and J. Lenard. 1977. Membrane asymmetry. *Science* **195**: 743–753.

Santos, E., J. R. Villanueva, and R. Sentandreu. 1978. The plasma membrane of *Saccharomyces cerevisiae*. Isolation and some properties. *Biochim. Biophys. Acta* **508**: 39–54.

Schibeci, A., J. B. M. Rattray, and D. K. Kidby. 1973. Isolation and identification of yeast plasma membrane. *Biochim. Biophys. Acta* **311**: 15–25.

Slayman, C. W., and C. L. Slayman. 1970. Potassium transport in *Neurospora crassa*: Evidence for a multisite carrier at high pH. *J. Gen. Physiol.* **55**: 758–791.

Stevens, B. J. 1977. Variation in number and volume of the mitochondria in yeast according to growth conditions: A study based on serial sectioning and computer graphics reconstruction. *Biol. Cellulaire* **28**: 37–56.

Suomalainen, H., T. Nurminen, and E. Oura. 1967. Isolation of the plasma membrne of yeast. *Acta Chem. Fenniae* **B40**: 323–326.

Taylor, I. E. P., and D. S. Cameron. 1973. Preparation and quantitative analysis of fungal cell walls: Strategy and tactics. *Ann. Rev. Microbiol.* **27**: 243–259.

Taylor, J. W., and M. S. Fuller. 1981. The Golgi apparatus, zoosporogenesis, and development of the zoospore discharge apparatus of *Chytridium confervae*. *Exp. Mycol.* **5**: 35–59.

Tu, J. C., and S. K. Malhotra. 1977. Plasma membrane of Phycomyces: Sequential changes in the structure during spore formation. *Cytobios* **20**: 121–132.

Urech, K., M. Durr, T. Boller, and A. Wiemkem. 1978. Localization of polyphosphate in vacuoles of *Saccharomyces cerevisiae*. *Arch. Microbiol.* **116**: 275–278.

van Dijken, J. P., M. Veenhuis, N. J. W. Kreger-van Rij, and W. Harder. 1975. Microbodies of methanol assimilating yeasts. *Arch. Microbiol.* **105**: 201–267.

Veenhuis, M., J. P. van Dijken, S. A. F. Pilon, and W. Harder. 1978. Development of crystalline peroxisomes in methanol-grown cells of the yeast *Hansenula polymorpha* and its relation to environmental conditions. *Arch. Microbiol.* **117**: 153–163.

Veenhuis, M., W. Harder, J. P. van Dijken, and F. Mayer. 1981. Substructure of crystalline peroxisomes in methanol-grown *Hansenula polymorpha*: Evidence for an in vivo crystal of alcohol oxidase. *Mol. Cell. Biol.* **1**: 949–957.

Wergin, W. P. 1973. Development of Woronin bodies from microbodies in *Fusarium oxysporum* f. sp. *lycopersici*. *Protoplasma* **76**: 249–260.

Wiebe, H. H. 1978. The significance of plant vacuoles. *Bioscience* **28**: 327–331.

3

CARBON NUTRITION

Raulin was one of the first fungal physiologists to demonstrate that a supply of organic carbon is critically needed for the growth of fungi. He observed that when sugar was omitted from a synthetic medium the growth of *Aspergillus niger* was less than 3 percent of that in a medium containing sugar. In retrospect one could have predicted the outcome of Raulin's experiment since at least half of the dry weight of a fungal thallus is carbon, and fungi do not fix significant CO_2. Inevitably the experimental demonstration of the need for organic carbon by fungi was followed by increased interest in the significance of carbon utilization for fungal growth and development.

Carbon sources serve two essential functions in the physiology of fungi and other heterotrophic organisms. (1) They supply the carbon needed for the synthesis of critical constituents such as carbohydrates, proteins, lipids, and nucleic acids. (2) Their oxidation provides a source of energy for proper functioning of the essential life processes of fungi.

Since fungi grow on a wide range of substrates the types of carbon sources that they are capable of using in the laboratory may provide clues to their preferred natural habitats. Moreover, an understanding of the physiological basis for the utilization of a preferred carbon source may be helpful to those who wish to exploit a species' capabilities and potentialities for research or industrial use.

Fungi as a group are capable of utilizing a tremendously large variety of carbon compounds for growth and/or development. These range from small molecules such as sugars, organic acids and alcohols to large polymers such as proteins, lipids, polysaccharides and lignin. Table 3.1 lists a sampling of the smaller organic molecules utilized by fungi; the list is far from complete. The early literature concerning carbon nutrition has been reviewed in detail by several authors including Cochrane (1958), Foster (1949), Lilly and Barnett (1951, 1953), and Perlman (1965). These authors provide details on the utilization of specific compounds. In this chapter we provide an overview of carbon nutrition with the emphasis placed on factors that control utilization.

Table 3.1 Some Compounds Utilized by Fungi as the Sole Source of Carbon

Carbohydrates and Related Compounds

Adonitol	Galactosamine	Kojic acid	Phytic acid
Amylose	Galactose	Lactose	Polygalacturonate
Arabinose	Gentiobiose	Lyxose	Raffinose
Arabitol	Gluconic acid	Maltose	Rhamnose
Cellobiose	Glucosamine	Mannitol	Ribose
Chitin	Glucose	Mannose	Salicin
Dextran	Glucose pentaacetate	Melibiose	Sorbitol
Dextrin	Glucuronic acid	Melizitose	Sorbose
Dulcitol	Glycerol	α-Methlgalactoside	Starch
Erythritol	Glycogen	α-Methylglucoside	Sucrose
Erythrose	Inositol	α-Methylmannoside	Trehalose
Fucose	Inulin	α-Methylxyloside	Turanose
Fructose	2-Ketogluconic acid	Panose	Xylan
Galactitol	5-Ketogluconic acid	Pectin	Xylose

Amino Acids and Related Compounds

Alanine	Betaine	Glutamic acid	Norleucine	Taurine
α-Aminoadipic acid	Citrulline	Glutamine	Norvaline	Threonine
	Creatine	Glycine	Ornithine	Tryptamine
α-Aminobutyric acid	Cysteine	Hippuric acid	Phenaceturic acid	Tryptophan
	Cystine	Histamine		Tyramine
Amygdalin	Diaminopimelic acid	Histidine	Phenylalanine	Tyrosine
Anthranilic acid		Isoleucine	Phenylglycine	Valine
Arginine	Djenkolic acid	Leucine	Proline	
Asparagine	Ephedrine	Lysine	Sarcosine	
Aspartic acid	Ethionine	Methionine	Serine	

Organic Acids and Related Compounds

Acetic acid	Formic acid	Mandelic acid	Pyruvic acid
Adipic acid	Fumaric acid	Methylricinoleate	Stearic acid
Arachidonic acid	Gentisic acid	Myristic acid	Stipitatic acid
Butyric acid	Glutaric acid	Oleic acid	Succinic acid
Capric acid	Glycerylricinoleate	Oxalic acid	Tartaric acid
Caproic acid	Homogentisic acid	Palmitic acid	Triacetin
Citric acid	α-Ketoglutaric acid	Phenylacetic acid	Tributyrin
Di-n-butyrl sebacate	Lactic acid	Phenylpyruvic acid	Triolein
Ethanol	Lauric acid	Pimelic aicd	Tristearin
Ethyl acetate	Linoleic acid	Propionic acid	Valeric acid

Polycyclic Compounds and Alkaloids

Androsterone	Digitonin	Ergosterol	Stigmasterol
Cholesterol	Digitoxigenin	Methylreserpate	Testosterone
Cholic acid	Diosgenin	Nicotine	Tetracycline
Cortisone	Ergotamine	Progesterone	Yohimbine

Source: Perlman (1965).

CARBOHYDRATES

The vast majority of fungi live as parasites on plants or grow saprophytically on their dead remains. Therefore, carbohydrates of particular interest as carbon sources are those which are constituents of plants. Classes of carbohydrates shown to be effective carbon sources for fungi include simple sugars, sugar acids, and sugar alcohols as well as short- and long-chain polymers of these subunits. However, fungi cannot utilize all carbohydrates equally well. Ever since the pioneering work of Raulin, fungal physiologists have noticed that fungi are selective in their ability to utilize sugars and other nutrients and that nutritional requirements differ among species and sometimes even among strains of the same species.

Reddy (1969) compared the growth of five species of *Helminthosporium* using various monosaccharides as carbon sources (Table 3.2). Notice that members of the same genus differ widely in their ability to utilize even those sugars that are closely related chemically. Glucose, mannose, and galactose are all aldohexoses, and yet glucose is a good carbon source for four of the five species and galactose is a poor one. Sorbose and fructose are both ketohexoses, but for each species fructose is a significantly better carbon source. The pentose arabinose is used only sparingly by each species, but the closely related xylose is a good source for some but not others. Rhamnose (a deoxysugar) gives the poorest growth of all.

Many common fungal species can utilize individual monosaccharides but not larger molecules composed of these same monosaccharide subunits. For example, *Rhizopus nigricans*, *Blakeslea trispora*, and *Sordaria fimicola* grow poorly on sucrose but well on glucose or fructose. This is due to the inability of these fungi to hydrolyze the disaccharide (Lilly and Barnett, 1951).

In addition, a fungal species may have the ability to utilize a particular carbon source for vegetative growth but be unable to use it for the production of specialized structures. This is illustrated in Table 3.3 where a comparison is made among the

Table 3.2   Growth of Five Species of *Helminthosporium* in Media Containing Different Monosaccharides as the Sole Carbon Source

Mono-saccharide	Dry Weight (mg)				
	H. spiciferum	*H. rostratum*	*H. hawaiiense*	*H. nodulosum*	*H. halodes*
D-Glucose	196	57	182	230	133
D-Mannose	150	89	110	142	103
D-Galactose	24	26	100	21	37
L-Rhamnose	14	17	18	21	31
D-Fructose	152	97	79	163	106
L-Sorbose	57	14	10	11	28
L-Arabinose	46	38	43	52	55
D-Xylose	158	33	108	119	57

Source: Reddy (1969).

Table 3.3   Mycelial Growth, Density of Conidium Formation, and Production and Dry Weight of Sclerotia for *Aspergillus niger* on Agar Media Containing Different Carbon Compounds

Carbon Compound	Mycelial Growth (mm)	Conidial Density[a]	No. of Sclerotia[b]	Dry Wt. of Sclerotia (mg)[b]
Ribose	31	+ + + +	0	0
Xylose	50	+ + +	248.2 ± 10.9	62.5 ± 3.78
Rhamnose	53	+ + + +	19.0 ± 3.6	6.5 ± 1.28
Sorbose	44	+ + +	60.0 ± 4.4	24.0 ± 1.82
Glucose	46	+ + +	134.2 ± 6.6	52.0 ± 2.1
Galactose	52	+ + + +	66.2 ± 4.8	28.5 ± 1.9
Fructose	52	+	289.5 ± 12.2	74.0 ± 2.9
Maltose	60	+ + + +	101.2 ± 10.0	51.0 ± 3.5
Sucrose	47	+ +	149.5 ± 8.41	52.2 ± 2.7
Melibiose	37	+ + +	62.6 ± 7.5	63.7 ± 4.7
Cellobiose	40	+ +	175.2 ± 4.6	34.0 ± 1.6
Starch	46	+ + + +	108.7 ± 10.0	40.7 ± 1.7
Dextrin	39	+ +	60.7 ± 8.4	23.0 ± 1.1
Glycogen	60	+ + + +	106.0 ± 10.5	44.0 ± 3.5
Dulcitol	46	+ + + +	91.5 ± 8.1	33.2 ± 2.6
Sorbitol	46	+ +	115.1 ± 9.3	42.5 ± 2.6
Mannitol	50	+ + + +	0	0
Malonic aicd	47	+ + + +	0	0
Fumaric acid	48	+ + + +	0	0
Citric acid	10	+ + + +	0	0
Control (no carbon)	28	+ +	0	0

Source: Agnihotri (1969).

[a]Index of conidial density: + poor, + + fair; + + + good; + + + + excellent.

[b]Based on four replications. Variation expressed as standard error.

effects of various carbohydrates and organic acids on mycelial growth, conidiation, and the production of sclerotia by *Aspergillus niger*. In this species ribose, mannitol, and organic acids can be utilized as carbon sources for mycelial growth, but the formation of sclerotia is inhibited. Fructose is a good carbon source for growth and sclerotia production but not for the production of conidia.

Not only the type but also the concentration of the carbon source is important in determining how effective it is in promoting growth. Notice in Figure 3.1 that sucrose is a very poor carbon source at low concentrations for growth of *Coprinus lagopus*. In contrast, concentrations of glucose and fructose as low as 10–20 mM markedly stimulate growth, but at approximately 100 mM a plateau is reached where increasing concentrations have little effect. Increasing the acetate concentration up to 100 mM results in increased growth, but at 300 mM growth is inhibited. The decline observed with acetate is typical of that obtained with excessively high con-

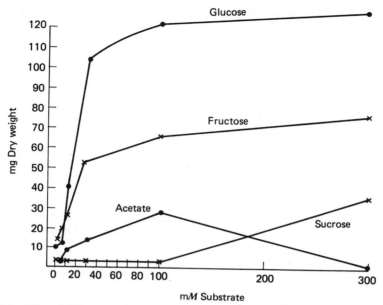

Figure 3.1 Effect of different initial concentrations of some potential carbon sources on the growth of *Coprinus lagopus*. (Redrawn from Moore, 1969.)

centrations of any nutrient. The most common causes for growth inhibition are (1) the depletion of other nutrients, (2) the accumulation of toxic metabolites, and (3) adverse osmotic effects.

What are the physiological bases for these differences in the utilization of carbon sources? In all fungi the extent to which a low molecular weight carbon compound can be used as a carbon source depends on the existence and proper functioning of (1) a specific carrier molecule to transport the compound across the plasma membrane and (2) biochemical pathways to metabolize the compound such that energy is released and organic building blocks are formed. Many factors, both chemical and physical, affect these two basic processes and thus regulate the patterns of growth and development in fungi. Let us examine the sugar transport systems in *Saccharomyces cerevisiae* as a model system for the uptake of carbon sources in general. The metabolic pathways involved in the utilization of these compounds are discussed in Chapter 10.

Sugar Transport Systems in *Saccharomyces cerevisiae*

CONSTITUTIVE VS. INDUCIBLE SYSTEMS

The principal sugar uptake systems in *S. cerevisiae* are summarized in Table 3.4. There is at least one transport system for the uptake of glucose and structurally

Table 3.4   The Principal Sugar Uptake Systems in *Saccharomyces cerevisiae*

Carrier System	Examples of Sugars Transported	Constitutive/ Inducible	References
Nonspecific pyranose system	D-Glucose L-Glucose 2-Deoxy-D-glucose D-Fructose D-Xylose D-Mannose D-Arabinose D-Lyxose L-Sorbose	Constitutive	Kotyk, 1967; Kotyk & Haskovec, 1968; Cirillo, 1968.
Galactose system	D-Galactose D-Fucose L-Xylose L-Arabinose D-Gulose	Induced by galactose	Kotyk & Haskovec, 1968; Van Steveninck & Dawson, 1968; Van Steveninck, 1972.
Disaccharide system	D-Glucose Maltose Trehalose	Inducible	Gorts, 1969; Kotyk & Michaljanicova, 1979.
Nonspecific α-methyl-D-glucoside system	D-Glucose Maltose Trehalose α-Methyl-D-glucosides α-Thioethyl-D-glucopyranoside	Inducible	Okada & Halvorson, 1964; Kotyk & Michaljanicova, 1979.

related sugars. This system is constitutive, meaning that the necessary carrier proteins are always present in the cell. The other three uptake systems are inducible; they become functional only after certain substrates called inducers are added to the medium. For example, when cells of a vigorously growing culture are transferred from a glucose medium to one containing galactose as the sole carbon source, growth slows appreciably and even ceases in some strains. This growth inhibition is due to the low affinity of the constitutive transport system for the galactose molecule. However, continued incubation with galactose for 4 hours (or longer in some strains) results in increased uptake of galactose and the resumption of vigorous growth. During this lag period, the genes coding for proteins involved in galactose transport and metabolism are turned on, and growth resumes once these newly synthesized proteins are present in sufficient concentration. This is called adaptive growth, characterized by an initial period of negligible growth followed by an increase in the growth rate. As indicated in Table 3.4, the induced galactose carrier enables the cell to take up other sugars in addition to galactose; however, galactose is the only effective inducer (Kotyk and Haskovec, 1968).

Table 3.5 illustrates the difference in transport parameters between uninduced

Table 3.5 **Transport Parameters of Sugar Transport in Uninduced and in Galactose-Induced Cells of *Saccharomyces cerevisiae* NCYC 240, at 25° C**

Substrate	K_m		V (mmoles/g/h)	
	Uninduced	Induced	Uninduced	Induced
Galactose	510	4.0	0.65	2.11
Glucose	5.1	5.1	2.13	2.14

Source: Van Steveninck (1972).

cells and those induced with galactose. Notice that both the affinity of the galactose carrier for galactose (as measured by the K_m) and the maximum uptake velocity change dramatically after induction and become equivalent to those for glucose. The glucose carrier, however, remains unaffected by the presence of galactose. Van Steveninck and Dawson (1968) have proposed that induction results in the synthesis not of a new carrier but rather of a "permease" that catalyzes the binding of galactose to the original carrier.

In order for induction of the galactose carrier or permease to occur, not only must galactose be present but glucose must be absent. If glucose is added to a culture of cells growing in a galactose medium, the utilization of galactose ceases and the uptake of glucose begins. This may be due to repression by glucose of the genes coding for galactose carrier proteins, in a manner analogous to the enzyme repression system described for bacteria by Jacob and Monod. However, in *S. cerevisiae* the mechanism by which glucose inhibits the utilization of galactose appears to be more a case of switching on additional genes than switching others off. There is evidence that glucose induces the synthesis of proteins that either degrade the galactose carrier molecule (Alonso and Kotyk, 1978) or decrease the affinity of the carrier for galactose (Matern and Holzer, 1977).

Inducible carrier proteins have a relatively short half-life (1–2 hours) compared with constitutive carriers (more than 24 hours) (Alonso and Kotyk, 1978). Thus, the fungal cell can respond relatively quickly to changes in the nutrient composition of the environment. Induction, repression, and activation/inactivation are all mechanisms for conserving cellular resources and increasing the energy efficiency of the cell. Whenever glucose or structurally related sugars are present in the environment they will be used preferentially by the constitutive uptake systems regardless of which other sugars are also present. Only when the substrates for the constitutive systems have been exhausted will the cell expend the energy necessary to synthesize and maintain the new carrier system. If glucose later becomes available, the cell will then "dismantle" the induced system in favor of the more efficient constitutive one.

ACTIVE VS. PASSIVE SYSTEMS

Uptake of sugars occurs by either facilitated diffusion or active transport; the type of transport depends on the species, the sugar, and the environmental conditions. In *S.*

cerevisiae, uptake of galactose is by facilitated diffusion in uninduced cells and by active transport in cells induced by galactose (Van Steveninck and Dawson, 1968). Glucose is also taken up by both facilitated diffusion and active transport, but in contrast to galactose the glucose is phosphorylated during active transport (Meredith and Romano, 1977). Uptake of maltose and trehalose is also active (Kotyk and Michaljanicova, 1979; Serrano, 1977). There is evidence that maltose is cotransported with hydrogen ions into the cell. The entry of one H^+ per maltose molecule is balanced by the exit of one K^+ or the entry of a permeable anion (Serrano, 1977). However, proton symport of sugars does not appear to be as common an uptake mechanism in *Saccharomyces* as in other fungi (Deak, 1978; Jennings, 1974).

FACTORS AFFECTING SUGAR UPTAKE

Sugar uptake is affected by many factors; one of the most striking is the presence of other sugars in the growth medium. Kotyk (1967) tested for such interactions by measuring the effect of unlabeled monosaccharides on the uptake of ^{14}C-labeled sugars. The results (Table 3.6) indicate that even though some sugars have no effect on the transport of others, many sugars are inhibitory. By the use of double-reciprocal plots Cirillo (1968) showed that three of the mutually inhibitory sugars all

Table 3.6   Inhibition of Uptake of ^{14}C-Labeled Monosaccharides by the Presence of Various Nonlabeled Sugars[a]

^{12}C-Sugar ╲ ^{14}C-Sugar	D-Glucose	D-Xylose	D-Arabinose	D-Galactose	D-Ribose	L-Arabinose
D-Glucose	+ +	+ +	+ +	+	+	+
2-Deoxy-D-glucose	+ +	+ +	+ +	+	+	?
D-Mannose	+ +	+ +	+ +	?	+	+
D-Galactose	—	—	+	+ +	+	+ +
L-Rhamnose	—	—	—	?	—	?
D-Xylose	+ +	+ +	+ +	—	—	?
D-Arabinose	+ +	+ +	+ +	—	—	?
D-Lyxose	+	+	+	—	—	?
D-Ribose	—	—	—	+	+	+
L-Arabinose	—	—	—	+	—	+
L-Xylose	—	—	—	+	—	—
D-Fructose	+ +	+ +	+ +	+	+	?
L-Sorbose	+ +	+ +	+ +	+	—	?

Source: Kotyk (1967).

[a]*Saccharomyces cerevisiae* was Incubated Anaerobically at 30° C, and Both Sugars were Added Simultaneously. The Number of Plus Signs Corresponds to the Intensity of Inhibition by the Unlabeled Sugar.

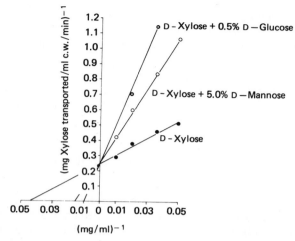

Figure 3.2 Competitive inhibition of xylose uptake by glucose and mannose in baker's yeast. (Cirillo, 1968)

compete for the same carrier molecule (Figure 3.2). Pairs of sugars that do not inhibit each other's uptake are assumed to be transported by different carriers.

In some cases, sugars actually stimulate the uptake of other sugars. Table 3.7 illustrates the maximum rate of uptake of D- and L-arabinose either alone or in the presence of D-galactose. The observation that the uptake of each sugar is high when galactose is also present suggests that the carrier induced by galactose can also transport both D- and L-arabinose, although not to the same extent.

A host of other factors have been shown to affect the uptake of sugars (as well as other nutrients). Because the carrier molecules are proteins, their activities are influenced by temperature, pH, and inhibitors just as enzymes are. The pH is particular important because of the involvement of H^+ in the symport of many molecules. Active transport can be influenced by chemical or physical factors that affect respi-

Table 3.7 Maximum Rate of Uptake of D- and L-Arabinose by *Saccharomyces cerevisiae* Before and After Induction by D-Galactose (In mg Sugar per ml Cell Water per Minute)

Sugar	Before	After
L-Arabinose	9.5	18
D-Arabinose	16	26

Source: Kotyk and Haskovec (1968).

ration and thus energy production in the cell. In addition, certain inorganic ions such as K^+ can affect uptake by functioning as transport cofactors, in maintaining electrical neutrality during proton symport, and by generally influencing the osmotic relations of the cell (Kotyk, 1967). Thus, the composition of the growth medium as well as various physical environmental factors determine not only which nutrients will be taken up but the rate of uptake as well. Such factors may thus be critical determinants for both the extent and type of growth and development that results.

Utilization of Large Carbohydrates

Many fungi can also utilize oligosaccharides and polysaccharides as carbon source, but in general these molecules are too large to be transported intact into the cell. Carrier systems for disaccharides such as maltose, trehalose, and α-methyl-D-glucosides have been described (Table 3.4), but other disaccharides and carbohydrates of longer chain length must first be hydrolyzed to their subunits before they can be taken up. The enzymes for hydrolyzing the disaccharide sucrose and the trisaccharide melibiose are localized in the cell wall of S. cerevisiae (Fries and Ottolenghi, 1959a,b). However, many fungi are capable of secreting into the environment a variety of enzymes that hydrolyze complex carbohydrates. These include cellulase, amylase, pectinase, chitinase, dextranase, xylanase, β-1,3-glucanase, and even DNAase and RNAase (e.g., Hankin and Anagnostakis, 1975; Simonson and Liberta, 1975).

Because cellulose is the most abundant polysaccharide in plant tissue, it represents a widespread potential source of carbon for fungi. Many pathogens and saprophytes produce a series of enzymes, collectively called "cellulase," which facilitate the degradation of cellulose to glucose units. "C_1-cellulase" acts on insoluble crystalline cellulose and converts it to a soluble form. This in turn is acted on by "C_x-cellulase," which hydrolyzes the soluble cellulose to monosaccharides. A third enzyme, cellobiase, which hydrolyzes the disaccharide cellobiose to glucose, may also

Table 3.8a   Effect of Starch Concentration on Mycelial Growth and Amylase Production by *Talaromyces emersonii* after 6 Days at 45° C on Starch–Yeast Extract Medium

Starch Concentration (%)	Dry Weight Mycelium mg/30 ml Medium (Mean of Five Flasks)	Amylase Units	pH Culture[a] Filtrate (Final)
0	7.0	0.0	6.0
0.25	95.2	2.5	5.6
0.5	108.1	3.1	5.9
0.75	119.3	3.9	5.4
1.0	133.6	4.5	5.5

Source: Oso (1979).

[a]Initial pH = 6.8.

Table 3.8b   Effect of Carbon Source on Mycelial Growth and Amylase Production by *Talaromyces emersonii* Grown at 45° C for 6 Days, and pH of the Medium.

Carbon Source[a]	Dry Weight Mycelium mg/30 ml Medium (Mean of Five Flasks)	Amylase Units	pH Culture[b] Filtrate (Final)
D(−)Glucose	43.6	1.3	5.9
D(−)Fructose	52.3	1.6	6.2
D(+)Galactose	49.1	1.1	5.8
Sucrose	21.4	0.0	5.9
Maltose	54.5	1.7	6.0
Cellobiose	89.8	1.4	6.1
Starch	105.2	3.1	5.8

Source: Oso (1979).

[a]Basal medium: Yeast extract, 2 g; NaNO$_3$, 3 g; K$_2$HPO$_4$, 1 g; MgSO$_4$·7H$_2$O, 0.5 g; distilled water, 1 liter.

[b]Initial pH = 6.8.

be present. The ability of certain fungi to synthesize and secrete this cellulase complex enables them to degrade a variety of cellulosic structures and gives them a great advantage in colonizing both living and nonliving plant materials (Bateman, 1976; Olutiola, 1976).

Starch is another plant polysaccharide that is utilized by many fungi. *Talaromyces emersonii* is a thermophilic ascomycete that can use starch as the sole source of carbon. Oso (1979) incubated this species for 6 days in a medium containing different concentrations of starch. The cultures were then filtered and the filtrate assayed for the presence of amylase, a major starch-degrading enzyme. As evident from Table 3.8a, both mycelial growth and the concentration of amylase in the medium increase with increasing starch concentration. If starch is replaced by other carbohydrates, amylase production decreases significantly; no amylase is detected when sucrose is the carbon source (Table 3.8b). In this organism amylase is an inducible enzyme; it is synthesized in the presence of only certain substrates and not others. Obviously, starch is the most effective inducer of amylase in *T. emersonii*.

LIGNIN

Lignin is a major component of perennial plants and is probably the second most abundant organic polymer on earth, next to cellulose. Lignin is thus a widespread potential source of carbon for both parasitic and saprophytic fungi. Lignin is the term given to a group of closely related, structurally complex, high molecular weight compounds that are branched polymers of three substituted alcohols: *p*-coumaryl alcohol, coniferyl alcohol, and sinapyl alcohol (Figure 3.3). The relative proportions of these subunits vary with the type of plant (Kirk, 1971).

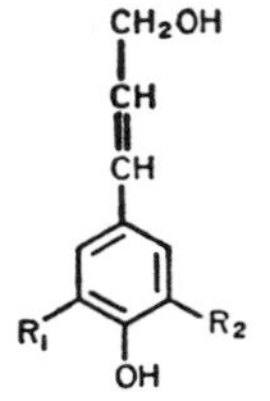

I $R_1 = R_2 = H$ = *p*—Coumaryl alcohol
II $R_1 = H, R_2 = OCH_3$ = Coniferyl alcohol
III $R_1 = R_2 = OCH_3$ = Sinapyl alcohol

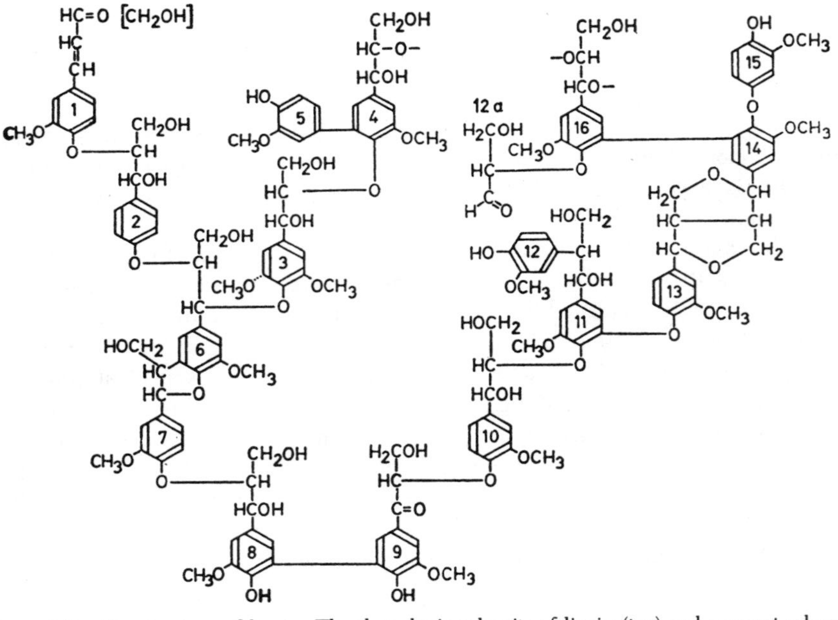

Figure 3.3 The structure of lignin. The three basic subunits of lignin (top) and a recent schematic formula for conifer lignin. (Kirk, 1971)

The chemical structure of lignin makes it resistant to attack by most microorganisms; however, several groups of fungi can utilize it as a source of carbon. The major groups of lignicolous fungi are the white-rot and litter-decomposing Basidiomycetes. In addition, the brown-rot Basidiomycetes and certain soft-rot Ascomycetes and Deuteromycetes can partially degrade lignin (Kirk et al., 1977). Only the white-rot fungi have been studied in detail.

The mechanisms by which fungi degrade lignin are only partially known. The evidence suggests that extracellular enzymes produced by the fungus oxidize both the aromatic rings and the aliphatic side chains to produce products of low molecular weight that can then be taken up. Phenoloxidizing enzymes such as laccase and peroxidase are involved in lignin degradation by most species and are produced in quantity by white-rot fungi (Kirk et al., 1977). Sopko (1967a,b) noted that the quantity of lignin degradation products released from beech wood during decay by *Pleurotus ostreatus* is correlated with the quantity of laccase produced. Also, a mutant strain of the white-rot fungus *Sporotrichum pulverulentum* that lacks the ability to synthesize phenoloxidases is unable to degrade lignin, but, degradation occurs if laccase is added to the lignin-containing medium (Ander and Eriksson, 1976).

Phenoloxidases do not appear to be the only enzymes involved in extracellular lignin degradation; in fact, some lignicolous species apparently do not produce any such enzymes (Scurti et al., 1972). Studies using ^{14}C-labeled synthetic lignin have shown that degradation can also proceed via demethylation of side chains as well as by cleavage of aromatic rings (Haider and Trojanowski, 1975; Kirk et al., 1978). Thus, there may be several enzyme systems involved in degrading the lignin molecule into subunits that have the necessary small size and chemical characteristics to be taken up.

On the basis of the few organisms studied, it seems that lignin cannot serve as the sole source of carbon for growth. No growth is observed when each of the two white-rot species *Phanerochaete chrysosporium* and *Coriolus versicolor* is incubated on a mineral salts medium containing ^{14}C-labeled lignin. However, both species grow well and release $^{14}CO_2$ when the medium is supplemented with cellulose, cellobiose, glucose or xylose (Kirk, 1971). This indicates that other carbohydrates, such as those naturally occurring with lignin in plants, facilitate lignin utilization. Other factors have been shown to be important as well. In *P. chrysosporium*, lignin degradation is found to be greatest at high oxygen partial pressure, at pH 4–4.5, and at low nitrogen concentrations (Kirk et al., 1976, 1978).

ALCOHOLS

Certain fungi can utilize alcohols as the sole source of carbon. Much research has been done using yeasts, a group which is industrially important because of their ability to produce alcohols via fermentation reactions. In Table 3.9 are the results obtained from culturing the yeast *Candida utilis* in a basal medium containing different alcohols or other compounds found as contaminants in synthetic ethanol. Of the compounds tested, ethanol and 1-propanol support the greatest amount of growth, but other alcohols are almost completely ineffective. Of the nonalcoholic compounds tested, some growth is obtained with ethyl acetate and acetic acid (Sestakova, 1976).

Several filamentous fungi can also utilize alcohols. *Aspergillus nidulans* can use ethanol as a carbon source for both growth and conidiation, and for the latter process

Table 3.9   Utilization of Major Impurities of Synthetic Ethanol and Some Metabolically
Significant Compounds by the Yeast *Candida utilis* 49

Substrate	Yeast Biomass Yield (%)[a]	Residual Substrate (%)	Unidentified Compounds	Medium pH	Ammonia Nitrogen in Medium (mg/100 ml)
Methanol	0	N	—	5.5	2.8
Acetaldehyde	0	0.024	—	3.1	0.1
Diethyl ether	0.3	Traces	x_1	6.2	2.0
Hexane	0	N	—	5.5	1.0
Ethanol	72.0	0.0	—	7.2	9.9
2-Propanol	1.0	0.016	x_2	7.8	17.4
Propanol	48.2	0.007	—	6.5	11.4
Acetone	0	N	—	5.5	2.1
Methyl ethyl ketone	0.2	0.23	—	5.6	2.1
Ethyl acetate	43.4	Traces	—	6.8	3.8
2-Methylpropan-2-ol	0	0.462	x_3	5.5	1.5
2-Butanol	2.6	0.149	x_4	7.8	12.7
2-Methylpropan-1-ol	1.1	0.24	x_5	3.8	1.5
Butanol	0.6	N	x_6, x_7, x_8	3.3	1.3
Butyraldehyde	0	N	x_9	3.4	1.1
Crotyl alcohol	0	N	x_{10}	4.9	1.2
Crotonaldehyde	0	0.261	—	4.5	0.9
Allyl alcohol	0	0.098	x_{11}	5.3	0.5
Crotonic acid	0	N	x_{12}	4.9	0.5
Acrolein	0	N	—	3.6	0.5
Acetic acid	34.2	Traces	—	7.9	3.7

Source: Sestakova (1976).

[a]Yield referred to the amount of substrate added.

ethanol is even more effective than glucose (Martinelli, 1976). Similarly, ethanol, 1-propanol, and 1-butanol are excellent carbon sources for the growth of *Armillaria mellea* (Weinhold and Garraway, 1966). However, *Schizophyllum commune* grows poorly on ethanol (Hoffman and Raper, 1974), and several species of *Colletotrichum* and *Gleosporium* fail to use ethanol at all (Ghosh and Tandon, 1967).

Polyhydric alcohols are also utilized as carbon sources by many fungi although glycerol is most commonly chosen for experimental work. Of a variety of alcohols tested as possible carbon sources for *Hansenula miso*, glycerol is the only polyhydric alcohol found to support growth (Table 3.10). In this study glycerol is much more effective than glucose and just as effective as ethanol (Harada and Hirabayashi, 1968).

In addition to their role as carbon sources, alcohols may promote the utilization of other carbon sources. Growth of the Basidiomycete *Armillaria mellea* cannot be initiated on a medium containing glucose (0.5%) as the sole carbon source. However, the addition of even a small amount (0.05%) of ethanol, 1-propanol, or 1-butanol to the glucose medium results in abundant growth and rhizomorph forma-

Table 3.10   Utilization of Various Alcohols
as Sole Sources of Carbon by
Hansenula miso

Alcohol 1%	Yield of Cells (mg dry wt./100 ml)
Methanol	0
Ethanol	580
n-Propanol	3
n-Butanol	1
1,2-Ethanediol	1
DL-1,2-Propanediol	2
1,3-Propanediol	1
DL-1,3-Butanediol	0
1,4-Butanediol	0
meso-2,3-Butanediol	1
Glycerol	560
DL-1,2,4-Butanetriol	0
Pentaerythritol	0
Glucose	108

Source: Harada and Hirabayashi (1968).

tion (Weinhold, 1963; Allerman and Sortkjaer, 1973). The growth of A. *mellea* on a glucose medium supplemented with ethanol is equivalent to that on a medium containing ethanol (0.5%) as the sole carbon source (Weinhold and Garraway, 1966). Analysis of the culture medium at various times during the incubation period shows that most of the growth occurs after the ethanol supplement has been depleted from the medium (Garraway and Weinhold, 1968). Thus, glucose can be utilized as a carbon source only after growth has been initiated on ethanol. In addition to its effect on glucose utilization by A. *mellea*, ethanol was also shown to stimulate the uptake of the nitrogen source (L-asparagine) and phosphate from the medium (Sortkjaer and Allerman, 1973).

Alcohols may also enhance the sensitivity of fungi to inhibitors. The growth of *Saccharomyces cerevisiae* var. *ellipsoideus* was shown to be unaffected by the inhibitor 2,6-dichloro-4-nitroanaline (DCNA) supplied at 1.2mg/liter when sucrose is the carbon source. However, growth is significantly reduced by DCNA even at concentrations as low as 0.25mg/liter when ethanol is added to this basal medium at concentrations similar to those found in fermentation vats (Michaels et al., 1977). The mechanism for this is unknown.

FATTY ACIDS, TRIGLYCERIDES, AND HYDROCARBONS

Many fungal groups can utilize a variety of relatively nonpolar molecules as carbon sources. As usual, most of the nutritional studies have been done using yeasts.

Fatty Acids and Triglycerides

The extent to which fatty acids can be utilized by fungi depends largely on the length of the hydrocarbon chain. In Figure 3.4 are the results obtained by incubating the yeast *Cunninghamella echinulata* on a nitrate–mineral salts medium containing various even-chain fatty acids as the sole carbon source. Long-chain fatty acids containing 14–18 carbons can be utilized by this species at each of the four pHs tested. However, pH is an important factor influencing the ability of the fungus to use those with shorter chain lengths. For example, the C_2–C_8 fatty acids are good carbon sources at pH 8.0 but not at pH 5.5; just the reverse is true of the C_{10}–C_{12} acids (Lewis and Johnson, 1967).

Lipids including triglycerides may also serve as sources of carbon. Many species secrete into the environment lipases that hydrolyze lipids into subunits which more readily enter the cell (e.g., Brockerhoff and Jensen, 1974; Das et al., 1979). The production of lipases can be controlled by nutritional and environmental factors. For example, in *Penicillium chrysogenum* maximum lipase production occurs at 30°C, pH 6.0, and in shake culture rather than stationary. Glucose is the most effective carbon source of those tested, followed by maltose, mannitol, galactose, sucrose, lactose, and fructose (Chander et al., 1977). The lipase produced by *Geotrichum candidum*, on the other hand, is an inducible enzyme whose synthesis is triggered by the presence of lipids in the medium (Iwai et al., 1973).

Even in those instances in which lipids or their components are not the sole carbon source, their presence in the medium may significantly affect the growth of

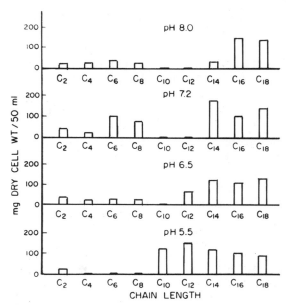

Figure 3.4 Effect of initial pH and fatty acid chain length on the growth of *Cunninghamella echinulata*. C_2, C_3, C_4, and so on refer to the fatty acid of the chain-length cited. (Lewis and Johnson, 1967)

fungi. A wide variety of effects have been reported. For example, various fatty acids fail to support growth of *Armillaria mellea* (Moody and Weinhold, 1972) and *Geotrichum candidum* (Steele, 1973) when supplied as the sole carbon source. However, they invariably enhance growth in a medium containing glucose. Wardle and Schisler (1969) showed that a wide variety of commercial lipids such as cottonseed oil, corn oil, and linseed oil stimulate mycelial growth of *Agaricus bisporus* when this species is incubated on a D-xylose + L-asparagine medium. In this study, the lipid-stimulated increase in growth is greater than could be explained merely by the added carbon in the lipids; perhaps the lipids are also functioning as growth factors or as aids to xylose utilization. The presence of as little as $0.1mM$ caproic acid (C_6) in a glucose-NH_4Cl medium is sufficient to inhibit growth of *Saccharomyces cerevisae* and *Candida utilis*. In the latter species, increasing the chain length of the fatty acid from C_2 to C_9 causes increasing growth inhibition, but longer chain lengths reverse the inhibition (Hunkova and Fencl, 1977). In seeking to understand the mechanisms underlying these inhibitory effects, Hunkova and Fencl (1978) found that high fatty acid concentrations decrease phosphate uptake and respiration but stimulate ethanol production. It was speculated that one of the key sites of fatty acid inhibition may be the mitochondrial membrane.

Lipids have also been shown to have striking effects on developmental processes. Long chain and short chain fatty acids induce rhizomorph formation in *A. mellea* (Moody and Weinhold, 1972); the formation of fruiting bodies in *Ceratocystis* sp. is stimulated by oleic, linoleic, and linolenic acids (Dalphe and Neumann, 1977). These unsaturated fatty acids also alter the circadian rhythm of spore formation in *Neurospora crassa* (Brody and Martins, 1979); the effect of linolenic acid can be reversed by the addition of the saturated fatty acid palmitate. Unsaturated fatty acids also inhibit the differentiation of *Dictyostelium discoideum* but have no effect on growth (Das and Weeks, 1979).

Sterols can be hydroxylated and subsequently degraded by fungi. However, sterols are not usually thought of as carbon sources but instead in terms of their growth factor or hormonal properties (Hendrix, 1970). These are discussed in detail in Chapter 9.

Hydrocarbons

The widespread utilization of hydrocarbons as carbon sources by fungi has received increasing attention in recent years because of the widespread use of petroleum-base products. For example, fungi from the genera *Cladosporium*, *Gliomastix*, and *Candida* are typical contaminants of aircraft fuels (Neihof and Bailey, 1978), and other genera have been found growing on substrates as unlikely as diesel fuel, kerosene, and paraffin (Klug and Markovetz, 1971). Crude oil was utilized by 40 out of 60 fungal isolates tested by Davies and Westlake (1979), and suggestions have been made for using fungi to degrade oil spills (Karrick, 1977). In addition, yeast cells grown on paraffin have been suggested as an inexpensive source of digestible protein (Shannon and McNab, 1973).

Although some filamentous fungi can utilize hydrocarbons as small as butane, most fungi able to grow on hydrocarbons require a chain length of nine carbon atoms or more (Klug and Markovetz, 1971). Among 55 strains (36 species) of *Candida* tested, most species grew most abundantly on hydrocarbons containing either 11–12 carbons or 18 carbons (Klug and Markovetz, 1967).

The effectiveness with which hydrocarbons are used as carbon sources is frequently enhanced by a fatty acid supplement to the growth medium. For example, a strain of *Rhodotorula glutinis* growing on 2% paraffin shows a twofold increase in growth when the medium is supplemented with 0.25% oleic acid (Vaskivnyuk et al., 1974). *Candida tropicalis* is able to grow on hydrocarbons as a carbon source, but there is a lag period before the maximal growth rate is obtained. This lag period can be reduced considerably by the addition of fatty acids at concentrations of 0.005–0.01% (Gololobov et al., 1969).

Hydrocarbons are thought to enter cells by dissolving directly in the lipids of the plasma membrane. There does, however, appear to be some specificity involved because only those hydrocarbons that support growth are able to enter the cell (Klug and Markovetz, 1971). Cells of *C. tropicalis* transferred from a glucose medium to one containing alkanes as the carbon source develop a mannan–fatty acid complex on the surface of the cell wall to which the alkanes adsorb (Kappeli and Fiechter, 1977; Kappeli et al., 1978). This inducible complex appears to facilitate the transport of the nonpolar hydrocarbons from the medium to the cell surface (Miura et al., 1977). Thus, the stimulatory effect of fatty acids on growth of this species (Gololobov et al., 1969) may be due to the increased production of this mannan–fatty acid complex.

CARBON SOURCES CONTAINING NITROGEN

Other organic compounds that are potential sources of carbon for fungi are those that contain nitrogen. The most common organic nitrogen encountered by fungi growing parasitically or saprophytically are proteins and their constituent amino acids. Many species of fungi secrete proteases into the environment, and the resulting amino acids can then be utilized as nutrients if the necessary uptake and metabolic systems are present (McIntyre and Hankin, 1978; Hankin and Anagnostakis, 1975; Das et al., 1979). Depending on the species, amino acids may serve as a source of carbon, a source of nitrogen, or both.

Gleason and co-workers (1970a,b) compared several water molds with regard to their capacity to use individual amino acids as sources of carbon. Their results for *Aphanomyces laevis* and *Pythium ultimum* are summarized in Table 3.11. Some amino acids such as alanine and glutamate are very good sources of carbon for both species tested. Others, such as aspartate, proline, and serine can be used by *Pythium* but not *Aphanomyces*. Use of ^{14}C-labeled amino acids indicated that *Aphanomyces* can readily transport all of the tested amino acids into the cell, even those that did not permit growth! Moreover, they all can be metabolized to $^{14}CO_2$. This suggests that various amino acids trigger different metabolic pathways and that some of these pathways are more conducive to growth than others.

Table 3.11   Utilization of Twelve Carbon Sources by *Aphanomyces laevis* and *Pythium ultimum*

| Compound | Aphanomyces laevis (107–52) | | Pythium ultimum (67–1) | |
	Dry Weight mg/flask	Days	Dry Weight mg/flask	Days
Glucose	76	9	77	5
Sucrose	4	14	81	5
Maltose	0	14	81	5
Cellobiose	0	14	85	5
Alanine	72	9	52	9
Glutamate	73	15	65	9
Aspartate	0	14	43	8
Proline	0	14	25	21
Leucine	0	14	0	21
Arginine	8	21	2	21
Phenylalanine	0	14	0	21
Serine	0	14	25	21
Control	0	14	4	7

Source: Gleason et al. (1970b).

Gleason and colleagues (1970b) suggested that differences in the ability of the fungi to utilize amino acids might be useful characteristics for distinguishing these genera taxonomically. However, Faro (1971) points out that even though proline and glutamate support good vegetative growth when supplied as carbon sources for *Achlya heterosexualis* as a representative species, the fungus is unable to form sexual reproductive structures. Thus, because the fungal life cycle cannot be completed on media containing amino acids as the sole carbon source, Faro concluded that amino acid utilization preferences are inappropriate for taxonomic purposes.

Certain other fungi are incapable of using amino acids as the sole carbon source. For example, the marine fungi *Culcitalna achrara*, *Humicola alopallonella*, and *Halosphaeria mediostegera* are unable to utilize any of the 24 amino acids provided as the sole source of carbon, even though many of these compounds are effective as nitrogen sources (Sguros et al., 1973).

CARBON DIOXIDE FIXATION

Although carbon dioxide is not usually considered an important carbon source for fungi, many species have the capacity of fixing CO_2 into organic compounds (Tabak and Bridge-Cooke, 1968). Most species fix CO_2 only in the presence of an additional, more complex, carbon source. However, a few such as *Cephalosporium* spp. and *Fusarium* spp. have been shown to utilize CO_2 as the sole source of carbon. These two organisms were grown for five years on a completely inorganic medium and were able to utilize $^{14}CO_2$ supplied as either a gas or as HCO_3^- (Mirocha and

DeVay, 1971). During the initial stages of growth there is a linear relationship between the growth rate and the concentration of CO_2 supplied. When the supplied air is "scrubbed" by being passed through sulfuric acid to remove any trace contaminants (but not CO_2), the growth rate of *Cephalosporium* decreases. This suggests that even though this species can indeed utilize CO_2 as the sole source of carbon, it also has the ability to use trace organics present in the air.

Some fungi not only fix CO_2 during the utilization of other carbon sources but appear to have an obligate requirement for it. The growth of *Verticillium albo-atrum* is inhibited by more than 90% when the fungus is incubated in a medium containing glucose or glycerol (1%) but lacking CO_2; no inhibition occurs if the CO_2-free medium contains succinate or acetate (Hartman et. al., 1972). *Aqualinderella fermentans* (Emerson and Held, 1969) and *Byssochlamys nivea* (Yates et al., 1967) require CO_2 when grown under anaerobic conditions. Other genera such as *Curvularia*, *Cochliobolus*, and *Chaetomium* do not have an absolute requirement for CO_2, but its presence significantly enhances growth (Macauley and Griffin, 1969). In *Phycomyces blakesleeanus* (Hilgenberg and Sandmann, 1977) and *Blastocladiella emersonii* (Cantino and Horenstein, 1956) more $^{14}CO_2$ is assimilated in the light than in the dark, indicating a light-driven fixation process.

Carbon dioxide also has widespread effects on fungal morphology. For example, in *B. emersonii*, CO_2 stimulates the production of thick-walled, resistant sporangia (Cantino, 1966). In several species such as *Cladosporium wernickii* (Houston et al., 1969) and *Mucor rouxii* (Bartnicki-Garcia and Nickerson, 1962) high CO_2 partial pressures cause the fungus to assume a yeastlike morphology instead of a filamentous one.

SUMMARY

Carbon is an element basic to the structure and function of all cells and is thus crucial for the survival of living organisms. Being heterotrophic, almost all fungi are dependent on their environment for reduced, relatively complex forms of carbon. Being absorbers, fungi must move these complex molecules across the plasma membrane before they can be utilized for growth and development, but the plasma membrane is an effective barrier to the movement of almost all carbon sources. Consequently, several adaptive systems have evolved for taking up nutrients from the environment.

First, fungi secrete extracellular enzymes such as cellulase, amylase, protease, lipase, phenoloxidase, and others which degrade organic macromolecules into subunits of transportable size. Second, fungi have specific uptake systems for moving these subunits across the membrane. Some of these transport systems are constitutive and thus are always present; others are inducible and thus are capable of regulatory control. Uptake may be by simple diffusion, as in the case of certain lipids; but most carbon sources are taken up by either facilitated diffusion or active transport. In the latter process the organism is able to exploit a particular environment quite completely by transporting the carbon source against a gradient and thus concentrate it within the cell.

In this chapter we have mentioned only the general types of carbon sources utilized by fungi: sugars and polysaccharides, lignin, alcohols, fatty acids and lipids of various types, amino acids, and carbon dioxide. The utilization of each of these sources is affected by a host of factors that have been touched on throughout the chapter:

1.   The genotype determines the organism's potential for the carbon sources that can be utilized. The information encoded in the cell's DNA determines which extracellular enzymes and which carrier proteins will be synthesized. In addition, if the genotype permits the induction of these enzymes, the fungus must be able to survive during the lag period necessary for sufficient induction to occur. Lastly, the fungus must possess the enzymes necessary to metabolize the carbon source once it is inside the cell.

2.   The concentration of the carbon source determines the shape of the growth curve and thus whether the fungus will starve, thrive, or be inundated with toxic metabolites.

3.   The presence of other chemicals in the environment affects the utilization of the carbon source. Similar shaped substrates compete for the same carrier; and small ions such as H^+ and K^+ (as well as large molecules) may be required to act as co-substrates, to maintain electrical neutrality, or to activate the carrier.

4.   Physical factors in the environment also influence the rate at which uptake and metabolism of the carbon source occur. Factors such as pH and temperature may directly affect enzyme and carrier activity; other factors such as the oxygen supply may indirectly affect the utilization of the carbon source by influencing the rate of respiration and thus the cell's energy supply.

As we shall see, some of these factors apply equally well to nutrients other than carbon sources.

REFERENCES

Agnihotri, V. P. 1969. Some nutritional and environmental factors affecting growth and production of sclerotia by a strain of *Aspergillus niger. Can. J. Microbiol.* 15: 835–840.

Allerman, K., and O. Sortkjaer. 1973. Rhizomorph formation in fungi. II: The effect of 12 different alcohols on growth and rhizomorph formation in *Armillaria mellea* and *Clitocybe geotropa. Physiol. Plant.* 28: 51–55.

Alonso, A., and A. Kotyk. 1978. Apparent half-lives of sugar transport proteins in *Saccharomyces cerevisiae. Folia Microbiol.* 23: 118–125.

Ander, P., and K. E. Eriksson. 1976. The importance of phenol oxidase activity in lignin degradation by the white-rot fungus *Sporotrichum pulverulentum. Arch. Microbiol.* 109: 1–8.

Bartnicki-Garcia, S., and W. J. Nickerson. 1962. Induction of yeastlike development in *Mucor* by carbon dioxide. *J. Bacteriol.* 84: 829–840.

Bateman, D. F. 1976. Cell wall hydrolysis by pathogens. In J. Friend and D. R. Threlfall (Eds.), *Biochemical Aspects of Plant–Parasite Relationships*, pp. 79–103. New York: Plenum.

Brockerhoff, H., and R. G. Jensen. 1974. *Lipolytic Enzymes.* New York: Academic.

Brody, S., and S. A. Martins. 1979. Circadian rhythms in *Neurospora crassa*: Effects of unsaturated fatty acids. *J. Bacteriol.* 137: 912–915.

Cantino, E. C. 1966. Morphogenesis in aquatic fungi. In G. C. Ainsworth and A. S. Sussman (Eds.), *The Fungi*, Vol. 2, pp.283–337. New York: Academic.

Cantino, E. C., and E. A. Horenstein. 1956. The stimulatory effect of light upon growth and CO_2 fixation in *Blastocladiella*. I: The S.K.I. cycle. *Mycologia* 48: 777–799.

Chander, H., S. S. Sannabhadti, J. Elias, and B. Ranganathan. 1977. Factors affecting lipase production by *Penicillium chrysogenum*. *J. Food Sci.* 42: 1677–1682.

Cirillo, V. P. 1968. Relationship between sugar structure and competition for the sugar transport system in baker's yeast. *J. Bacteriol.* 95: 603–611.

Cochrane, V. W. 1958. *Physiology of Fungi*. New York: Wiley.

Dalphe, Y., and P. Neumann. 1977. L'induction chez *Ceratocystis* de fructifications de types Graphium et Leptographium par des acides insatures. *Can. J. Bot.* 55: 2159–2167.

Das, A., M. Chatterjee, and A. Roy. 1979. Enzymes of some higher fungi. *Mycologia* 71: 530–536.

Das, D. V. M., and G. Weeks. 1979. Effects of polyunsaturated fatty acids on the growth and differentiation of the cellular slime mold, *Dictyostelium discoideum*. *Exp. Cell Res.* 118: 237–243.

Davies, J. S., and W. S. Westlake. 1979. Crude oil utilization by fungi. *Can. J. Microbiol.* 25: 146–156.

Deak, T. 1978. On the existence of H^+-symport in yeasts. *Arch. Microbiol.* 116: 205–212.

Emerson, R., and A. A. Held. 1969. *Aqualinderella fermentans*. gen. et sp. n., a Phycomycete adapted to stagnant water. II: Isolation, cultural characteristics, and gas relations. *Amer. J. Bot.* 56: 1102–1120.

Faro, S. 1971. Utilization of certain amino acids and carbohydrates as carbon sources by *Achlya heterosexualis*. *Mycologia* 63: 1234–1237.

Foster, J. W. 1949. *Chemical Activities of Fungi*. New York: Academic.

Fries, J., and P. Ottolenghi. 1959a. Localisation of invertase in a strain of yeast. *C. R. Trav. Lab. Carlsberg* 31: 259–271.

Fries, J., and P. Ottolenghi. 1959b. Localisation of melibiase in a strain of yeast. *C. R. Trav. Lab. Carlsberg*. 31: 272–281.

Garraway, M. O., and A. R. Weinhold. 1968. Period of access to ethanol in relation to carbon utilization and rhizomorph initiation and growth in *Armillaria mellea*. *Phytopathology* 58: 1190–1191.

Ghosh, A. K., and R. N. Tandon. 1967. Utilization of alcohols by some anthracnose fungi. *Pathol. Microbiol.* 30: 64–70.

Gleason, F. H., C. R. Rudolph, and J. S. Price. 1970a. Growth of certain aquatic Oomycetes on amino acids. I. *Saprolegnia*, *Achlya*, *Leptolegnia* and *Dictyuchus*. *Physiol. Plant.* 23: 513–516.

Gleason, F. H., T. D. Stuart, J. S. Pierce, and E. T. Niebach. 1970b. Growth of certain aquatic Oomycetes on amino acids. II: *Apodachlya*, *Aphanomyces* and *Pythium*. *Physiol. Plant.* 23: 769–774.

Gololobov, A. D., E. R. Davidov, O. B. Razumov, and V. V. Rachinskii. 1969. Effect of fatty acids on the adaptation of yeasts to the hydrocarbon type of nutrition. *Prikl. Biokhim. Mikrobiol.* 5: 241–251.

Gorts, C. P. M. 1969. Effect of glucose on the activity and kinetics of the maltose uptake system of alpha-glucosidase in *Saccharomyces cerevisiae*. *Biochim. Biophys. Acta* 184: 299–305.

Haider, K., and J. Trojanowski. 1975. Decomposition of specifically ^{14}C-labelled phenols and dehydropolymers of coniferyl alcohol as models for lignin degradation by soft and white rot fungi. *Arch. Microbiol.* 105: 33–41.

Hankin, L., and S. L. Anagnostakis. 1975. The use of solid media for detection of enzyme production by fungi. *Mycologia* 67: 597–607.

Harada, T., and T. Hirabayashi. 1968. Utilization of alcohols by *Hansenula miso*. *Agr. Biol. Chem.* 32: 1175–1180.

Hartman, R. E., N. T. Keen, and M. Long. 1972. Carbon dioxide fixation by *Verticillium albo-atrum*. *J. Gen. Microbiol.* 73: 29–34.

Hendrix, J. W. 1970. Sterols in growth and reproduction of fungi. *Ann. Rev. Phytopathology* 8: 111–130.

Hilgenberg, W., and G. Sandmann. 1977. Light-stimulated carbon-dioxide fixation in *Phycomyces blakesleeanus*. *Exp. Mycol.* 1: 265–270.

Hoffman, R. M., and J. R. Raper. 1974. Genetic impairment of energy conservation in development of *Schizophyllum commune*. *J. Gen. Microbiol.* 82: 67–75.

Houston, M. R., K. H. Meyer, N. Thomas, and F. T. Wolf. 1969. Dimorphism in *Cladosporium werneckii*. *Sabouraudia* 7: 195–198.

Hunkova, Z., and Z. Fencl. 1977. Toxic effects of fatty acids on yeast cells: dependence of inhibitory effects on fatty acid concentration. *Biotechnol. Bioeng.* 19: 1623–1641.

Hunkova, Z., and Z. Fencl. 1978. Toxic effects of fatty acids on yeast cells: possible mechanisms of action. *Biotechnol. Bioeng.* 20: 1235–1248.

Iwai, M., Y. Tsujisaka, T. Okamoto, and J. Fukumoto. 1973. Lipid requirement for the lipase production of *Geotrichum candidum*. *Agric. Biol. Chem.* 37: 929–931.

Jennings, D. H. 1974. Sugar transport in fungi: An essay. *Trans. Brit. Mycol. Soc.* 62: 1–24.

Kappeli, O., and A. Fiechter. 1977. Component from the cell surface of the hydrocarbon-utilizing yeast *Candida tropicalis*, induced by hydrocarbon substrate. *J. Bacteriol.* 133: 952–958.

Kappeli, O., M. Muller, and A. Fiechter. 1978. Chemical and structural alterations at the cell surface of *Candida tropicalis*, induced by hydrocarbon substrate. *J. Bacteriol.* 133: 952–958.

Karrick, N. L. 1977. Alteration in petroleum resulting from physicochemical and microbiological factors. In D. C. Mains (Ed.), *Effects of Petroleum on Arctic and Subarctic Marine Environments and Organisms*, pp. 235–279. London: Academic.

Kirk, T. K. 1971. Effects of microorganisms on lignin. *Ann. Rev. Phytopathol.* 9: 185–210.

Kirk, T. K., W. J. Conners, and J. G. Zeikus, 1976. Requirement for a growth substance during lignin decomposition by two wood-rotting fungi. *Appl. Environ. Microbiol.* 32: 192–194.

Kirk, T. K., W. J. Conners, and J. G. Zeikus. 1977. Advances in understanding the microbiological degradation of lignin. In F. A. Loewus and V. C. Runeckles (Eds.), *Recent Advances in Phytochemistry* 11: 369–394.

Kirk, T. K., E. Schultz, W. J. Conners, L. F. Lorenz, and J. G. Zeikus. 1978. Influence of culture parameters on lignin metabolism by *Phanerochaete chrysosporium*. *Arch. Microbiol.* 117: 277–285.

Klug, M. J., and A. J. Markovetz. 1967. Degradation of hydrocarbons by members of the genus *Candida*. II: Oxidation of *n*-alkanes and 1-alkenes by *Candida lipolytica*. *J. Bacteriol.* 93: 1847–1853.

Klug, M. J., and A. J. Markovetz. 1971. Utilization of aliphatic hydrocarbons by microorganisms. *Adv. Microbiol. Physiol.* 4: 1–43.

Kotyk, A. 1967. Properties of the sugar carrier in baker's yeast. II: Specificity of transport. *Folia Microbiol.* 12: 121–131.

Kotyk, A., and C. Haskovec. 1968. Properties of the sugar carrier in baker's yeast. III: Induction of the galactose carrier. *Folia Microbiol.* 13: 12–19.

Kotyk, A., and D. Michaljanicova. 1979. Uptake of trehalose by *Saccharomyces cerevisiae*. *J. Gen. Microbiol.* 110: 323–332.

Lewis, H. L., and G. T. Johnson. 1967. Growth and oxygen-uptake response of *Cunninghamella echinulata* on even-chain fatty acids. *Mycologia* 59: 878–887.

Lilly, V. G., and H. L. Barnett. 1951. *Physiology of the Fungi*. New York: McGraw-Hill.

Lilly, V. G. and H. L. Barnett. 1953. The utilization of sugars by fungi. Agricultural Experiment Station, University of West Virginia, Bulletin 362T.

Macauley, B. J., and D. M. Griffin. 1969. Effect of carbon dioxide and the bicarbonate ion on the growth of some soil fungi. *Trans. Brit. Mycol. Soc.* 53: 223–228.

Martinelli, S. D. 1976. Conidiation of *Aspergillus nidulans* in submerged culture. *Trans. Brit. Mycol. Soc.* 67: 121–128.

Matern, H., and H. Holzer. 1977. Catabolite inactiviation of the galactose uptake system in yeast. *J. Biol. Chem.* 252: 6399–6402.

McIntyre, J. L., and L. Hankin. 1978. An examination of enzyme production by *Phytophthora* spp. on solid and liquid media. *Can. J. Microbiol.* 24: 75–78.

Meredith, S. A., and A. H. Romano. 1977. Uptake and phosphorylation of 2-deoxy-D-glucose by wild type and respiration-deficient baker's yeast. *Biochim. Biophys. Acta* 497: 745–759.

Michaels, G. E., N. L. Wolfe, and D. L. Lewis. 1977. Fermentation inhibition by 2,6-dichloro-4-nitroanaline (DCNA). *J. Agric. Food Chem.* 25: 419–420.

Mirocha, C. J., and J. E. DeVay. 1971. Growth of fungi on an inorganic medium. *Can. J. Microbiol.* 17: 1373–1378.

Miura, Y., M. Okazaki, S. Hamada, S. Murakawa, and R. Yugen. 1977. Assimilation of liquid hydro-carbon by microorganisms. I: Mechanism of hydrocarbon uptake. *Biotechnol. Bioeng.* **19**: 701–714.

Moody, A. R., and A. R. Weinhold. 1972. Fatty acids and naturally occurring plant lipids as stimulants of rhizomorph production in *Armillaria mellea*. *Phytopathology* **62**: 264–267.

Moore, D. 1969. Sources of carbon and energy used by *Coprinus lagopus* sensu Buller. *J. Gen. Microbiol.* **58**: 49–56.

Neihof, R. A., and C. A. Bailey. 1978. Biocidal properties of anti-icing additives for aircraft fuels. *Appl. Environ. Microbiol.* **35**: 698–703.

Okada, H., and H. O. Halvorson. 1964. Uptake of α-thioethyl-D-gluco-pyranoside by *Saccharomyces cerevisiae*. II: General characteristics of an active transport system. *Biochim. Biophys. Acta* **82**: 547–555.

Olutiola, P. O. 1976. Cellulase enzymes in culture filtrates of *Ceratocystis paradoxa*. *Mycologia* **68**: 1083–1092.

Oso, B. A. 1979. Mycelial growth and amylase production by *Talaromyces emersonii*. *Mycologia* **71**: 520–536.

Perlman, D. 1965. The chemical environment for fungal growth. 2: Carbon sources. In G. C. Ainsworth and A. S. Sussman (Eds.), *The Fungi*, Vol. 1, pp. 479–490. New York: Academic.

Reddy, S. M. 1969. Utilization of monosaccharides by five species of *Helminthosporium*. *Path. Microbiol.* **33**: 185–190.

Scurti, J. C., N. Fuissello, and R. Jodice. 1972. (The role of fungi in humic substance formation. III: Utilization of lignin, lignosulfonate, humic and fulvic acids by fungi and phenoloxidase occurrence). *Allionia* (Turin) **18**: 117–128.

Serrano, R. 1977. Energy requirements for maltose uptake in yeast. *Eur. J. Biochem.* **80**: 97–102.

Sestakova, M. 1976. Assimilation spectrum of the yeast *Candida utilis* 49 used for producing Fodder yeast from synthetic ethanol. *Folia Microbiol.* **21**: 384–390.

Sguros, P. L., J. Rodrigues, and J. S. Simms. 1973. Role of marine fungi in the biochemistry of the oceans. V: Patterns of constituitive nutritional growth responses. *Mycologia* **65**: 161–174.

Shannon, D. W. F., and J. M. McNab. 1973. The digestibility of the nitrogen, amino acids, lipid, carbohydrates, ribonucleic acid and phosphorus of *n*-paraffin grown yeast when given to colostomised laying hens. *J. Sci. Food Agric.* **24**: 27–34.

Simonson, L. G., and A. E. Liberta. 1975. New sources of fungal dextranase. *Mycologia* **65**: 845–851.

Sopko, R. 1967a. Ligninolytic activity of the white-rot fungus *Pleurotus ostreatus*. I: The influence of some nutrient solutions on the decomposition of lignin. *Drev. Vysk.* **2**: 81–91.

Sopko, R. 1967b. Ligninolytic activity of the white-rot fungus *Pleurotus ostreatus*. II: The effect of temperature on the decomposition of lignin. *Drev. Vysk.* **3**: 121–130.

Sortkjaer, O., and K. Allerman. 1973. Rhizomorph formation in fungi. III: The effect of ethanol on the synthesis of DNA and RNA and the uptake of asparagine and phosphate in *Armillaria mellea*. *Physiol. Plant.* **29**: 129–133.

Steele, S. D. 1973. Self-inhibition in *Geotrichum candidum*. *Can. J. Microbiol.* **19**: 943–947.

Tabak, H. H., and W. M. Bridge-Cooke. 1968. The effect of gaseous environments on the growth and metabolism of fungi. *Bot. Rev.* **34**: 126–252.

Van Steveninck, J. 1972. Transport and transport-associated phosphorylation of galactose in *Saccharomyces cerevisiae*. *Biochim. Biophys. Acta* **274**: 575–583.

Van Steveninck, J., and E. C. Dawson. 1968. Active and passive galactose transport in yeast. *Biochim. Biophys. Acta* **150**: 47–55.

Vaskivnyuk, V. T., E. I. Kvasnikov, T. P. Ostapchenko, T. A. Grinberg, and V. Y. Masumyan. 1974. (Effect of oleic acid on the growth and carotene formation of *Rhodotorula gracilis* K 1/48 on hydrocarbon containing media). *Prikl. Biokhim. Mikrobiol.* **10**: 563–566.

Wardle, K. S., and L. C. Schisler. 1969. The effects of various lipids on growth of mycelium of *Agaricus bisporus*. *Mycologia* **61**: 305–314.

Weinhold, A. R. 1963. Rhizomorph production by *Armillaria mellea* induced by ethanol and related compounds. *Science* **142**: 1065–1066.

Weinhold, A. R., and M. O. Garraway. 1966. Nitrogen and carbon nutrition of *Armillaria mellea* in relation to growth promoting effects of ethanol. *Phytopathology* **56**: 108–112.

Yates, A. R., A. Seaman, and M. Woodbine. 1967. Growth of *Byssochlamys nivea* in various carbon dioxide atmospheres. *Can. J. Microbiol.* **13**: 1120–1123.

4
NITROGEN NUTRITION

A source of nitrogen is indispensible for fungal growth and development. Fungi require nitrogen for the synthesis of a variety of critically important cellular constituents, including amino acids and proteins; purines, pyrimidines, and nucleic acids; glucosamine and chitin; and various vitamins. Most fungi are able to utilize simple inorganic nitrogen sources as well as organic sources such as amino acids. In the cell, amino acids usually serve as the nitrogen donors for synthesizing more complex molecules. Thus, the particular nitrogen source in the culture medium often must first be converted intracellularly to an amino acid before it is actually utilized by the organism. In this chapter we discuss the principal types of nitrogen sources used by fungi and the factors affecting their utilization. Nitrogen metabolism in filamentous fungi has been reviewed by Pateman and Kinghorn (1976).

NUTRITIONAL CLASSES OF FUNGI
BASED ON NITROGEN UTILIZATION

Fungi have been divided into three and possibly four nutritional classes depending on their ability to utilize molecular nitrogen (N_2), nitrate, nitrite, ammonium, and organic nitrogen compounds (Table 4.1). There is no unequivocal evidence that fungi can fix molecular nitrogen. Early reports suggesting that fungi possess this capability were later found to be erroneous due to bacterial contamination. On the other hand, attempts have been made to transfer nitrogen-fixing ability to a fungus via a symbiotic relationship with bacteria. For example, Giles and Whitehead (1975, 1976) induced the uptake of the nitrogen-fixer *Azotobacter vinelandii* into protoplasts of *Rhizopogon*, a mycorrhizal fungus. This fungal–bacterial association contained nitrogenase, the principal enzyme involved in nitrogen fixation, and was able to grow on a culture medium exposed to N_2 in the air but lacking any other nitrogen source. However, nitrogenase activity decreased over a period of months, which indicates that nitrogen fixation was not possible over the long term.

A survey of reports of nitrogen nutrition suggests that the majority of fungi can

Table 4.1 Classification of Fungi Based on the Type of Nitrogen Source Utilized

Class	Nitrogen Source				
	N_2	NO_3^-	NO_2^-	NH_4^+	Organic-N
1	?	+	+	+	+
2		+	+	+	+
3				+	+
4					+

Source: Lilly (1965).

grow efficiently when supplied with either an inorganic or an organic source of nitrogen and thus belong to Class 2 (Table 4.1). However, some types of nitrogen compounds within this class may be used more effectively than others by a particular species. Notice in Table 4.2 that the growth of isolate 2 of *Alternaria tenuis* is poorest on ammonium compounds and best on amino acids. Nitrate and nitrite produce intermediate growth, but nitrate (the calcium salt) results in the greatest sporulation. Variations in patterns of utilization may be present even among different isolates of the same species. Isolate 2 is the only one of the three tested that can

Table 4.2 Average Dry Weight and Sporulation of Three Isolates of *Alternaria tenuis* on Different Sources of Nitrogen

Nitrogen Source	Isolate 1		Isolate 2		Isolate 3	
	Dry Weight (mg)	Sporu- lation	Dry Weight (mg)	Sporu- lation	Dry Weight (mg)	Sporu- lation
None	00.0	−	00.0	−	00.0	−
$NaNO_2$	00.0	−	56.6	+	00.0	−
KNO_3	78.5	+	69.6	+	70.5	+ + +
$NaNO_3$	62.6	+ +	67.0	+ +	74.0	+ +
$Ca(NO_3)_2$	97.5	+ + +	81.3	+ + + +	85.0	+ +
NH_4NO_3	45.0	+	42.6	+	35.0	+
$(NH_4)_2SO_4$	51.6	−	50.3	+	29.3	+
NH_4Cl	55.0	+	44.0	+	28.3	+ +
Glycine	67.3	+	79.0	+	74.5	+
DL-Valine	108.6	+	102.0	+	96.0	−
L-Glutamic acid	93.9	+ + + +	99.5	+	97.3	+
L-Asparagine	84.0	+ +	72.3	+	60.5	+ +
Peptone	105.5	+	91.6	+ +	88.0	+ +
Urea	66.3	+	39.0	+	62.0	+ + +
Thiourea	00.0	−	00.0	−	00.0	−

Source: Singh and Tandon (1970).

utilize nitrite. The best nitrogen source for growth of all three isolates is DL-valine, but in isolate 1 the greatest sporulation is achieved with L-glutamic acid and in isolate 3 with KNO$_3$ and urea.

A few known fungi are capable of using ammonium and organic nitrogen but not nitrate and thus belong to Class 3. The *Rhizoctonia* endophytes of the orchid *Arundia chinensis* belong to this class, as well as *Linderina pennispora* and *L. macrospora* (Stephen and Chan, 1970; Stephen and Fung, 1971). However, the latter two organisms grow much more poorly on ammonium than on amino acids.

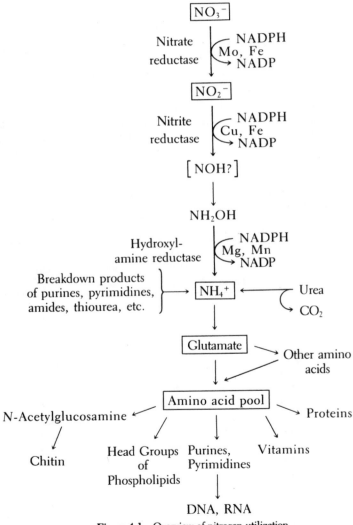

Figure 4.1 Overview of nitrogen utilization.

Apparently, only a very few species can utilize only organic nitrogen. One of these Class 4 species, *Teighemiomyces parasiticus*, has an absolute requirement for L-cysteine and a near total requirement for L-valine and L-leucine (Binder and Barnett, 1974). Other fungi with a requirement for one or more amino acids include *Blastocladiella emersonii*, *Catenaria anguillulae*, and *Leptomitus lacteus* (Whitaker, 1976).

Although fungi as a group can utilize many different compounds as nitrogen sources, most of the experimental work has been done with nitrate, nitrite, ammonium, amino acids, and urea. These compounds are interconvertible and are utilized by fungi via the generalized pathway illustrated in Figure 4.1.

NITRATE UTILIZATION

Most fungi can utilize nitrate. The only known metabolic route for the assimilation of nitrate nitrogen is by reduction to nitrite via the enzyme nitrate reductase and then to ammonia (Pateman and Kinghorn, 1976). Nitrate reductase is an enzyme that catalyzes the transfer of electrons from NADPH to nitrate and requires molybdenum and iron as cofactors (Figure 4.1). Certain groups of fungi such as the higher Basidiomycetes, the Saprolegniaceae, and the Blastocladiales are unable to use nitrate. This is thought to be due to their inability to synthesize nitrate reductase (Whitaker, 1976).

219824

Nitrate uptake studies have been done using various species such as *Saccharomyces cerevisiae*, *Aspergillus nidulans*, *Neurospora crassa*, and *Penicillium chrysogenum*. Although there is some confusion in the literature about the mechanism of nitrate uptake, the strongest evidence indicates that uptake is directly linked to nitrate reductase activity. Pateman and Kinghorn (1976) compared the uptake of ^{15}N-nitrate by a normal strain of *A. nidulans* with uptake by a mutant strain that lacked nitrate reductase activity. Both strains were grown on a urea medium, transferred to media containing ^{15}N-nitrate, incubated for various periods of time, and the cells assayed for the percentage enrichment of ^{15}N. The normal strain shows a 40-minute lag before nitrate is detected within the cell (Figure 4.2). This lag is roughly coincidental with the appearance of nitrite reductase activity and is characteristic of an inducible enzyme system. When the fungus is transferred from the urea medium to one containing nitrate plus ammonium, the normal strain fails to take up nitrate and also lacks nitrate reductase activity. Thus, ammonium is an effective repressor of nitrate reductase. On the other hand, the mutant strain lacking nitrate reductase activity fails to take up nitrate regardless of the ammonium concentration.

On the basis of these data Pateman and Kinghorn conclude that there is no indication of a nitrate transport system independent of nitrate reduction. They hypothesize that nitrate entering the cell binds immediately to nitrate reductase in a manner similar to that observed for the binding of a substrate to a carrier protein. In this manner, nitrate enters the cell by moving down the nitrate concentration gradient created by the action of nitrate reductase. Thus, nitrate may enter via simple diffusion or facilitated diffusion linked to nitrate reductase (Pateman and Kinghorn, 1976).

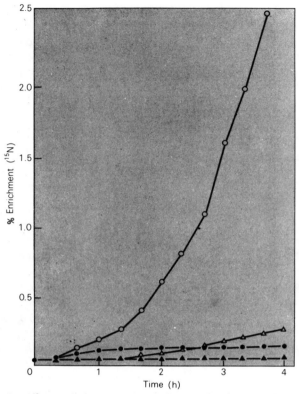

Figure 4.2 Uptake of ¹⁵N-enriched potassium nitrate by normal and nitrate reductase deficient cells of *Aspergillus nidulans*. (○) Normal cells on nitrate medium. (△) Normal cells on nitrate plus ammonium. (●) Nitrate reductase deficient cells on nitrate. (▲) Nitrate reductase deficient cells on nitrate plus ammonium. (Pateman and Kinghorn, 1976)

If nitrate reductase activity is indeed associated with nitrate uptake, then this enzyme should be localized at or near the plasma membrane. Immunohistochemical studies by Roldan and co-workers (1982) have shown this to be the case. Nitrate reductase is localized in the cell wall–plasmalemma region and in the vacuolar membranes as well.

NITRITE UTILIZATION

Although nitrite is utilized as a nitrogen source by some fungi it is toxic to many species. Figure 4.3 illustrates the inhibitory effect of high nitrite concentrations on growth of *Neurospora crassa*. Nitrite toxicity may be due to its ability to deaminate amino acids and to its interference with sulfur metabolism because of its similarity to the sulfite ion (Pateman and Kinghorn, 1976). Poor growth on nitrite may be

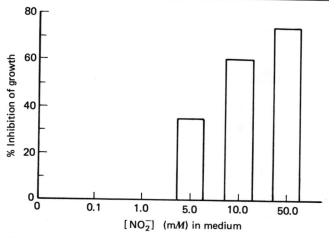

Figure 4.3 Inhibition of growth of *Neurospora crassa* by nitrite. (From Nicholas, 1965.)

related in part to an effect of pH. For example, growth of *Aspergillus niger* on nitrite is accelerated by adjusting the pH of the growth medium from 5.6 to 7.5 (Agnihotri, 1968).

Nitrite is converted to ammonium in a series of steps that are not clearly understood (Figure 4.1). Nitrite first reacts with nitrite reductase, an enzyme containing copper and iron as cofactors that catalyzes the transfer of electrons from NADPH to nitrite. The product of this reaction is not known although nitroxyl (NOH), hyponitrite (N_2O_2), and hydroxylamine (NH_2OH) have been implicated. Eventually hydroxylamine is formed and is acted upon by hydroxylamine reductase (an enzyme closely associated with nitrite reductase) to form ammonium (Pateman and Kinghorn, 1976).

In a set of experiments similar to those with nitrate, Pateman and Kinghorn (1976) measured the uptake of ^{15}N-nitrite using normal and nitrite reductase-deficient strains of *A. nidulans*. The two strains were grown on urea and then incubated with labeled nitrite with and without ammonium chloride. Nitrite uptake was invariably associated with nitrite reductase activity and occurred only when the normal strain was incubated on a medium with nitrite and not in significant amounts when ammonium was present. Also, the mutant lacking nitrite reductase failed to take up significant amounts of nitrite when incubated for up to 4 hours on a medium with nitrite alone or with nitrite plus ammonium. Thus, in a manner reminiscent of nitrate uptake, nitrite transport is closely associated with nitrite reductase activity, and ammonium represses enzyme synthesis.

AMMONIUM UTILIZATION

The majority of fungi can use ammonium as the sole source of nitrogen. In addition to its effect on growth, ammonium ion has also been implicated as a regulator of several developmental systems in fungi. In *Calonectria camelliae* the presence of ammonium in the medium is essential for the formation of fertile perithecia; infertile perithecia are formed in the presence of nitrate or nitrite (Table 4.3). This process appears to be mediated by calcium. With NH_4NO_3 and $(NH_4)_2PO_4$, increasing concentrations of calcium carbonate result in decreased perithecia production, but with NH_4Cl and $NaNO_3$ calcium carbonate is stimulatory (Shipton, 1977).

In *Saccharomyces cerevisiae*, ascus formation is completely inhibited by relatively low levels (2–4 m*M*) of ammonium, either in the form of ammonium sulfate or the nonmetabolizable analog methylammonia (CH_3NH_2) (Table 4.4). Apart from glutamine, which may serve as an intracellular source of ammonium, no other nitrogen sources are inhibitory. The principal site of ammonium action in this species is the DNA. Ammonium inhibits meiosis and pre-meiotic DNA synthesis, and continued incubation of the cells with NH_4^+ causes massive DNA degradation (Pinon, 1977). In addition, ammonium inhibits both protein synthesis and protein degradation in this organism at the time of ascus formation (Croes et al., 1978).

Ammonium ion is also produced by amoebae of *Dictyostelium discoideum* before and during aggregation and stimulates the formation of a maturing slug instead of a fruiting body (Schindler and Sussman, 1977). There is evidence that ammonium-induced slug formation is caused by ammonium inhibition of cyclic-AMP synthesis and/or release (Sussman and Schindler, 1978).

Table 4.3   The Development and Fertility of Perithecia of *Calonectria camelliae* in the Presence of Different Nitrogen Sources and Different Concentrations of Calcium Carbonate.

Treatment	pH value		Perithecia per 100 mm²	Fertility of Perithecia[a]
	Initial	Final		
NH_4Cl	6.7	6.6	3.1	+
NH_4Cl + $CaCO_3$, 0.5 g/liter	6.8	6.4	3.7	+
NH_4Cl + $CaCO_3$, 1.0 g/liter	6.9	6.6	6.9	+
NH_4NO_3	6.7	7.3	8.2	+
NH_4NO_3 + $CaCO_3$, 0.5 g/liter	6.6	7.3	7.0	+
NH_4NO_3 + $CaCO_3$, 1.0 g/liter	6.7	7.3	4.6	+
NH_4NO_3, Ca^{2+} not added	6.6	7.2	7.2	+
$(NH_4)_2PO_4$	6.4	7.2	5.4	+
$(NH_4)_2PO_4$ + $CaCO_3$, 0.5 g/liter	6.6	7.2	4.3	+
$(NH_4)_2PO_4$ + $CaCO_3$, 1.0 g/liter	6.4	7.2	2.3	+
$NaNO_3$	6.7	7.9	1.6	−
$NaNO_3$ + $CaCO_3$, 0.5 g/liter	6.8	7.7	1.8	−
$NaNO_3$ + $CaCO_3$, 1.0 g/liter	6.7	7.7	2.5	−
$NaNO_2$	6.6	8.3	0.3	−
$NaNO_2$ + $CaCO_3$, 0.5 g/liter	6.6	8.3	0.4	−
$NaNO_2$ + $CaCO_3$, 1.0 g/liter	6.7	8.4	0.4	−

Source: Shipton (1977).

[a]Fertility assessed at harvest and again 4 weeks later.

Table 4.4   Effect of Nitrogen Sources on Sporulation of
Strain 131 of *Saccharomyces cerevisiae*

Nitrogen Source	Sporulation at 24 h (%)
No Nitrogen	75
1 μM(NH$_4$)$_2$SO$_4$	76
10 μM (NH$_4$)$_2$SO$_4$	68
0.1 mM (NH$_4$)$_2$SO$_4$	67
1.0 mM (NH$_4$)$_2$SO$_4$	48
2.0 mM (NH$_4$)$_2$SO$_4$	0 <0.001
5.0 mM NH$_4$Cl	0 <0.001
2.0 mM NaNO$_2$	72
10.0 mM NaNO$_2$	72
10.0 mM NaNO$_3$	69
1 mM CH$_3$NH$_2$	30
2 mM CH$_3$NH$_2$	0
4 mM glutamine	0.25
8 mM glutamine	0
4 mM glutamate	79
8 mM glutamate	76
6 mM glutamate + 6 mM (NH$_4$)$_2$SO$_4$	0.2
8 mM adenine	73
8 mM uracil	69
4 mM histidine	61
4 mM arginine	50
8 mM arginine	32
6 mM urea	60

Source: Adapted from Pinon (1977).

The utilization of ammonia by most fungi is accompanied by a drop in pH of the culture medium. As illustrated in Table 4.5, the pH of the medium for *Scopulariopsis brevicaulis* decreases to as low as 2.7 when ammonium is present compared with pH 8.0–8.1 when KNO$_3$ is the nitrogen source. Note that this low pH is associated with minimal nitrogen assimilation and poor growth, but the addition of organic acids such as tartrate and malate lessens the pH drop and permits good growth of this species. A wide variety of ammonium compounds including ammonium sulfate, ammonium chloride, ammonium phosphate, and ammonium tartrate support perithecial production by *Venturia inaequalis* but only when calcium carbonate is added to the medium to keep the pH from dropping. No such pH decreases are observed when ammonium carbonate or ammonium oxalate is used (Ross and Bremner, 1971).

In many fungi ammonium is taken up preferentially to nitrate. Indirect evidence for this comes from pH studies (Table 4.5) where the final pH of an *S. brevicaulis* growth medium is just as low with NH$_4$NO$_3$ as with (NH$_4$)$_2$SO$_4$. A direct assay of

Table 4.5 Growth of *Scopulariopsis brevicaulis* on Different Nitrogen Sources with Glucose as the Carbon Source

Parameter	Nitrogen Source				
	KNO₃	NH₄NO₃	(NH₄)₂SO₄	(NH₄)₂SO₄ + Tartrate	(NH₄)₂SO₄ + Malate
Mean dry matter (mg/flask)	200	18	12	160	225
Nitrogen assimilated (% of initial)	100	15–20	15–20	100	100
Range of final pH	8.0–8.1	3.0–3.9	2.7–3.8	3.4–4.4	7.5–8.6

Source: Morton and MacMillan (1954).

ammonium and nitrate in the medium (Figure 4.4) indicates that the addition of ammonia to a nitrate medium causes the immediate cessation of nitrate uptake, but nitrate has no effect on ammonium uptake. This is understandable from what we know about the repression of nitrate reductase by ammonium.

Studies of ammonium uptake have been facilitated by the use of ¹⁵N-ammonium and more commonly the ¹⁴C-labeled analog methylammonia. Evidently these two compounds are sufficiently similar so that they are transported by the same uptake system. Roon and co-workers (1975a) demonstrated that ammonium is a strong competitive inhibitor of methylammonia transport and that mutants of *Saccharomyces cerevisiae* selected on the basis of their reduced ability to transport methylammonia simultaneously exhibit a decreased ability to transport ammonium.

In *S. cerevisiae* methylammonia uptake is pH and temperature dependent and shows characteristics of an active transport system. This transport system exhibits

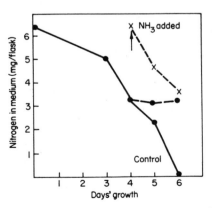 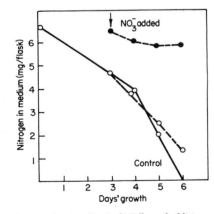

Figure 4.4 Uptake of nitrate and ammonium by *Scopulariopsis brevicaulis*. (Left) Effect of addition of ammonia on the uptake of nitrate from the medium. —— Culture in nitrate only; ––– Culture in nitrate and ammonia; ● nitrate N; × ammonia N. (Right) Effect of addition of nitrate on the uptake of ammonia from the medium. —— Culture in ammonia; ––– Culture in ammonia and nitrate; ● nitrate N; ⊙ ammonia N. (From Morton and MacMillan, 1954.)

maximal activity in ammonium-grown cells and is repressed 60–70% when gluta-mine or asparagine is added to the growth medium. Uptake of methylammonia is inhibited noncompetitively by L-amino acids, and inhibition persists until the exter-nal amino acid supply is depleted (Roon et al., 1975a,b).

The relationships between methylammonia and ammonium transport in filamen-tous fungi are similar to those in yeasts. An active transport system for ammonium and methylammonia in *Aspergillus nidulans* has been characterized using both ^{15}N-ammonium and ^{14}C-methylammonia. The system has a K_m of less than 50 μM for ammonia and 20 μM for methylammonia, and it can accumulate methylammonia against a concentration gradient. A number of mutations result in low transport rates and altered K_m values for both substrates (Pateman et al., 1974; Cook and Anthony, 1978).

One complicating factor in the elucidation of ammonium uptake mechanisms was the discovery of mutants of both A. *nidulans* (Arst and Cove, 1969) and S. *cerevisiae* (Middlehoven, 1977) that can transport ammonium as effectively as the wild type but are unable to transport methylammonia. Thus, further study of the mechanisms of ammonium transport are needed.

Once inside the cell, ammonium can be incorporated into amino acids by either of two basic reactions. By far the most common is reductive amination involving a reaction between ammonium and α-ketoglutarate catalyzed by glutamate dehydro-genase:

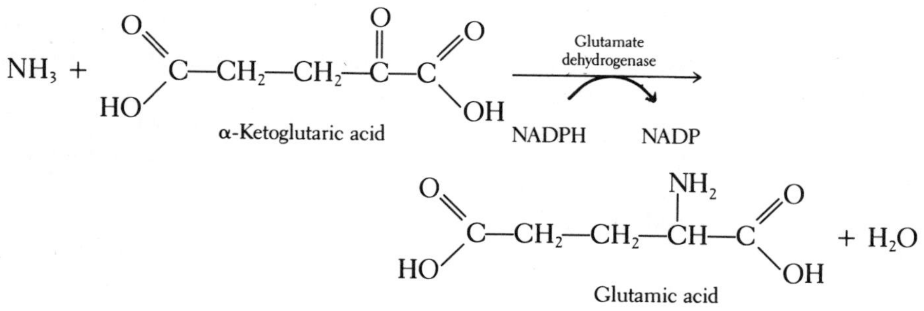

The glutamate formed in this reaction can then react with another ammonium ion in the presence of glutamine synthetase to yield glutamine:

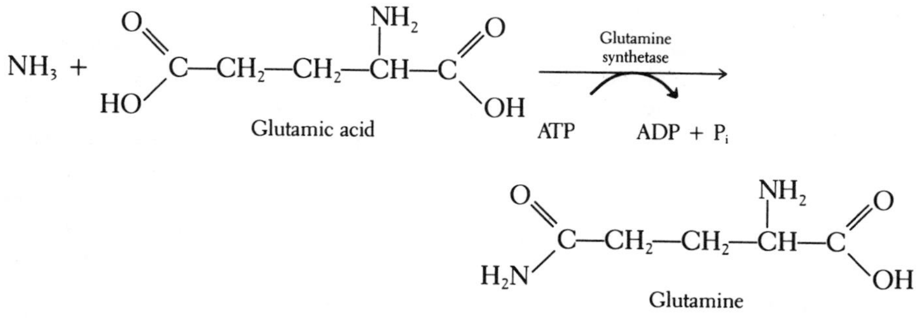

In addition, glutamate can act as the amino donor in the synthesis of most of the other cellular amino acids in a transamination reaction. In this process glutamate (or sometimes aspartate) transfers its amino group to an α-keto acid:

$$\underset{\text{Glutamic acid}}{\underset{\text{HO}}{\overset{\text{O}}{\diagup}}\hspace{-0.3em}\text{C}\!-\!\text{CH}_2\!-\!\text{CH}_2\!-\!\underset{}{\overset{\text{NH}_2}{\text{CH}}}\!-\!\text{C}\!\underset{\text{OH}}{\overset{\text{O}}{\diagup}}} + \underset{\alpha\text{-Keto acid}}{\text{R}\!-\!\overset{\text{O}}{\text{C}}\!-\!\text{C}\!\underset{\text{OH}}{\overset{\text{O}}{\diagup}}} \xrightarrow{\text{Transaminase}}$$

$$\underset{\alpha\text{-Ketoglutaric acid}}{\underset{\text{HO}}{\overset{\text{O}}{\diagup}}\hspace{-0.3em}\text{C}\!-\!\text{CH}_2\!-\!\text{CH}_2\!-\!\overset{\text{O}}{\text{C}}\!-\!\text{C}\!\underset{\text{OH}}{\overset{\text{O}}{\diagup}}} + \underset{\text{Amino acid}}{\text{R}\!-\!\underset{}{\overset{\text{NH}_2}{\text{CH}}}\!-\!\text{C}\!\underset{\text{OH}}{\overset{\text{O}}{\diagup}}}$$

The carbon skeletons for most of the amino acids are siphoned off from glycolytic and Krebs Cycle intermediates. Other amino acids are synthesized directly from glutamate or aspartate. Once amino acids are formed they may be used in the synthesis of a wide variety of molecules, such as proteins, vitamins, nucleic acids, chitin, and certain phospholipids (Figure 4.1).

Although a discussion of the pathways of amino acid biosynthesis is beyond the scope of this book, we will briefly mention one metabolic route that is of evolutionary interest. The amino acid lysine cannot be synthesized by animals, but among other organisms two different biosynthetic pathways exist. In the Oomycetes (as well as bacteria, various algae, and higher plants), lysine is synthesized from a condensation of pyruvate and aspartate semialdehyde. This pathway is the diaminopimelic (DAP) pathway, named after one of the intermediates. In the rest of the fungi (and the euglenids) lysine is synthesized from α-ketoglutarate and acetylCoA via the α-aminoadipic acid (AAA) pathway. These two pathways have no enzymes or intermediates in common, and thus they probably evolved independently of each other. Because the Oomycetes are the only fungal group to have the DAP pathway, they probably had a phylogenetic origin that is independent of the rest of the fungi. It has been hypothesized that the presence of cellulose instead of chitin in the cell walls of Oomycetes may be related to the presence of the DAP pathway. There is evidence that the DAP path evolved before the AAA path and that cellulosic walls evolved before chitinous walls. One hypothesis suggests that as chitinous walls were selected for, the DAP path was selected against because it interfered in some way with chitin biosynthesis. Thus, natural selection would favor the evolution of a new lysine pathway that was compatible with cell walls of chitin (Vogel, 1965; LeJohn, 1971).

In contrast to the pathway in Figure 4.1 for the reduction of nitrate to organic nitrogen, some fungi are able to carry out the reverse reactions: to oxidize organic nitrogen to nitrate. This process, once thought to be carried out principally by bacteria such as *Nitrosomonas* and *Nitrobacter,* has been found to occur in soil-borne fungi including several *Aspergillus* species and a species of *Penicillium* and *Cephalosporium.* However, the rate of oxidation by these fungi is considerably less than in bacteria. Peptone and casein are the best nitrogen sources for oxidation (Eylar and Schmidt, 1959). In cell-free extracts of A. *wentii*, even inorganic sources such as ammonia, hydroxylamine, and nitrite can be oxidized to nitrate (Alleem et al., 1964). These oxidation steps release energy that can be used for cellular activities.

UREA UTILIZATION

Urea is produced by fungi upon degradation of such compounds as purines, pyrimidines, and arginine. The urea thus formed can be used as a nitrogen source for growth as can urea supplied in the medium. The use of urea as a nitrogen source for growth of fungi involves the prior intracellular conversion to ammonium (Pateman and Kinghorn, 1976). Growth of *Saccharomyces cerevisiae* on a medium containing urea is comparable to that on a medium containing ammonium (Sumrada et al., 1976). Urea also supports good growth, good conidial production, and moderate sclerotial development by *Aspergillus niger* (Agnihotri, 1968) and good growth and sporulation of isolates of *Alternaria tenuis* (Table 4.2). However, some fungi grow poorly on the concentrations of urea used in synthetic media. For example, mycelial growth of *Botrytis convoluta* on a medium containing urea is less than 10% of that on a medium containing casein hydrolysate (Maas and Powelson, 1972).

Uptake of ^{14}C-urea has been demonstrated using both yeasts and filamentous fungi. In S. *cerevisiae*, two urea transport systems are present (Cooper and Sumrada, 1975; Sumrada et al., 1976). The first is an active transport system that is effective at low urea concentrations. This system is induced by urea and repressed by good nitrogen sources such as asparagine and glutamine. Urea can also be taken up by a facilitated diffusion system that is effective only at high urea concentrations (above 500 μM). The carrier for this system is neither inducible nor repressible. Urea is also taken up via active transport by *Geotrichum candidum* (Shorer et al., 1972) and A. *nidulans* (Dunn and Pateman, 1972). In the latter species the same carrier transports both urea and its toxic analog thiourea.

AMINO ACID UTILIZATION

Most fungi can utilize amino acids as sources of nitrogen, but not all are used equally well. In Figure 4.5 are the results obtained from incubating *Bipolaris maydis* race T on a basal glucose or glucose-xylose medium containing different nitrogen sources. Notice that there is considerable variation in the extent to which amino acids can be utilized for growth (Figure 4.5a) and sporulation (Figure 4.5b). In general there is a positive correlation between mycelial dry weight and sporulation per unit weight. For example, tryptophan, leucine, and methionine are poor nitrogen sources for growth as well as for sporulation; serine, γ-aminobutyrate, and glutamate are effective nitrogen sources for both processes. This generalization is not always true, however. Ornithine and arginine are good sources for growth but poor

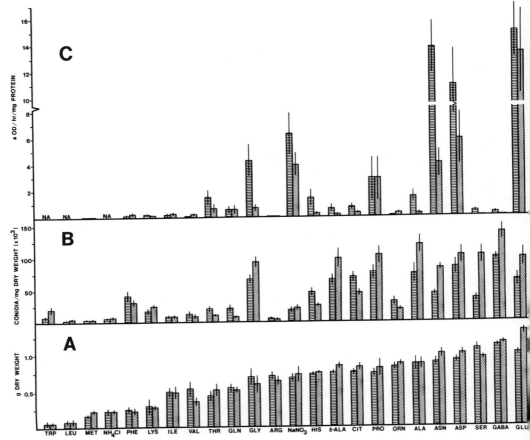

Figure 4.5   (A) Growth, (B) sporulation, (C) polyphenoloxidase activity in *Bipolaris maydis* incubated 6 days on a basal glucose medium containing different nitrogen sources at 1 g N/liter with (right columns) or without (left columns) a supplement of 2 g/liter xylose. (Evans and Black, 1981)

sources for sporulation. In general, amino acids with acidic side chains are the best sources for both growth and sporulation; those with uncharged polar side chains are moderately good, and those with nonpolar side chains larger than a methyl group are the poorest. The addition of xylose to the media has various effects depending on the nitrogen source present. Xylose stimulates growth slightly when β-alanine, citrulline, asparagine, or glutamate is the nitrogen source and inhibits growth when added with valine or serine. On the other hand, xylose stimulates sporulation in the presence of sources such as tryptophan, glycine, alanine, and asparagine but inhibits sporulation with sources such as glutamine, threonine, and others. With still other nitrogen sources, xylose has no effect on either growth or sporulation.

In addition, the effect of nitrogen sources on the activity of polyphenoloxidase (PPO) in *B. maydis* was measured to see if there was a correlation between PPO activity and sporulation. As indicated in Figure 4.5c the activity of PPO varies considerably depending on the nitrogen compound. Cultures grown on glucose media have particularly high PPO activity in the presence of glutamate, asparagine, or aspartate; with most of the other compounds, however, enzyme activity is quite low. It appears that there is little correlation between sporulation and PPO activity. Notice that low PPO activity is associated with high sporulation with GABA and serine, but there is high PPO activity and high sporulation with glutamate, aspartate, and asparagine. Clearly the type of nitrogen source strongly affects the physiology of this species, and those sources which are optimal for mycelial growth may be different from those which are optimal for differentiation. This generalization has been borne out for a wide variety of species (Evans and Black, 1981).

Effective utilization of amino acids by fungi depends on the amount and type of carbon source. In the presence of certain plant growth hormones, *Armillaria mellea* will utilize L-asparagine and several L-amino acids more effectively for growth when ethanol is the carbon source compared with glucose. Figure 4.6 illustrates that ethanol stimulates rhizomorph formation by this species when L-alanine, L-asparagine, or casein hydrolysate is the nitrogen source but not in the presence of glycine, phenylalanine, or methionine. Evidently the carbon source promotes growth, in part, by influencing the capacity of the fungus to utilize the available nitrogen source

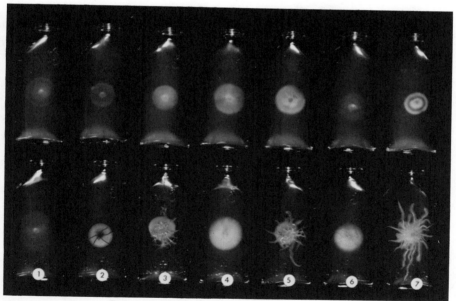

Figure 4.6 Growth and rhizomorph production by *Armillaria mellea* on a solid medium with several amino acids and casein hydrolysate as sources of nitrogen in the presence of 500 ppm ethanol (bottom row) and without ethanol (top row). (1) Control, no nitrogen. (2) Glycine. (3) L-Alanine. (4) DL-Phenylalanine. (5) L-Asparagine. (6) DL-Methionine. (7) Casein hydrolysate. (Weinhold and Garraway, 1966)

(Weinhold and Garraway, 1966). Along these same lines, Aube and Gagnon (1969) tested the response of three isolates of *Trichoderma viride* to different carbon and nitrogen concentrations using glucose and L-asparagine, respectively (Table 4.6). With 6 g carbon per liter a threefold increase in the L-asparagine concentration from 93 to 280 mg nitrogen per liter produced no significant increase in mycelial growth. In contrast, with 10 or 14 g carbon per liter, increasing the L-asparagine concentration causes a dramatic increase in growth. Results such as these have focused attention on the importance of the carbon-to-nitrogen ratio in the culture medium in controlling physiological processes.

Amino Acid Uptake

Amino acids constitute one of the largest and most diverse groups of substances transported across the fungal membrane. Thus it is not surprising that a variety of different transport systems have evolved for the uptake of these important nitrogen sources. Reviews of work in the field of amino acid uptake and accumulation point overwhelmingly to active transport as the major mechanism involved (Pateman and Kinghorn, 1976; Whitaker, 1976). In most natural environments, nutrients that include amino acids are present in relatively low concentrations. Therefore, a process by which these nutrients can be accumulated against a concentration gradient is of obvious importance. Indeed, it has been suggested that the appearance of transport systems for this purpose may have occurred very early in the evolution of fungi and other microorganisms (Pardee and Palmer, 1973).

Whitaker (1976) has described four categories of transport systems in fungi: (1) a single system of broad specificity, (2) systems of broad but differing specificities, (3) systems specific for acidic, neutral, or basic amino acids, and (4) specific systems for

Table 4.6 Mean Dry Weight of Two Isolates (1500, 1503) of *Trichoderma viride* in a Basal Medium Containing Different Concentrations of Carbon (as Glucose) and Nitrogen (as L-Asparagine)

Amount of Carbon and Nitrogen per Liter	Dry Weight (mg/isolate no.)	
	1500	1503
6 g C + 93 mg N	273	204
6 g C + 280 mg N	312	204
10 g C + 93 mg N	263	265
10 g C + 280 mg N	510[a]	322[a]
14 g C + 93 mg N	286	247
14 g C + 280 mg N	487[a]	361[a]

Source: Adapted from Aube and Gagnon (1969).

[a]Values showed a significant increase in growth, at the 1% level, in response to increase in nitrogen concentration.

single amino acids. These general categories are based mostly on work with yeasts, *Neurospora crassa*, *Penicillium chrysogenum*, and *Aspergillus nidulans*, hence, our discussion will emphasize these organisms. The designation of transport systems into categories is in part a matter of interpretation because of differences in technique among investigators, the species used, and the stage of development of the organism.

SINGLE SYSTEM OF BROAD SPECIFICITY

One of the amino acid transport systems in *Saccharomyces cerevisiae* is a general amino acid "permease" (Grenson et al., 1970). This carrier system has a maximal level of activity that is fivefold to fiftyfold higher than that found with any of the specific amino acid transport systems (Larimore and Roon, 1978). Although this general uptake system was first thought to transport only neutral and basic L-amino acids, subsequent studies indicated that it transports acidic (Darte and Grenson, 1975) and D-amino acids (Rytka, 1975) as well. Proline, which is more precisely an imino acid, is not transported by this system.

Evidence for the existence of this transport system came initially from genetic studies in which a mutant strain was obtained that was deficient in the general amino acid permease and that showed a reduced capability to take up a wide variety of amino acids. In addition, Lineweaver–Burke plots of uptake experiments when pairs of amino acids were present in the growth medium indicated that these amino acids all compete for the same carrier. The addition of ammonium causes a decrease in the amino acid uptake rate, but there is no evidence that this inhibition is due to a repression of the permease. It appears, therefore, that the general transport system is constitutive, with ammonium probably affecting the carrier directly in an allosteric manner (Grenson et al., 1970; Darte and Grenson, 1975; Rytka, 1975).

Similar types of nonspecific amino acid transport systems have also been described for filamentous fungi such as *N. crassa* (Pall, 1969), *P. chrysogenum* (Benko et al., 1967, 1969), *Arthobotrys conoides* (Gupta and Pramer, 1970), and *Botrytis fabae* (Jones, 1963).

SYSTEMS OF BROAD BUT DIFFERING AFFINITIES

The separation of this transport system from the general system discussed above is somewhat arbitrary. The designation of this catagory is based largely on the results of competition studies. The uptake of ^{3}H-phenylalanine by *Geotrichum candidum* is inhibited by 50% or more when this amino acid is added in combination with unlabeled L-alanine, L-cysteine, L-histidine, L-leucine, L-tryptophan, or L-valine. In contrast, the inhibition of uptake is 40% or less with L-arginine, L-aspartate, L-glutamate, L-lysine, L-threonine, or L-tyrosine. These broad differences in competition with phenylalanine transport are taken to mean that two or more systems, each capable of transporting several amino acids and differing in their affinities for individual amino acids, are involved (McEnvoy and Murray, 1972).

With *Penicillium griseofulvum*, Whitaker and Morton (1971) noted varying degress of inhibition of uptake of either ^{14}C-L-leucine or ^{14}C-L-lysine when tested in competition experiments with individual amino acids. They observed that most of the amino acids studied, including L-lysine, compete with L-leucine for uptake. This indicates that L-leucine is taken up by a transport system which also transports several other amino acids. On the other hand, out of 22 amino acids tested only L-arginine, L-histidine, L-phenylalanine, and L-leucine reduce L-lysine uptake to any great extent. This latter observation indicates that one of the transport systems associated with L-lysine uptake has a narrower affinity range for amino acids than the other.

SYSTEMS SPECIFIC FOR ACIDIC, NEUTRAL, OR BASIC AMINO ACIDS

Several fungal species, including *S. cerevisiae*, *N. crassa*, *Aspergillus niger*, and *P. chrysogenum* have carrier systems for transporting specific classes of amino acids. In *N. crassa*, acid, basic, and neutral amino acids each have a separate transport system (Pall, 1969, 1970a,b). A comparison of these systems with the general transport system in *Neurospora* is given in Table 4.7. Notice the different affinities of the same amino acid for the different systems. The four systems also respond differently to changes in the environment. For example, the onset of conidial germination is correlated with a decrease in the activity of System III; during germination and subsequent hyphal growth Systems I and II increase in activity (Railey and Kensey, 1976).

Table 4.7   The Major Amino Acid Transport Systems in *Neurospora*[a]

	System 1: L-Neutral Amino Acids	System II: D- or L-Basic, Neutral and Acidic Amino Acids	System III: L-Basic Amino Acids	System IV: D- or L-Acidic Amino Acids
Amino acids transported and affinity constants (μM)	L-Tryptophan (60) L-Leucine (110) L-Phenylalanine (50)	L-Arginine (6.2) L-Phenylalanine (2) D-Phenylalanine (25) Glycine (7) L-Aspartic acid (1200)	L-Arginine (2.4) L-Lysine (4.8)	L-Cysteic acid (7) L-Aspartic acid (13) L-Glutamic acid (16)
Other amino acids showing affinity	L-Valine L-Alanine Glycine L-Histidine L-Serine	L-Lysine L-Leucine α-Aminoisobutyric acid β-Alanine L-Histidine	L-Ornithine L-Canavanine L-Histidine (low affinity)	D-Aspartic acid D-Glutamic acid

Source: Pall (1970b).

[a]The amino acids in this table are only a partial listing of the amino acids having affinity for the different transport systems. In most cases, amino acids with similar properties to those listed will also have affinity. Affinity constants (K_m or K_i) are all expressed in μM.

Table 4.8 Properties of Some Transport Systems Present in *Penicillium*

Transport System	Conditions for Development	V_{max} (μmoles/g-min)	K_m (M)
L-Methionine	Sulfur deficiency	1	10^{-5}
L-Cystine	Sulfur deficiency	1.2	2×10^{-5}
L-Cysteine	Constitutive	2.1	1.4×10^{-4}
Choline-O-sulfate	Sulfur deficiency	3	2×10^{-4}
Sulfate (thiosulfate, selenate, molybdate)	Sulfur deficiency	1.5	$1-7 \times 10^{-5}$
Tetrathionate	Sulfur deficiency	3	2.5×10^{-5}
Glutathione (reduced)	Sulfur deficiency	2.3	1.7×10^{-5}
Ammonium	Nitrogen deficiency	10	2.5×10^{-7}
L-α-Neutral and basic amino acids	Nitrogen or carbon deficiency	10	2×10^{-5}
L-α-Acidic amino acids	Nitrogen or carbon deficiency	10	$0.2-1 \times 10^{-4}$
L-α-Basic amino acids	Constitutive	1	5×10^{-6}
L-Arginine	Constitutive	N	N
L-Lysine	Constitutive	N	N
L-Proline	Nitrogen or carbon deficiency	N	N

Source: Hunter and Segel (1971).

N: Not determined, or undeterminable in wild-type mycelium.

SPECIFIC SYSTEMS FOR SINGLE AMINO ACIDS

Many fungi possess a carrier that is specific for a single amino acid. In yeasts these systems are not usually inhibited by ammonium and have a higher affinity for their substrate than do the more general transport systems (Beckerich and Heslot, 1978). *P. chrysogenum* has a variety of uptake systems which are quite specific (Table 4.8). For example, the uptake of L-methionine and L-cystine occurs by separate systems which develop as a result of sulfur starvation; a specific carrier for L-cysteine is constitutive (Benko et al., 1967; Skye and Segel, 1970). Also, in addition to the basic amino acid transport system there are separate specific systems for the uptake of L-arginine and L-lysine. A carrier for proline becomes active under conditions of nitrogen or carbon deficiency (Hunter and Segel, 1971).

FACTORS AFFECTING AMINO ACID UPTAKE AND ACCUMULATION

Many factors affect uptake and accumulation of amino acids. As with sugar uptake (Chapter 3), the response of the organism to a particular factor depends on the chemical structure of the molecule to be transported and the genetic makeup of the organism.

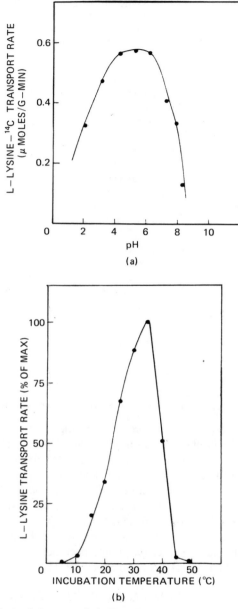

Figure 4.7 Factors affecting L-lysine uptake by *Penicillium*. (a.) pH dependence. (b.) Effect of temperature. (c.) Concentration dependence. (d.) Effect of ionic strength. (Hunter and Segel, 1971)

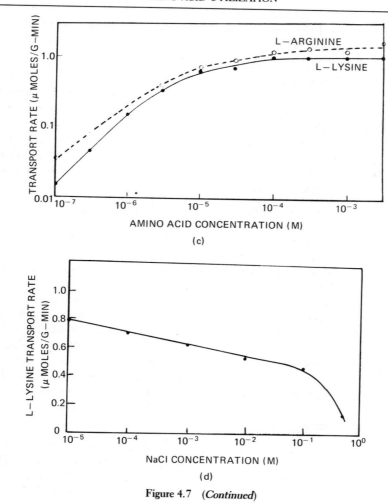

Figure 4.7 (*Continued*)

pH. The pH of the medium influences uptake by determining the charge of the amino acid and the charge of the carrier protein, and thus the extent of binding between the two. In addition, proton symport is a common mechanism of amino acid transport in fungi (Whitaker, 1976), and thus an adequate hydrogen ion concentration is required for sufficient uptake. Although the pH optimum varies for individual amino acids, in general uptake occurs most rapidly at pH 5–6. Figure 4.7a illustrates the typical pH dependence of amino acid uptake in fungi (Hunter and Segel, 1971).

Temperature. In addition to influencing the rates of chemical reactions within the cell, temperature affects the rates at which nutrients are taken up and thus made available for utilization. Figure 4.7b illustrates a typical temperature profile for amino acid uptake. Increasing the incubation temperature from 5° to 30° C results in increasing uptake of ^{14}C-lysine by *P. chrysogenum*; however, the uptake rate drops sharply at higher temperatures (Hunter and Segel, 1971).

Concentration. Because amino acid transport is carrier-mediated, a graph of the uptake rate at different substrate concentrations gives a curve demonstrating that saturation kinetics are in effect Figure 4.7c). Thus, increasing the amino acid concentration may result in increased uptake at low concentrations but may have no effect at high concentrations. In addition, it is usual for several amino acids to be transported by the same carrier. When two such amino acids are present together in the incubation medium, the extent to which one is utilized over the other will be determined by their relative concentrations and their affinities for the carrier.

Energy Requirements. Active transport appears to be the principal mode of amino acids uptake in fungi, and proton symport is an important mechanism in the few species studied (e.g., Hunter and Segel, 1973; Seaston et al., 1973, 1976). Thus, chemical or physical factors which affect energy production will consequently influence amino acid uptake. For example, the uptake of glycine and γ-aminoisobutyric acid by *S. cerevisiae* increases substantially when the cells are preincubated with a metabolizable carbon source; the increase in uptake rates is coincident with an increase in the intracellular supply of high-energy phosphates (Kotyk and Rihova, 1972a,b). Similarly, inhibitors of energy production such as iodoacetate (which inhibits glycolysis), azide, and 2,4-dinitrophenol (which accelerate the leakage of protons across the membrane), and anaerobic conditions all decrease or eliminate amino acid uptake (Seaston et al., 1973, 1976; Whitaker and Morton, 1971). However, care should be taken when interpreting the results of inhibitor studies; Horak and co-workers (1978) found that many metabolic inhibitors actually *stimulate* amino acid uptake at low concentrations!

Other Nutrients. We have seen examples in which certain other nutrients (or the lack thereof) affect amino acid uptake. For example, ammonium ion inhibits uptake in many fungi, and several uptake systems become active during periods of nitrogen, carbon, or sulfur deficiency (Table 4.8). In addition, any molecule which alters the ionic strength of the medium may affect uptake. Increasing NaCl concentrations result in decreased uptake of ^{14}C-lysine in *P. chrysogenum* (Figure 4.7d); however, the uptake of L-methionine is unaffected by NaCl in the same concentration range (Hunter and Segel, 1971). In *S. cerevisiae* certain monosaccharides appear to interfere indirectly with amino acid uptake (Kotyk and Rihova, 1972b). However, in most fungi sugars are stumulatory because of the requirement for an energy source (Whitaker and Morton, 1971).

Concentration of the Intracellular Amino Acid Pool. The concentration of free amino acids within the cell may inhibit the uptake of additional amino acids from the medium. This phenomenon is called **transinhibition** and is a mechanism by which the cell prevents an extended accumulation of a nutrient (Horak et al., 1977). For example, cells of *S. cerevisiae* that have a high lysine content exhibit a low rate of lysine uptake; however, transport is rapid when the intracellular lysine content is low (Morrison and Lichstein, 1976). Moreover, the effect of the intracellular amino acid pool is usually nonspecific. Apparently, uptake is affected most directly by the size of the pool rather than the types of amino acids present (Woodward and Cirillo, 1977). Horak and co-workers (1977) have suggested that each separate amino acid transport system in *S. cerevisiae* may share a common component that is sensitive to transinhibition.

In some fungi, on the other hand, the extent of transinhibition varies with specific amino acids. For example, amino acids having a high specificity for the L-methionine carrier protein in *N. crassa* show strong transinhibition of L-methionine uptake, those which have a low affinity show weak transinhibition, and those with no affinity show no transinhibition (Pall, 1971). Similar phenomena have been reported for tryptophan transport in *N. crassa* (Wiley and Matchett, 1968) and with L-methionine transport in *P. chrysogenum* (Benko et al., 1967).

Even growing the organism with an inorganic nitrogen source such as ammonium results in the synthesis of sufficient amino acids to "load" the intracellular pool to the extent that transinhibition can occur. Whitaker and Morton (1971) conducted an experiment whereby *P. griseofulvum* was grown on a medium containing low (0.08%), normal (0.16%), or high (0.32%) concentrations of ammonium sulfate. They then transferred the mycelia to a minimal medium containing leucine as the sole nitrogen source and measured the accumulation of leucine within the tissue. As indicated in Figure 4.8, a preincubation in a high ammonium medium results in

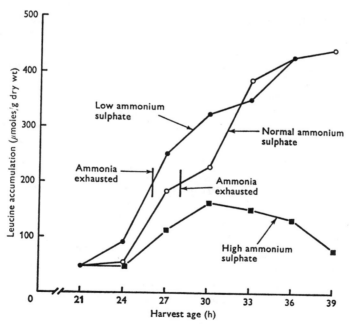

Figure 4.8 Accumulation of leucine by *Penicillium griseofulvum* following transfer to minimal medium containing leucine, after harvesting from culture medium containing different initial concentrations of ammonium sulfate as a nitrogen source. (Whitaker and Morton, 1971)

low accumulation of leucine; preincubation in a lower ammonium concentration has no inhibitory effect.

MISCELLANEOUS NITROGEN SOURCES

Fungi have been reported to utilize a wide range of nitrogen sources for growth in addition to those discussed above. They include proteins and peptides, amides, D-amino acids, purines and pyrimidines, imidiazoles, amines, and thiourea.

Polyporus versicolor grows well on proteins and peptides as nitrogen sources (Levi et al., 1968). The ability of fungi to utilize proteins and shorter polypeptides is dependent on the secretion of proteases that hydrolyze these macromolecules into smaller subunits (Cohen 1973a,b). Any peptide resulting from this hydrolysis may be cleaved into individual amino acids by the action either of extracellular proteases (Wolfinbarger and Marzluf, 1974, 1975a) or proteases covalently bound to the cell wall (Frey and Roehm, 1978). In addition, some fungi possess transport systems for the uptake of short-chain peptides. For example, strains of *Saccharomyces cerevisiae* and *Candida albicans* have been identified which transport a variety of di- and tripeptides (Davies, 1980; Marder et al., 1977). *Neurospora crassa* can transport oligopeptides that have a volume no greater than trileucine, but it is unable to take up dipeptides (Wolfinbarger and Marzluff, 1975a,b). In each case, the incorporated peptides are cleaved by intracellular peptidases.

Purines and pyrimidines can be used as nitrogen sources by fungi. This has been shown in several species including *Aspergillus nidulans* (Kinghorn and Pateman, 1976), *N. crassa* (Tsao and Marzluf, 1976), *Penicillium chrysogenum* (Allam and Elzainy, 1970b), *Fusarium moniliforme* (Allam and Elzainy, 1970a), and a large number of yeasts (LaRue and Spencer, 1968). The latter workers showed that purines and pyrimidines used by yeasts include adenine, hypoxanthine, guanine, xanthine, uric acid, cytosine, uracil, dihydrouracil, thymine, and dihydrothymine. In addition, β-alanine and β-isobutyric acid, which are degradation products of cytosine and thymine, respectively, are also utilized, as are allantoin and allantoic acid, which are breakdown products of purines. The ultimate nitrogenous degradation product of all purines, pyrimidines, and their derivatives is ammonium, which is readily utilized for growth. Distinct transport systems for purines and pyrimidines are known. Tsao and Marzluf (1976) reported a transport system for uric acid and xanthine and a second one for hypoxanthine, adenine, and guanine. They suggest that these may be distinct from yet another transport system for adenosine and uracil.

D-Amino acids are considered by Pateman and Kinghorn (1976) to be extremely poor nitrogen sources for A. *nidulans* and *N. crassa*. This is not surprising considering that the D-isomers are much less common in nature than the L-isomers. The utilization of D-amino acids appears to be dependent on the presence of D-amino acid oxidase, an enzyme that catalyzes the removal of amino groups. Since many fungal strains contain this enzyme (Kishore et. al., 1976; Onishi et al., 1962; Rosenfeld and Leiter, 1977) they probably use D-amino acids to some extent. In this regard LaRue and Spencer (1967a) tested 91 strains of yeast from 19 genera for their ability to utilize D-amino acids as nitrogen sources. Eighty strains from 16 genera were able to grow on at least one of 17 D-amino acids tested. Many of the yeast strains grew as vigorously on D-amino acids as on the corresponding L-amino acids.

Some of the other compounds shown to be utilized as nitrogen sources include amides (Hynes and Pateman, 1970), imidazoles such as histamine and histidinol (LaRue and Spencer, 1967b), thiourea, and amines. Thiourea, a compound often associated with sewage, was rapidly metabolized by a *Penicillium* species isolated from soil and resulted in the excretion of ammonium into the medium (Lashen and Starkey, 1970). Of 461 yeast strains tested, 86% were able to use one or more amines as a nitrogen source. The ability to utilize ethylamine, propylamine, butylamine, and benzylamine is generally greater than the ability to utilize various methylated amines (van Dijken and Bos, 1981).

The compounds discussed briefly above are among the wide range of nitrogen sources fungi can use as substrates. In addition, fungi appear capable of recycling their own nitrogen-containing constituents for use in further growth. In this regard Levi, Merrill, and Cowling (1968), on the basis of studies with mycelial fractions and model nitrogen compounds, concluded that wood-destroying fungi conserve their nitrogen supply by a process of autolysis and reuse of such fungal constituents as proteins, peptides, amino acids, nucleic acids, nucleotides, and amino sugars. Most of the nitrogenous compounds discussed above are broken down, at least in part, to ammonium before being assimilated.

SUMMARY

In this chapter we have presented a sampling of the ways in which fungal growth and development are regulated by the type of nitrogen source in the culture medium. We have focused on nitrate, nitrite, ammonium, urea, and amino acids as the sources most widely used, although fungi can obtain nitrogen from a wide variety of other compounds with the exception of N_2.

Except for proteins, most of the common nitrogen sources are small enough to enter the cell directly, without prior extracellular degradation. Thus, for these small molecules the plasma membrane represents the first point at which fungi can exercise control over their incorporation. For this reason, we have placed emphasis on mechanisms of uptake of nitrogen sources. To a large extent, these mechanisms provide important clues for factors that influence the utilization of nitrogen compounds as nutrients.

Because nitrate and nitrite apparently enter fungal cells by diffusion, the principal factors affecting their utilization will be those involving metabolism, not transport. For example, the inability of a species to grow on nitrate is often due to the absence of nitrate reductase; and the inability to grow on nitrite may be due to the lack of nitrite reductase or because the nitrite is inhibitory to certain metabolic systems. On the other hand, the utilization of actively transported compounds such as ammonium, urea, and amino acids is more intricately regulated because utilization can be controlled both at the cell surface and intracellularly. Thus, factors affecting the shape and charge of the carrier molecule (such as temperature and pH), factors affecting the supply of ATP necessary for active transport (such as oxygen levels, pH, and carbon source), and intracellular feedback systems (such as the concentrations of sulfur and nitrogen in amino acid pools) all affect the uptake and thus the utilization of these nitrogen sources.

The variety of transport systems involved in amino acid uptake illustrates the precision with which the utilization of these nitrogen sources is regulated. In nutrient-poor media the general uptake system is usually operating, resulting in a scavenging of any amino acids that become available. Once the environment becomes nutrient rich, the fungus can "afford" to be more selective, and individual transport systems are switched on.

REFERENCES

Agnihotri, V. P. 1968. Effects of nitrogenous compounds on sclerotium formation in *Aspergillus niger*. *Can. J. Microbiol.* **14**: 1253–1258.

Allam, A. M., and T. A. Elzainy. 1970a. Purine catabolism in *Fusarium moniliforme*. *J. Gen. Microbiol.* **63**: 183–187.

Allam, A. M., and T. A. Elzainy. 1970b. Utilization and deamination of adenine by *Penicillium chrysogenum*. *J. Chem. U.A.R.* **13**: 253–255.

Alleem, M. I. H., H. Lees, and R. Lyric. 1964. Ammonium oxidation by cell-free extracts of *Aspergillus wentii*. *Can. J. Biochem.* **42**: 989–998.

Arst, H. N., and D. J. Cove. 1969. Methylammonium resistance in *Aspergillus nidulans*. *J. Bacteriol.* **98**: 1284–1293.

Aube, C., and C. Gagnon. 1969. Effect of carbon and nitrogen nutrition on growth and sporulation of *Trichoderma viride* Pers. ex Fries. *Can. J. Microbiol.* **15**: 703–706.

Beckerich, J. M., and H. Heslot. 1978. Physiology of lysine permeases in *Saccharomyces lipolytica*. *J. Bacteriol.* **133**: 492–498.

Benko, P. V., T. C. Wood, and I. H. Segel. 1967. Specificity and regulation of methionine transport in filamentous fungi. *Arch. Biochem. Biophys.* **122**: 783–804.

Benko, P. V., T. C. Wood, and I. H. Segel. 1969. Multiplicity and regulation of amino acid transport in *Penicillium chrysogenum*. *Arch. Biochem. Biophys.* **129**: 498–508.

Binder, F. L., and H. L. Barnett. 1974. Amino acid requirement for axenic growth of *Teighemiomyces parasiticus*. *Mycologia* **66**: 265–271.

Cohen, B. L. 1973a. The neutral and alkaline proteases of *Aspergillus nidulans*. *J. Gen. Microbiol.* **77**: 521–528.

Cohen, B. L. 1973b. Regulation of intracellular and extracellular neutral and alkaline proteases in *Aspergillus nidulans*. *J. Gen. Microbiol.* **79**: 311–320.

Cook, R. J., and C. Anthony. 1978. The ammonia and methylamine active transport system of *Aspergillus nidulans*. *J. Gen. Microbiol.* **109**: 265–274.

Cooper, T. G., and R. A. Sumrada. 1975. Urea transport in *Saccharomyces cerevisiae*. *J. Bacteriol.* **121**: 571–576.

Croes, A. F., J. M. J. M. Steijns, G. J. M. L. De Vries, and T. M. J. A. van der Putte. 1978. Inhibition of meiosis in *Saccharomyces cerevisiae* by ammonium ions: Interference of ammonia with protein metabolism. *Planta* **141**: 205–209.

Darte, C., and M. Grenson. 1975. Evidence for three glutamic acid transporting systems with specialized physiological functions in *Saccharomyces cerevisiae*. *Biochim. Biophys. Res. Commun.* **67**: 1028–1033.

Davies, M. B. 1980. Peptide uptake in *Candida albicans*. *J. Gen. Microbiol.* **121**: 181–186.

Dunn, E., and J. A. Pateman. 1972. Urea and thiourea in *Aspergillus nidulans*. *Heredity* **29**: 129.

Evans, R. C., and C. L. Black. 1981. Interactions between nitrogen sources and xylose affecting growth, sporulation, and polyphenoloxidase activity in *Bipolaris maydis* race T. *Can. J. Bot.* **59**: 2102–2107.

Eylar, O. R., and E. L. Schmidt. 1959. A survey of heterotrophic microorganisms from soil for ability to form nitrite and nitrate. *J. Gen. Microbiol.* **20**: 473–481.

Frey, J., and K. Roehm. 1978. Subcellular localization and levels of amido peptidases and dipeptidase in *Saccharomyces cerevisiae*. *Biochim. Biophys. Acta* **527**: 31–41.

Giles, K. L., and H. C. M. Whitehead. 1975. The transfer of nitrogen fixing ability to a eukaryote cell. *Cytobios* **14**: 49–61.

Giles, K. L., and H. C. M. Whitehead. 1976. Uptake and continued metabolic activity of *Azotobacter* within fungal protoplasts. *Science* **193**: 1125–1126.

Grenson, M., C. Hou, and M. J. Crabeel. 1970. Multiplicity of the amino acid permeases in *Saccharomyces cerevisiae*. 4: Evidence for a general amino acid permease. *J. Bacteriol.* **103**: 770–777.

Gupta, R. K., and D. Pramer. 1970. Amino acid transport by the filamentous fungus *Arthrobotrys conoides*. *J. Bacteriol.* **103**: 120–130.

Horak, J., A. Kotyk, and L. Rihova. 1977. Specificity of transinhibition of amino acid transport in baker's yeast. *Folia Microbiol.* **22**: 360–362.

Horak, J., A. Lotyk, and L. Rihova, 1978. Stimulation of amino acid transport in *Saccharomyces cerevisiae* by metabolic inhibitors. *Folia Microbiol.* **23**: 286–291.

Hunter, D. R., and I. H. Segel. 1971. Acidic and basic amino acid transport systems of *Penicillium chrysogenum*. *Arch. Biochem. Biophys.* **144**: 168–183.

Hunter, D. R., and I. H. Segel. 1973. Effect of weak acids on amino acid transport by *Penicillium chrysogenum*: Evidence for a proton or charge gradient as a driving force. *J. Bacteriol.* **113**: 1184–1192.

Hynes, M. J., and J. A. Pateman. 1970. The use of amides as nitrogen sources by *Aspergillus nidulans*. *J. Gen. Microbiol.* **63**: 317–324.

Jones, O. T. G. 1963. The accumulation of amino acids by fungi with particular reference to the plant parasitic fungus *Botrytis fabae*. *J. Exp. Bot.* **14**: 399–411.

Kinghorn, J. R., and J. A. Pateman. 1976. Nitrogen metabolism. In J. E. Smith and J. A. Pateman (Eds.), *Genetics and Physiology of Aspergillus*, pp. 147–202. New York: Academic.

Kishore, G., M. Sugumaran, and C. S. Vaidyanathan. 1976. Metabolism of DL-(+)-phenylalanine by *Aspergillus niger*. *J. Bacteriol.* **128**: 182–191.

Kotyk, A., and L. Rihova. 1972a. Transport of alpha-aminobutyric acid in *Saccharomyces cerevisiae*: Feedback control. *Biochim. Biophys. Acta* **288**: 380–389.

Kotyk, A., and L. Rihova. 1972b. Energy requirement for amino acid uptake in *Saccharomyces cerevisiae*. *Folia Microbiol.* **17**: 353–356.

Larimore, F. S., and R. J. Roon. 1978. Possible site specific reagent for the general amino acid transport system of *Saccharomyces cerevisiae*. *Biochemistry* **17**: 431–436.

LaRue, T. A., and J. T. F. Spencer. 1967a. The utilization of D-amino acids by yeasts. *Can. J. Microbiol.* **13**: 777–788.

LaRue, T. A., and J. T. F. Spencer. 1967b. The utilization of imidazoles by yeasts. *Can. J. Microbiol.* **13**: 789–794.

LaRue, T. A., and J. T. F. Spencer. 1968. The utilization of purines and pyrimidines by yeasts. *Can. J. Microbiol.* **14**: 79–86.

Lashen, E. S., and R. L. Starkey. 1970. Decomposition of thiourea by a *Penicillium* species and soil and sewage-sludge microflora. *J. Gen. Microbiol.* **64**: 139–150.

LeJohn, H. B. 1971. Enzyme regulation, lysine pathways and cell wall structures as indicators of major lines of evolution in fungi. *Nature* **231**: 164–168.

Levi, M. P., W. Merrill, and E. B. Cowling. 1968. Role of nitrogen in wood deterioration. VI: Mycelial fractions and model nitrogen compounds as substrates for growth of *Polyporus versicolor* and other wood-destroying and wood-inhabiting fungi. *Phytopathology* **58**: 626–634.

Lilly, V. G. 1965. The chemical environment for fungal growth. In G. C. Ainsworth and A. S. Sussman (Eds.) *The Fungi*, Vol. 1, pp. 465–478. New York: Academic.

Maas, J. L., and R. L. Powelson. 1972. Growth and sporulation of *Botrytis convoluta* with various carbon and nitrogen sources. *Mycologia* **64**: 897–903.

Marder, R., J. M. Becker, and F. Naider. 1977. Peptide transport in yeast: Utilization of leucine- and lysine-containing peptides by *Saccharomyces cerevisiae*. *J. Bacteriol.* **131**: 906–916.

McEnvoy, J. J., and J. R. Murray. 1972. Amino acid transport in germinated arthrospores of *Geotrichum candidum*. *Arch. Microbiol.* **86**: 101–110.

Middlehoven, W. J. 1977. Isolation and characterization of methylammonium-resistant mutants of *Saccharmyces cerevisiae* with relieved nitrogen metabolite repression of allantoinase, arginase and ornithine transaminase synthesis. *J. Gen. Microbiol.* **100**: 257–269.

Morrison, C. E., and H. C. Lichstein. 1976. Regulation of lysine transport by feedback inhibition in *Saccharomyces cerevisiae*. *J. Bacteriol.* **125**: 864–871.

Morton, A. G., and A. MacMillan. 1954. The assimilation of nitrogen from ammonium salts and nitrate by fungi. *J. Exp. Bot.* **5**: 232–252.

Nicholas, D. J. D. 1965. Utilization of inorganic nitrogen compounds and amino acids by fungi. In G. C. Ainsworth and A. S. Sussman (Eds.), *The Fungi*, Vol. 1., pp. 349–376. New York: Academic.

Onishi, E., H. Macleod, and M. H. Horowitz. 1962. Mutants of *Neurospora* deficient in D-amino acids. *J. Biol. Chem.* **237**: 138–142.

Pall, M. L. 1969. Amino acid transport in *Neurospora crassa*: I. Properties of two amino acid transport systems. *Biochim. Biophys. Acta* **173**: 113–127.

Pall, M. L. 1970a. Amino acid transport in *Neurospora crassa*: II. Properties of a basic amino acid transport system. *Biochim. Biophys. Acta* **203**: 139–149.

Pall, M. L. 1970b. Amino acid transport in *Neurospora crassa*: III. Acidic amino acid transport. *Biochim. Biophys. Acta* **211**: 513–520.

Pall, M. L. 1971. Amino acid transport in *Neurospora crassa*: IV. Properties and regulation of a methionine transport system. *Biochim. Biophys. Acta* **233**: 201–214.

Pardee, A. B., and L. M. Palmer. 1973. Regulation of transport systems: A means of controlling metabolic rates. In D. D. Davies (Ed.), *Rate Control of Biological Processes*, pp. 133–144. Cambridge: Symposia of the Society for Experimental Biology, Vol. 27.

Pateman, J. A., and J. R. Kinghorn. 1976. Nitrogen metabolism. In J. E. Smith and D. R. Berry (Eds.), *The Filamentous Fungi*, Vol. 2, pp. 159–237. New York: Wiley.

Pateman, J. A., J. R. Kinghorn, and E. Dunn. 1974. Regulatory aspects of L-glutamate transport in *Aspergillus nidulans*. *J. Bacteriol.* **119**: 534–542.

Pinon, R. 1977. Effects of ammonium ions on sporulation of *Saccharomyces cerevisiae*. *Exp. Cell. Res.* **105**: 367–378.

Railey, R. M., and J. A. Kensey. 1976. Development of amino acid uptake activity in *Neurospora*. *Can. J. Microbiol.* **22**: 115–120.

Roldan, J. M., J. Verbelen, W. L. Butler, and K. Tokuyasu. 1982. Intracellular localization of nitrate reductase in *Neurospora crassa*. *Plant Physiol.* **70**: 872–874.

Roon, R. J., H. L. Even, P. Dunlop, and F. L. Larimore. 1975a. Methylamine and ammonia transport in *Saccharomyces cerevisiae*. *J. Bacteriol.* **122**: 502–509.

Roon, R. J., F. Larimore, and J. S. Levy. 1975b. Inhibition of amino acid transport by ammonium in *Saccharomyces cerevisiae*. *J. Bacteriol.* **124**: 325–331.

Rosenfeld, M. G., and E. H. Leiter. 1977. Isolation and characterization of a mitochondrial D-amino oxidase from *Neurospora crassa*. *Can. J. Biochem.* **55**: 66–74.

Ross, R. G., and F. D. J. Bremner. 1971. Effect of ammonium nitrogen and amino acids on perithecial formation of *Venturia inaequalis*. *Can. J. Plant Sci.* **51**: 29–33.

Rytka, J. 1975. Positive selection of general amino acid permease mutants in *Saccharomyces cerevisiae*. *J. Bacteriol.* **121**: 562–570.

Schindler, J., and M. Sussman. 1977. Ammonia determines the choice of morphogenetic pathways in *Dictyostelium discoideum*. *J. Mol. Biol.* **116**: 161–170.

Seaston, A., G. Carr, and A. A. Eddy. 1976. The concentration of glycine by preparations of the yeast *Saccharomyces carlsbergensis* depleted of adenosine triphosphate. *Biochem. J.* **154**: 669–676.

Seaston, A., C. Inkson, and A. A. Eddy. 1973. The absorption of protons with specific amino acids and carbohydrates by yeast. *Biochem. J.* **134**: 1031–1043.

Shipton, W. A. 1977. Some nutritional factors regulating formation of fertile perithecia of *Calonectria camelliae*. *Trans. Brit. Mycol. Soc.* **69**: 59–62.

Shorer, J., I. Zelmanowicz, and I. Barash. 1972. Utilization and metabolism of urea during spore germination by *Geotrichum candidum*. Phytochemistry **11**: 595–605.

Singh, B. P., and R. N. Tandon. 1970. Nitrogen requirements of certain isolates of *Alternaria tenuis*. *Mycopathol. Mycol. Appl.* **42**: 33–38.

Skye, G. E., and I. H. Segel. 1970. Independent regulation of cysteine and cystine transport in *Penicillium chrysogenum*. *Arch. Biochem. Biophys.* **138**: 306–318.

Stephen, R. C., and C. Chan. 1970. Nitrogen requirements of the genus *Linderina*. *Can. J. Bot.* **48**: 695–698.

Stephen, R. C., and K. K. Fung. 1971. Nitrogen requirements of the fungal endophytes of *Arundina chinensis*. *Can. J. Bot.* **49**: 407–410.

Sumrada, R., M. Gorski, and T. Cooper. 1976. Urea transport-defective strains of *Saccharomyces cerevisiae*. *J. Bacteriol.* **125**: 1048–1056.

Sussman, M., and J. Schindler. 1978. A possible mechanism of morphogenetic regulation in *Dictyostelium discoideum*. Differentiation **10**: 1–6.

Tsao, T. F., and G. A. Marzluf. 1976. Genetic and metabolic regulation of purine base transport in *Neurospora crassa*. *Mol. Gen. Genet.* **149**: 347–355.

van Dijken, J. P., and P. Bos. 1981. Utilization of amines by yeasts. *Arch. Microbiol.* **128**: 320–324.

Vogel, H. J. 1965. Lysine biosynthesis and evolution. In V. Bryson and H. J. Vogel (Eds.), *Evolving Genes and Proteins*. New York: Academic.

Weinhold, A. R., and M. O. Garraway. 1966. Nitrogen and carbon nutrition in *Armillaria mellea* in relation to growth promoting effects of ethanol. *Phytopathology* **56**: 108–112.

Whitaker, A. 1976. Amino acid transport in fungi: An essay. *Trans. Brit. Mycol. Soc.* **67**: 365–376.

Whitaker, A., and A. G. Morton. 1971. Amino acid transport in *Penicillium griseofulvum*. *Trans. Brit. Mycol. Soc.* **56**: 353–369.

Wiley, W. R., and W. H. Matchett. 1968. Tryptophan transport in *Neurospora crassa*: II. Metabolic control. *J. Bacteriol.* **95**: 959–966.

Wolfinbarger, L., and G. A. Marzluf. 1974. Peptide utilization by amino acid auxotrophs of *Neurospora crassa*. *J. Bacteriol.* **119**: 371–378.

Wolfinbarger, L., and G. A. Marzluf. 1975a. Size restrictions on utilization of peptides by amino acid auxotrophs of *Neurospora crassa*. *J. Bacteriol.* **122**: 949–956.

Wolfinbarger, L., and G. A. Marzluf. 1975b. Specificity and regulation or peptide transport in *Neurospora crassa*. *Arch. Biochem. Biophys.* **171**: 637–644.

Woodward, J. R., and V. P. Cirillo. 1977. Amino acid transport and metabolism in nitrogen starved cells of *Saccharomyces cerevisiae*. *J. Bacteriol.* **130**: 714–723.

5

INORGANIC NUTRIENTS

An inorganic element is considered an essential nutrient if its absence from an otherwise complete medium leads to reduced growth or reduced reproduction. Inorganic nutrients are needed by fungi in considerably lower quantities than are carbon and nitrogen sources. In experimental media for culturing fungi the latter nutrients are usually supplied in gram/liter quantities. Inorganic nutrients, however, are provided in quantities which range from several hundred mg/liter in the case of phosphates to less than one mg/liter for compounds containing copper.

The amount of a specific inorganic nutrient needed for maximum growth varies with the species, the medium composition, and the environment. The results of extensive research show that inorganic nutrients may be grouped into two broad categories on the basis of the quantities needed to promote maximum growth. Macronutrients are nutrients that are required in relatively large quantities; they include magnesium, phosphorus, potassium, sulfur, and often calcium. Micronutrients, also called trace elements, are required in much smaller amounts; they include copper, iron, manganese, zinc, and often molybdenum. Other elements such as boron, cobalt, and sodium may also be required by certain species, but trace metal contaminants in reagents and water may make it difficult to demonstrate their necessity. Also, there are many pitfalls associated with media preparation, including precipitation of certain inorganic nutrients due to pH or autoclaving.

In the following sections we shall briefly discuss the major macronutrients and micronutrients used by fungi, emphasizing their mode of uptake, their biochemical roles in the cell, and the patterns of growth and development that they affect.

MAGNESIUM

Magnesium is required by all fungi and has a wide variety of regulatory roles. The intracellular magnesium concentration varies widely depending on the species, the stage of growth, and the composition of the culture medium. A comparison of the total and osmotically free (i.e., unbound) magnesium and phosphate in three different genera grown under different cultural conditions is illustrated in Table 5.1. Notice that the total magnesium concentration of the cells increases with increasing Mg^{2+} concentration of the medium. In most cases, the concentration of unbound magnesium increases as well. However, the internal concentration is surprisingly stable; a 14,000-fold increase in external magnesium results in only a 1.5–3.3-fold increase in intracellular levels (Okorokov et al., 1975). Notice also that the cellular Mg^{2+} concentration is influenced by the rate of growth. Rapidly growing cells of *Endomyces magnusii* and *Penicillium chrysogenum* have significantly less magnesium than cells that are growing more slowly.

In *P. chrysogenum*, phosphorus is essential for the accumulation and retention of magnesium. Transferring the fungus to a medium lacking phosphorus results in a sharp drop in the total magnesium concentration. In this species, magnesium has been found bound to polymers of orthophosphate. The resulting complex is large, having a molecular weight of more than 400,000 and a molar phosphorus/magnesium ratio of 0.1–2.0. These polyphosphate–magnesium complexes are thought to provide a storehouse for cellular Mg^{2+}. Through selective binding and release these complexes may help maintain the intracellular Mg^{2+} concentration at a relatively constant level, even when external concentrations change (Okorokov et al., 1974, 1975).

This effect of phosphorus on magnesium levels suggests that phosphorus may be involved in magnesium uptake. Studies by Rothstein using yeast demonstrate that phosphorus and low concentrations of potassium stimulate the uptake of magnesium

Table 5.1 Content of Magnesium and Phosphate in Various Fungi (Mg and P Expressed as mg/g dry wt)

Microorganism	Magnesium Concentration in the Medium (mM)	Total Content		Osmotically Free	
		Mg	P	Mg^{2+}	PO_4^{3-}
Saccharomyces cerevisiae	0.082	1.24	26.4	0.9	2.7
(12 h)	2.86	3.1	26.4	1.66	2.7
	36.2	4.02	27.1	2.03	2.3
Endomyces magnussii	0.26	12.8	33.9	0.63	12.7
(9 h slow growth)	13.9	16.2	28.6	0.70	7.3
	79	19.8	30.5	0.42	6.8
E. magnusii	0.082	1.45	22.0	0.95	1.1
(9 h rapid growth)	2.95	2.54	15.9	1.55	1.3
	37	2.93	21.5	1.63	1.5
Penicillium chrysogenum	0.0025	2.39	24.2	0.39	2.5
140A (slow growth)	2.5	2.7	9.83	0.64	1.6
P. chrysogenum	32	1.54	9.0	0.78	1.4
(rapid growth)					

Source: Okorokov et al. (1975).

(Rothstein et al., 1958; Fuhrmann and Rothstein, 1968). Other divalent cations, notably cobalt and zinc and to a lesser extent manganese, nickel, and strontium, are taken up by the same carrier. Transport requires metabolic energy and involves the secretion of 2 K^+ or 2 Na^+ for each divalent cation taken up.

However, neither phosphorus nor potassium affects the uptake of Mg^{2+} by *Neocosmospora vasinfecta*. This uptake system transports zinc and manganese as well as magnesium and is inhibited by anaerobiosis and to a lesser extent by azide and dinitrophenol. In this species, the intracellular Mg^{2+} concentration exerts a strong control over uptake. There is an inverse relationship between the cytoplasmic levels of Mg^{2+} and the rate of uptake; as the external magnesium concentration increases, Mg^{2+} uptake decreases to almost zero (Budd, 1979).

Once inside the cell, magnesium participates in a large array of metabolic processes. Both ATP and ADP have an affinity for divalent cations because of their negatively charged phosphate groups. Magnesium and manganese are the two most common ions to form a complex with these nucleotides, and thus most cellular ATP and ADP are present as either the magnesium or manganese salts. In addition, almost all metabolic reactions involving the transfer of phosphate groups usually require either Mg^{2+} or Mn^{2+} as a cofactor. When bound to phosphate groups, these ions facilitate phosphate transfer by altering the distribution of charge on the molecule, thus making it more susceptible to attack by certain reactants (Mahler and Cordes, 1971).

Enzymes that are not involved in phosphate transfers may also require magnesium as a cofactor. Enolase, for example, is a glycolytic enzyme that becomes fully active only after the binding of four Mg^{2+} atoms. At high magnesium concentrations, however, enolase is inactivated due to the binding of additional Mg^{2+} at inhibitory sites (Faller et al., 1977). The Krebs cycle enzyme isocitrate dehydrogenase also requires Mg^{2+} or Mn^{2+} even though it does not catalyze a phosphate transfer. There are two forms of this enzyme: one is activated by Mg^{2+} and the other by Mn^{2+}. The Mg^{2+}-enzyme is more sensitive to metabolic regulation, but the Mn^{2+}-enzyme has a greater affinity for its substrates. Thus the relative concentrations of Mg^{2+} and Mn^{2+} determine which form of the enzyme is more active and in this way help regulate the functioning of the Krebs cycle (Barratt and Cook, 1978). Some of the other important Mg^{2+}-requiring enzymes in fungi are phosphofructokinase, DNA polymerase, glutamine synthetase, chitin synthetase, and cellulase.

Magnesium also has an effect on membrane structure and function. In *Neurospora crassa*, Mg^{2+} is required for the synthesis of membrane glycoproteins (Gold and Hahn, 1979). It also stabilizes membrane preparations by causing the component proteins to be most closely packed together (Lewis and Patel, 1978). This Mg^{2+}-induced dense packing appears to decrease membrane permeability and thus inhibits transport. In yeast, for example, magnesium partially blocks the exchange of K^+ for H^+ and K^+ for K^+ between the medium and the cell (Rodriguez-Navarro and Sancho, 1979).

In view of the broad range of biochemical processes influenced by magnesium, it is not surprising that this ion has been shown to affect many aspects of growth and development. Work with the fission yeast *Schizosaccharomyces pombe* has provided evidence for the involvement of magnesium in cell division, a process of central importance in the life cycle of fungi. Duffus and co-workers showed that the magnesium content of yeast cells decreases during the cell cycle and reaches a low point at the time the cell divides (Duffus and Paterson, 1974b). EDTA, which chelates Mg^{2+}, completely inhibits DNA synthesis after 30 minutes and RNA synthesis after 40 minutes (Ahluwalia et al., 1978). Also, the ionophore A23187, which promotes passive transport of divalent cations across the membrane and thus prevents the active accumulation of ions within the cell, alters the intracellular concentration of Mg^{2+} and rapidly inhibits cell division (Duffus and Paterson, 1974a). It is thought that cells of S. *pombe* require a high Mg^{2+} concentration in order for nucleic acid synthesis to occur. However, work with animal cells has shown that the microtubules making up the spindle fibers are unstable at high Mg^{2+} concentrations (Borisy et al., 1974). Thus, the formation of spindle fibers in S. *pombe* may occur only at the end of the cell cycle when the magnesium concentration has reached a critically low level (Duffus and Paterson, 1974a; Ahluwalia et al., 1978). Once cell division has occurred the high magnesium content is restored and the spindle fibers break down.

PHOSPHORUS

Most if not all cellular phosphorus occurs as phosphate (PO_4^{3-}). In this form phosphorus is an integral component of important macromolecules such as DNA, RNA, and phospholipids as well as smaller molecules such as NAD, FAD, thiamine-pyrophosphate, pyridoxal phosphate, vitamin B_{12}, and coenzyme A. In addition, as a component of nucleotides such as ATP, GTP, CTP, and UTP, phosphates are intimately involved with the storage and transfer of energy in the cell.

The structure of the phosphate molecule is responsible for many of its special properties. The four oxygen atoms are arranged at the apices of a tetrahedron with the phosphorus atom in the center (Figure 5.1a). This structure results in the four phosphorus–oxygen bonds being equivalent, with each single bond having some double-bond character and the double bond having some single-bond character (Figure 5.1b). This situation is termed **resonance,** and phosphate is a resonance hybrid molecule. The capability of resonance results in lower energy and thus greater stability for the phosphate molecule compared with molecules that are incapable of resonance. However, when phosphates are linked together in a chain as in an ATP molecule there is interference between the adjacent phosphate groups, and resonance is decreased. Thus the products of ATP hydrolysis have more resonance stability than the reactants. In addition, at pH 7.0 the polyphosphate portion of ATP has four closely spaced negative charges (Figure 5.1c). These similarly charged groups repel each other strongly, and some of this electrostatic stress is relieved when ATP is hydrolyzed. Thus, both electrostatic repulsion and resonance stability contribute to the large free energy difference between reactants and products and make nucleotide triphosphates so important in the energy economy of the cell (Lehninger, 1975).

Some of the "phosphate bond energy" present in ATP can be transferred to other

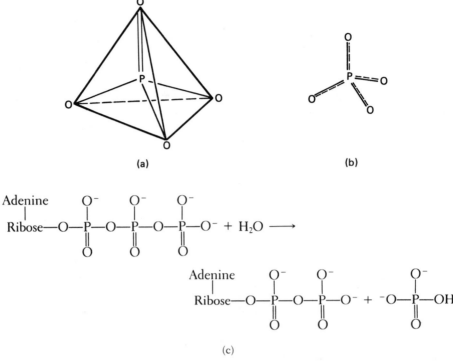

(a)   (b)

(c)

Figure 5.1 (a.) The tetrahedral structure of orthophosphate. (b.) The hybrid nature of the phosphorus–oxygen bonds as a result of resonance. (c.) The hydrolysis of ATP.

molecules via the terminal phosphate group. Such phosphorylation reactions play an important role in energizing a wide variety of cellular constituents. However, phosphorylation may result in increased activity in some cases and decreased activity in others. For example, the intermediates in glycolytic pathways are all phosphorylated and thus are quite reactive. On the other hand, phosphorylation of certain enzymes such as glutamate dehydrogenase from *Candida utilis* results in a tenfold reduction in enzyme activity (Hemmings, 1978).

Phosphorus is taken up by fungal cells usually as the orthophosphate ion (HPO_4^{2-}). Many species of fungi secrete phosphatases (phosphomonoesterases) into the environment which cleave the phosphate group from phosphorylated compounds (Hankin and Anagnostakis, 1975). The synthesis of extracellular phosphatase by *Penicillium brevi-compactum* is repressed by elevated levels of inorganic phosphate in the medium as well as by phosphorylated compounds such as AMP and even RNA (Ezhov, 1978; Ezhov et al., 1978).

In the few species studied, the uptake of phosphate is an active process. In *Neurospora crassa* there are two uptake systems, one of which is constitutive and the other derepressed by phosphate starvation (Lowendorf et al., 1975). *Saccharomyces cerevisiae* also has two systems. The first is an Na^+-dependent carrier that co-transports two Na^+ (or Li^{2+}) with each phosphate. The second carrier is Na^+-independent; each molecule of phosphate is co-transported with two H^+, and transport is accompanied by the efflux of one K^+. This system is inhibited by divalent cations at pH 7.2 but not at pH 4.5 (Roomans et al., 1977; Roomans and Borst-Pauwels, 1979).

Phosphate stimulates the transport of divalent cations such as calcium (Roomans et al., 1979) and magnesium (Okorokov et al., 1975). However, the uptake of certain monovalent cations is inhibited by phosphate (Roomans and Borst-Pauwels, 1977).

Once inside the cell, excess phosphate may be polymerized into long chains and stored as polyphosphate in vacuoles (Urech et al., 1978; Aiking et al., 1977). In *S.*

Table 5.2. Influence of $MgCl_2$ on Polyphosphate Accumulation, pH of Reaction Mixture, and Cell Mass of *Saccharomyces mellis*.

Determination	Concentration of Mg^{2+} (M)	Hours of Incubation				
		0	4	8	12	24
Total polyphosphate[b]	0	30	110	105	85	60
	10^{-4}	30	145	152	130	125
	10^{-3}	30	210	220	210	185
	0.1	30	225	252	250	215
	0.3	30	250	280	250	235
Total orthophosphate[b]	0	15	50	55	45	42
	10^{-4}	15	43	40	35	40
	10^{-8}	15	38	35	35	40
	0.1	15	38	35	35	40
	0.3	15	30	38	35	40
pH of reaction mixture	0	6.1	5.6	5.0	4.8	4.8
	10^{-4}	6.1	5.4	4.9	4.7	4.8
	10^{-3}	6.0	4.8	3.9	3.5	3.3
	0.1	5.5	4.5	3.7	3.4	3.1
	0.3	5.5	4.3	3.7	3.3	3.0
Cell mass[c]	0	1.36	1.36	1.48	1.44	1.40
	10^{-4}	1.44	1.60	1.60	1.60	1.64
	10^{-3}	1.48	1.68	1.96	2.16	2.78
	0.1	1.40	1.64	1.92	2.15	2.75
	0.3	1.44	1.56	1.88	2.04	2.28

Source: Weimberg (1975).

[a]Resting phosphate-starved cells of *S. mellis* at a concentration of 0.015 g/ml were incubated aerobically at 30°C in reaction mixtures containing 5% glucose, 0.3 M KCl, 0.015 M sodium orthophosphate (pH 6.5), and $MgCl_2$ at several concentrations. At intervals, volumes of 25 ml were removed, pH was measured, and cells were harvested by centrifugation, weighed, and assayed for ortho- and polyphosphates.

[b]Micromoles of phosphate per gram of cells.

[c]Grams per 100 ml.

cerevisiae polyphosphate accounts for 35–40% of the cell dry weight. As illustrated in Table 5.2, magnesium stimulates polyphosphate accumulation in S. *mellis* grown in low-phosphate medium but has no effect on cellular orthophosphate levels. In cells satiated with phosphate, magnesium has no effect on polyphosphate accumulation (Weimberg, 1975). Polyphosphate has been shown to play a role in the storage of cations in fungi. For example, large quantities of arginine become firmly bound to polyphosphate in vacuoles of S. *cerevisiae*. Several enzyme systems have been identified that regulate the binding of arginine in the vacuoles and its subsequent release into the cytoplasm (Durr et al., 1979).

Alterations in the phosphorus concentration of the medium often result in striking changes in many biochemical processes. For example, studies with S. *cerevisiae* have shown that decreasing the phosphate concentration from 3 g/liter to 82.1 mg/liter causes a decrease in the protein and glucan content of cell walls and a decrease in cellular sterols and triglycerides (Ramsey and Douglas, 1979). Lowering the phosphate content also decreases the rate of glycolysis in S. *cerevisiae* (Borst-Pauwels, 1967) and alters the relative participation of the Embden–Meyerhof pathway and the hexose-monophosphate shunt in C. *utilis* (Dawson and Steinhauer, 1977).

As can be expected, the addition of phosphate to a phosphorus deficient medium results in increased growth of fungi. Growth of *Candida* sp. is increased by a variety of organic and inorganic sources of phosphate (Shamis et al., 1968). Similarly, growth of *Alternaria tenuis* on a medium lacking phosphorus is strongly enhanced when calcium phosphate, sodium phosphate, potassium phosphate, or ammonium phosphate is added (Singh and Tandon, 1967a,b). However, increasing the concentration of KH_2PO_4 from 125 to only 150 ug/ml results in more than a 50% decrease in sporulation by *Claviceps purpurea* (Kybal et al., 1968). Conceivably, high levels of phosphate are inhibitory to enzymes crucial to the sporulation process in this organism.

Thus, concentrations of phosphorus that are optimal for certain processes are inhibitory for others. Most microorganisms grow best in media with phosphate concentrations of 0.3 to 300 μM. However, these concentrations are inhibitory for the production of many metabolites (Weinberg, 1974). In *Penicillium digitatum*, decreasing the phosphate content of the culture medium from 100 to 0.1 mM results in an 80-fold increase in ethylene production (Figure 5.2) (Chalutz et al., 1978). In addition, the activities of protein kinase, an enzyme catalyzing the phosphorylation of proteins, and phosphatase are also higher in phosphate-deficient cultures than in those that are phosphate satiated. This suggests that ethylene production may be regulated by phosphorylation and dephosphorylation reactions (Mattoo et al., 1979).

POTASSIUM

All fungi require potassium for growth, but this requirement is often difficult to demonstrate because of trace contaminants in the medium. For example, Slayman and Tatum (1964) found that even their "potassium-free" medium contained con-

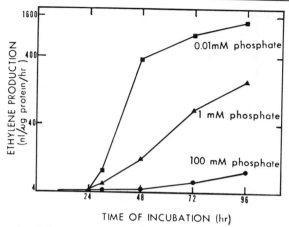

Figure 5.2 Effect of phosphate content of the medium on ethylene production by shake cultures of *Penicillium digitatum*. (Chalutz et al., 1978)

taminating potassium at the relatively high concentration of 16 μM. Figure 5.3a shows growth curves for *Neurospora crassa* when incubated in this medium alone or with potassium supplements. It is clear that although this species grows moderately well on the potassium contaminants, higher K^+ concentrations result in both a faster rate of growth and a delay in the onset of the stationary phase of growth. Figure 5.3b indicates that growth increases with increasing K^+ concentrations up to 1 mM; concentrations above 10 mM are inhibitory.

An assay of intracellular potassium showed that *N. crassa* accumulates significant amounts of potassium; by the end of the log phase of growth the intracellular potassium concentration is 160–200 mmoles/kg cell water in a medium containing only 0.3 mmoles potassium per liter. Figure 5.4 shows that once the external potassium concentration is decreased below 1 mM the drop in intracellular K^+ is balanced by a rise in intracellular Na^+. This suggests that at low K^+ concentrations there is an exchange of internal K^+ for external Na^+. Hydrogen ions and even extracellular K^+ can also substitute for Na^+ in this exchange (Slayman and Tatum, 1964; Slayman and Slayman, 1968).

The exchange of K^+ for K^+, K^+ for Na^+, and K^+ for H^+ are all energy dependent and follow Michaelis–Menton saturation kinetics. Competition experiments indicate that rubidium is transported by the same carrier as K^+ (Slayman and Slayman, 1968; Slayman and Tatum, 1965). These exchange mechanisms operate at pH 4–6 and appear to involve a single carrier transporting K^+ into the cell in exchange for K^+, Na^+ or H^+, depending on their relative concentrations. At pH 8, potassium uptake no longer exhibits Michaelis–Menton kinetics but rather gives a sigmoidal curve (Slayman and Slayman, 1970). This suggests that the carrier has several binding sites, some of which, when filled, induce a conformational change in the carrier. The Michaelis–Menton kinetics observed at low pH and the sigmoidal kinetics

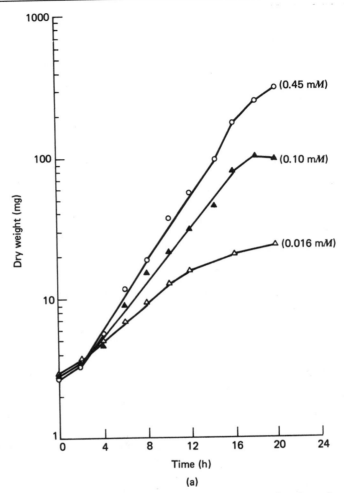

Figure 5.3 (a.) Growth curves for *Neurospora crassa* in "K⁺-free" minimal medium plus 0.45, 0.10, and 0.016mM K⁺. The lowest concentration represents trace contaminants in the "K⁺-free" medium alone.

observed at high pH can be explained by the presence of a single carrier with two binding sites, one for K^+ and the other for a modifier (H^+ at low pH and K^+ at high pH). The carrier can also have multiple subunits each with a site for K^+; at high pH's, H^+ acts as an allosteric effector causing interactions among the subunits.

The accumulation of large quantities of potassium within the cell is crucial for the proper functioning of the organism. Polyene antibiotics such as amphotericin B, candicidin, and nystatin are effective in treating mycotic infections such as vaginitis because they bind to membrane sterols and cause rapid leakage of 90% of the cellular

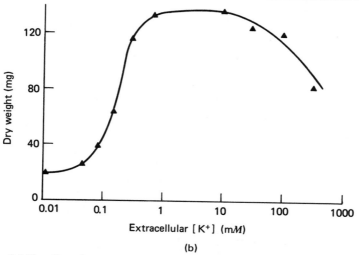

Figure 5.3 (b.) The effect of external K^+ concentration on the dry weight of N. *crassa* from 15 hour shake cultures. (Slayman and Tatum, 1964)

K^+ (Gale, 1974; Hammond et al., 1974). The loss of potassium results in the inhibition of a variety of metabolic processes, including glycolysis and respiration. The effects of these antibiotics can be reversed by raising the external K^+ concentration, thus decreasing the gradient for K^+ leakage, or by raising the external concentration of Ca^{2+} or Mg^{2+}, which reduces membrane damage thus preventing K^+ leakage (Hammond and Kliger, 1976).

A major role of potassium in fungi is regulating the cellular osmotic potential. The osmotic potential of the cell relative to that of the medium determines whether or not water will enter and thus bring about the turgor pressure necessary for growth.

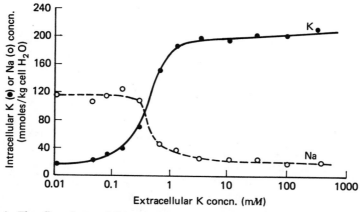

Figure 5.4 The effect of external K^+ concentration on the concentrations of intracellular K^+ and Na^+ in N. *crassa* from 15 hour shake cultures. (Slayman and Tatum, 1964)

The ability of marine fungi to grow in relatively high concentrations of NaCl is a problem of osmoregulation that is solved by K^+ transport. When exposed to high salt concentrations these fungi selectively take up K^+ and secrete either Na^+ or H^+ (Jennings, 1972). The influx of K^+ lessens the difference in osmotic potential between media and cytoplasm and also prevents Na^+ from building to inhibitory levels. Salt-tolerant fungi such as the yeast *Debaryomyces hansenii* have a higher intracellular K^+/Na^+ ratio than less tolerant species such as *Saccharomyces cerevisiae*. A comparison of these two species indicates that the former has a more efficient Na^+-extrusion and K^+-uptake system than the latter (Norkrans and Kylin, 1969).

Many of the physiological effects of K^+ are related to its role in transport processes. We have already seen that one of the phosphate uptake systems in *S. cerevisiae* involves proton symport coupled with the efflux of K^+. Glycine uptake in *S. carlsbergensis* occurs by a similar mechanism. In this organism the rate of glycine uptake decreases fourfold when the intracellular K^+ concentration is depleted by 20% (Eddy and Nowacki, 1971; Seaston et al., 1976). On the other hand, high extracellular K^+ inhibits glycine uptake in *S. carlsbergensis* (Eddy et al., 1970) and purine uptake in *S. cerevisiae* (Reichert et al., 1975).

Potassium also binds to proteins within the cell and activates enzymes such as aldolase, aldehyde dehydrogenase, and pyruvate kinase.

SULFUR

Sulfur is a required macronutrient because of its role as a component of proteins and vitamins. The amino acids cysteine and methionine each contain sulfur, and the free sulfhydryl group in cysteine makes it particularly important in stabilizing a protein's structure. This sulfhydryl group is very reactive, and two cysteine molecules can be oxidized and linked together via a disulfide bridge forming a molecule of cystine:

$$
\begin{array}{ccccccc}
\text{COOH} & & \text{COOH} & & & \text{COOH} & \text{COOH} \\
| & & | & \xrightarrow{\ 2H\ } & & | & | \\
\text{H}_2\text{N}-\text{C}-\text{H} & + & \text{H}_2\text{N}-\text{C}-\text{H} & & & \text{H}_2\text{N}-\text{C}-\text{H}\quad \text{H}_2\text{N}-\text{C}-\text{H} \\
| & & | & & & | & | \\
\text{CH}_2 & & \text{CH}_2 & & & \text{CH}_2-\text{S}-\text{S}-\text{CH}_2 \\
| & & | & & & \\
\text{SH} & & \text{SH} & & & \\
\text{Cysteine} & & \text{Cysteine} & & & \text{Cystine}
\end{array}
$$

Disulfide bridges in a protein help to maintain the proper folded shape and thus influence the ways that proteins function in cells. Sulfur is also a constituent of the coenzymes thiamine, biotin, lipoic acid, and coenzyme A; and certain enzymes contain complexes of iron and sulfur at the active site. Aconitase from *Candida lipolytica* is one of these "iron-sulfur proteins" and loses activity if these atoms are removed (Suzuki et al., 1976). Sulfur, as sulfate, also serves as an enzyme activator of yeast phosphoglycerate kinase (Scopes, 1978).

Sulfate (SO_4^{2-}) is the most common inorganic form of sulfur utilized by fungi. Studies with *Saccharomyces cerevisiae* (McReady and Din, 1974), *Penicillium chrysogenum* (Yamamoto and Segel, 1966), *Neurospora crassa* (Valee and Segel, 1971), and *Aspergillus nidulans* (Tweedie and Segel, 1970) all indicate that sulfate enters the mycelium by active transport. In *N. crassa* a low-affinity carrier is predominantly in the conidia, and a high-affinity system is mainly in the mycelium (Marzluf, 1973). In *P. chrysogenum* and *A. nidulans*, reciprocal inhibition studies have shown that thiosulfate ($S_2O_3^{2-}$) is also transported by the sulfate carrier system as are structural analogs of sulfate such as molybdate (MoO_4^{2-}) and selenate (SeO_4^{2-}) (Tweedie and Segel, 1970). Chromate is also a competitive inhibitor of sulfate uptake in *N. crassa* and *A. nidulans* (Lukasziewicz and Pieniazek, 1972; Roberts and Marzluf, 1971). In contrast, there are separate carriers for sulfite (SO_3^{2-}) and tetrathionate ($S_4O_6^{2-}$). The uptake of sulfide (S^{2-}) is apparently not carrier mediated (Tweedie and Segel, 1970).

A novel sulfate uptake systems has been reported by Reinert and Marzluf (1974). They found that *N. crassa* can utilize glucose-6-SO_4 as a sole source of sulfur although it cannot serve as a carbon source as well. The glucose-6-SO_4 molecule is transported intact into the cell via an energy-dependent carrier system that is distinct from the permeases responsible for the uptake of glucose and sulfate.

Once inside the cell, sulfate is rapidly converted into organic molecules and then into proteins. When *P. chrysogenum* is incubated in a medium containing $^{35}SO_4^{2-}$, the rapid depletion of label from the medium is matched by a rapid increase in labeled organic sulfur in the mycelium (Figure 5.5). Within five minutes the label appears in protein, and the amount of label in this fraction increases steadily during the experiment.

The incorporation of sulfur into protein can be divided into two phases. In the first phase sulfate is initially activated by reacting with ATP (Figure 5.6). The resulting adenosine phosphosulfate (APS) is then phosphorylated, the sulfur moiety reduced to sulfite, and the adenosine released. Finally, the free sulfite is reduced to

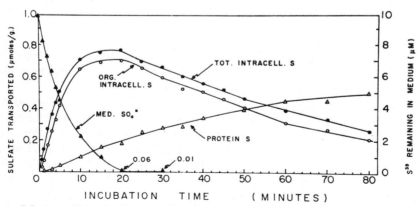

Figure 5.5 Intracellular distribution of ^{35}S in pulse-labeled mycelium of *Penicillium chrysogenum*. (Yamamoto and Segel, 1966)

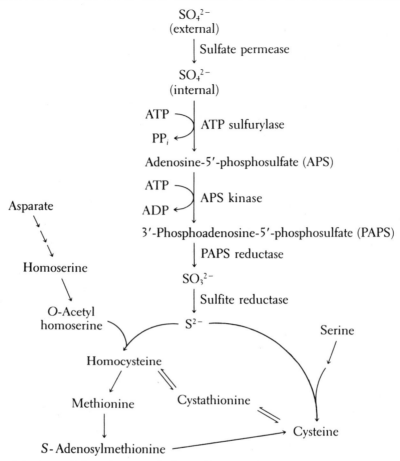

Figure 5.6 Some of the pathways involved in sulfate assimilation. (Modified from Morzycha and Paszewski, 1979)

sulfide. In the second phase sulfide is converted to organic sulfur, and cysteine and methionine are formed. Figure 5.6 shows that these two amino acids can be synthesized from either aspartate or serine, depending on the organism.

Even though sulfate uptake can occur independently of sulfate reduction and organic sulfur formation, the pathways from extracellular sulfate to intracellular cysteine and methionine are closely regulated by both precursors and endproducts (Yamamoto and Segel, 1966). For example, high intracellular levels of either sulfate or methionine repress the synthesis of sulfate permeases in *N. crassa* (Marzluf, 1973); in *S. cerevisiae* the permeases are regulated by the levels of sulfate, APS, methionine, and S-adenosylmethionine (Breton and Surdin-Kerjan, 1977). Other steps in the pathways from sulfate to protein are regulated in most cases by cysteine, methionine, and/or S-adenosylmethionine (Morzycha and Paszewski, 1979).

Table 5.3 Results of Tests for the Presence of Elemental Sulfur in Different Species of Fungi

Species	Medium or Natural Substrate	Structures Tested	Method of Analysis	
			TLC	H_2S
Physarum polycephalum	Medium 1	Sclerotia	+	+
Spongospora subterranea	Potato tubers	Spore balls	+	+
Albugo candida	Shepherd's purse	Sporangia	+	+
Pythium ultimum	Medium 2	Oogonia	+	+
Botrytis cinerea	Medium 3	Conidia	+	+
B. cinerea	Medium 3	Sclerotia	+	+
Erysiphe convolvulii	Leaves of small bindweed	Perithecia	–	–
Podosphaera leucotricha	Leaves of apple trees	Oidia	+	+
Gnomonia comari	Medium 3	Sporulating pycnidia	+	+
Pestalozzia sp.	Medium 3	Sporulating acervuli	+	+
Nectria cinnabarina	Deadwood	Stromata	+	+
Cytospora sp.	Medium 3	Sporulating pycnidia	+	+
Sclerotinia sclerotiorum	Medium 3	Sclerotia	+	+
Phomopsis viticola	Medium 3	Sporulating pycnidia	+	+
P. sclerotioides	Medium 3	Sclerotia and pycnidia	+	+
Monilia fructigena	Medium 3	Microconidia	+	+
Diatrype sp.	Medium 3	Perithecia	+	+
Eutypa armeniacae	Medium 3	Sporulating pycnidia	+	+
Claviceps purpurea	Rye	Ergot (sclerotia)	+	+
Melanconis sp.	Shoots of grapevine	Perithecia	+	+
Coniella diplodiella	Medium 3	Sporulating pycnidia	+	+
Septoria nodorum	Sterilized wheat	Sporulating pycnidia	+	+
Neurospora crassa	Medium 3	Conidia	+	+
N. crassa (fluffy)	Medium 3	Microconidia	–	–
Gliocladium roseum	Medium 3	Conidia	+	+
Trichoderma viride	Medium 3	Conidia	+	+
Rhizoctonia solani	Medium 3	Sclerotia	+	+
Typhula ishikariensis	Sterilized wheat	Sclerotia	+	+
Ustilago hordei	Barley	Chlamydospores	+	+
Tilletia caries	Wheat	Chlamydospores	+	+
Puccinia graminis	Barberry	Aeciospores	+	+
Gymnosporangium juniperi	Pine	Teliospores	–	–
Schizophyllum commune	Medium 3	Sporocarps and basidiospores	+	+
Agaricus bisporus	Commercial fungi	Young sporocarps	–	–

Source: Pezet and Pont (1977).

Fungi can store excess sulfur in various forms. Pools of soluble compounds such as methionine, cysteine, and homocysteine from the sulfur-assimilation pathways function as reserves in many species. In addition, choline-O-sulfate is a major storage form in higher fungi (Harada and Spencer, 1960). In *N. crassa* the choline sulfate ranges from 10% of total soluble sulfur in mycelia to 80% in ascospores (McGuire and Marzluf, 1974).

Elemental sulfur (S_8) has been detected in a large number of fungal groups (Table 5.3). This form of sulfur is especially prevalent in self-inhibited and dormant structures such as spores, chlamydospores, and sclerotia. Its presence in spores of *Phomopsis viticola* has been shown to be responsible for their failure to germinate under normal conditions. Because elemental sulfur has been shown to accept electrons from the middle of the electron transport chain, the element in this form may contribute to the low respiratory rate and thus the extent of dormancy in these structures (Pezet and Pont, 1975, 1977).

Fungi vary considerably in their ability to utilize different sources of sulfur. *Aureobasidium pullulans*, for example, possesses an oxidation pathway for enzymatically converting elemental sulfur (S_0) to dithionate ($S_2O_3^{2-}$) to tetrathionate ($S_4O_6^{2-}$) to sulfate, and each of these compounds can be utilized as a sulfur source (Killham et al., 1981). *Pestalotiopsis funera* can use a variety of organic and inorganic sources for growth and sporulation (Table 5.4). $MgSO_4$ is the best source for both these processes, and thiourea is not utilized at all. $NaSO_3$ results in good growth but is inhibitory to sporulation (Upadhyay and Dwivedi, 1977). Many lower fungi, including all members of the Blastocladiales, are unable to use sulfate as a sulfur source, but many of these species can grow on reduced forms of inorganic sulfur as well as organic sources. An isolate of *Puccinia graminis* is particularly fastidious in its sulfur requirements in that it cannot utilize any of the inorganic sources tested; only

Table 5.4 The Average Hyphal Dry Weight, Final pH, and Sporulation of *Pestalotiopsis funerea* on Different Sulfur Sources

Sulfur Compounds	Hyphal Dry Wt (in mg)	Final pH	Sporulation
Sodium sulfate	253	5.8	Poor
Sodium thiosulfate 5 hydrate	425	5.8	Poor
Potassium sulfate	346	5.8	Fair
Sodium disulfite	470	4.0	Fair
Sodium sulfite	310	5.0	Absent
Potassium peroxydisulfate	384	5.8	Good
Methionine	405	4.0	Fair
Thiourea	0	5.8	Absent
Magnesium sulfate	549	5.5	Excellent
Ammonium sulfate	471	4.0	Good
L-Cystine	488	4.0	Fair
No sulfur (control)	43	5.8	Absent

Source: Upadhyay and Dwivedi (1977).

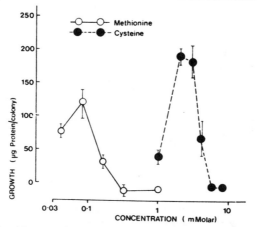

Figure 5.7 Effect of methionine and cysteine on growth of *Puccinia graminis*. (Howes and Scott, 1972)

homocysteine, cysteine, and methionine support growth. The latter two sources are effective at widely different concentrations (Figure 5.7) (Howes and Scott, 1972). Experiments using ^{35}S-labeled sulfide and sulfate showed that even though neither of these sources can be utilized for growth of *P. graminis*, both are taken up by the fungus. Sulfide can even be converted to cysteine and methionine. However, more than 70 percent of the label from sulfide incorporated into amino acids is released into the culture medium as cysteine, methionine, glutathione, cystinylglycine, and S-methylcysteine. Thus, sulfide cannot be utilized for growth of this isolate, evidently because the rate of conversion of sulfide to amino acids is not sufficient to replenish the leakage of organic sulfur from the fungus. More cysteine than methionine leaks from the mycelium, and this may explain why this species requires higher concentrations of cysteine than methionine for optimal growth (Figure 5.7). Sulfate cannot be utilized for growth of this isolate because even though it can be taken up it cannot be reduced to sulfide (Howes and Scott, 1973).

CALCIUM

The growth and reproduction of many species of fungi are enhanced when calcium is added to the culture medium. Calcium at 500 mg/liter causes almost a threefold increase in the growth of *Saccharomyces carlsbergensis* (Figure 5.8); of all the other divalent and trivalent cations tested, only strontium can duplicate the calcium effect (Lotan et al., 1976). Calcium also stimulates the production of sporangia in *Phytophthora fragariae* (Maas, 1976), *P. cactorum* (Elliot, 1972), and *Saprolegnia diclina* (Fletcher, 1979). Several species of *Pythium* require calcium for oospore formation (Lenny and Klemmer, 1966). The extent to which calcium is required often depends on the composition of the growth medium. For example, *Phytophthora parasitica* has an absolute requirement for calcium when cultured on a nitrate

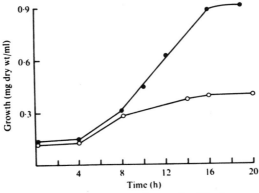

Figure 5.8 Effect of CaCl₂ and incubation time on the growth of *Saccharomyces carlsbergensis* 21. ○ control (no CaCl₂ added); ● with 500 μg/ml CaCl₂. (Lotan et al., 1976)

medium; when L-asparagine is the nitrogen source, calcium enhances growth but is not required (Hendrix and Guttman, 1970). The Oomycete *Achlya* has been used for several studies of calcium nutrition. In this organism calcium is required for the production of sporangia (Griffin, 1966). As seen in Table 5.5, 0.1 and 1.0 mM Ca^{2+} are the most effective concentrations for sporangial production.

Calcium uptake in *Achlya* exhibits saturation kinetics and is unaffected by respiratory inhibitors; accordingly, transport is assumed to occur by facilitated diffusion (LeJohn et al., 1974). In contrast, uptake of Ca^{2+} by *Schizosaccharomyces pombe* requires metabolic energy and appears to involve proton symport. This system has a

Table 5.5 Calcium Requirement for the Formation of Spores by *Achlya*

Treatment	mM	No. of Spore Balls after 24 h
CaCl₂	0.001	0
CaCl₂	0.01	0
CaCl₂	0.1	80
CaCl₂	1	80
CaCl₂	10	30
EDTA	1	0
EDTA	0.1	0
EDTA	0.01	0
EDTA	0.001	0
EDTA, CaCl₂	1.1	16
EDTA, CaCl₂	0.1,1	101
Water	—	0
Dilute pond water	1:3	78

Source: Griffin (1966).

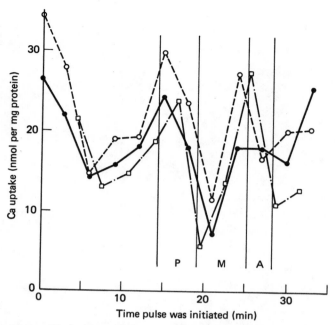

Figure 5.9 Uptake of $^{45}Ca^{2+}$ during mitosis in *Physarum polycephalum*. (Holmes and Stewart, 1977)

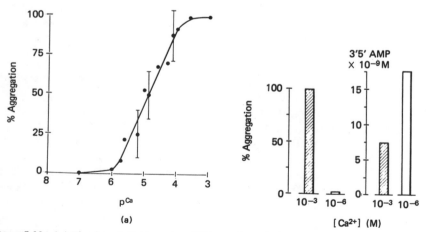

Figure 5.10 (a.) The degree of aggregation of *Dictyostelium discoideum* as a function of the negative log of the calcium ion concentration (pCa) in the medium. (b.) Cyclic 3'5' AMP production and aggregation when the calcium ion concentration was either 10^{-6} or 10^{-3} M. (After Mason et al., 1971)

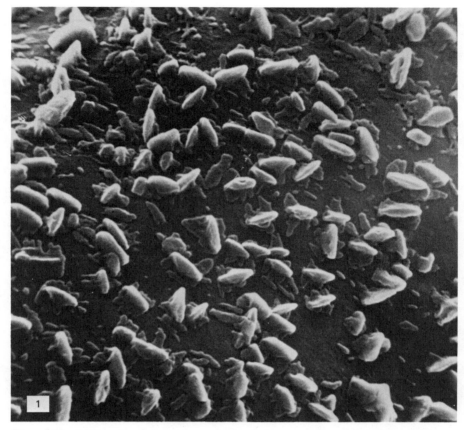

Figure 5.11   (1.) Scanning electron micrograph of washed sporangiophore wall of *Mucor mucedo* covered with tetragonal platelets of calcium oxalate crystals with their thin edges pointing outward. X12,500.

high affinity for calcium but also transports (with decreasing effectiveness) Sr^{2+}, Mn^{2+}, Co^{2+}, and Mg^{2+} (Boutry et al., 1977).

Calcium transport also occurs by an active process in the Myxomycete *Physarum polycephalum*, although some Ca^{2+} is taken up by pinocytosis. In this organism the internal Ca^{2+} concentration is quite stable, varying only tenfold when the external concentration varies 500-fold. There is considerable movement of Ca^{2+} across the membrane in both directions, and these fluxes occur in a cyclical manner (Holmes and Stewart, 1979). Figure 5.9 shows that the cyclical nature of $^{45}Ca^{2+}$ uptake is correlated with the stages of mitosis. There are accumulations of calcium during prophase and anaphase and a substantial outflow during metaphase. Paradoxically, these striking changes in uptake do not cause significant changes in the intracellular Ca^{2+} content, as measured by atomic absorption spectrometry. This suggests that the $^{45}Ca^{2+}$ taken up during mitosis accumulates in a small intracellular pool that is separate from a large pool containing the major portion of the total cellular calcium

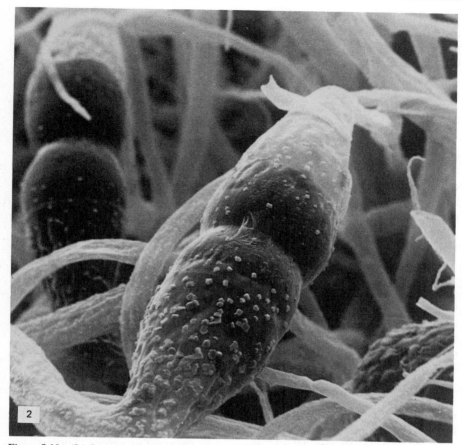

Figure 5.11   (2.) Scanning electron micrograph of fused gametangia of *M. mucedo* with suspensors. Note the increasing density of crystals on the older parts of the structures. X2200. (Urbanus et al., 1978)

(Holmes and Stewart, 1977). Because Ca^{2+} causes depolymerization of microtubule subunits in animal cells (Weisenberg, 1972), it is thought that perhaps the cyclical influx of Ca^{2+} during mitosis triggers a release of stored calcium near the nuclei. Thus, calcium may regulate mitosis through an effect on microtubule (and thus spindle fiber) formation (Holmes and Stewart, 1977).

Cyclical changes in intracellular calcium concentration have also been observed during cytoplasmic streaming in *P. polycephalum*; contraction of the cytoplasm occurs in those regions where Ca^{2+} accumulates (Ridgway and Durham, 1976). Calcium regulates the interaction between actin and myosin in this organism, and thus cytoplasmic streaming may result from reactions of these contractile filaments with calcium (Kato and Tonomura, 1975a,b).

In many fungi there is also evidence that calcium produces many effects at the plasma membrane without even entering the cell. Aggregation of *Dictyostelium discoideum* amoebae into a pseudoplasmodium is triggered by the secretion of cyclic

AMP (cAMP) from certain cells, which in turn is dependent on the presence of Ca^{2+} in the medium (Mason et al., 1971). There is no aggregation at Ca^{2+} concentrations below $10^{-6} M$, and the percentage of aggregation increases with increasing Ca^{2+} concentrations up to $10^{-4} M$ (Figure 5.10). The rapidity with which this calcium effect occurs (within 4 seconds) suggests that Ca^{2+} binds directly to the membrane and causes new cAMP binding sites to be uncovered (Juliani and Klein, 1977). Even D. discoideum cell "ghosts," which consists primarily of intact plasma membranes, are capable of Ca^{2+}-mediated adhesion (Sussman and Boschwitz, 1975).

Calcium is also a component of various cellular structures. Calcium in the form of calcium oxalate crystals have been observed on sporangia, gametangia, and zygophores of various Mucorales (Jones et al., 1976; Urbanus et al., 1978). The crystals in the latter structures are firmly attached to the cell wall and in some cases may be covered with wall material (Figure 5.11). Calcium is also a constituent of the gamma particle in zoospores of Blastocladiella emersonii and is released from this structure when the zoospore encysts (Hutchinson et al., 1977).

COPPER

Copper is a required micronutrient primarily because of its role as a metal activator of several fungal enzymes, particularly oxidases. However, at supraoptimal concentrations copper is a potent inhibitor of fungal growth and is a key component of several fungicides. Thus, the concentration of copper in growth media is of critical concern.

In most species studied, copper is taken up by active transport. For example, uptake of Cu^{2+} by Candida utilis obeys Michaelis–Menton kinetics, is inhibited by metabolic inhibitors, and is stimulated by the addition of a metabolizable energy source. There is evidence that Cu^{2+} is rapidly adsorbed to the cell surface prior to entry; treatment with 1 N HCl is sufficient to remove this copper (Khovrychev, 1973). In Dactylium dendroides, however, copper uptake is unaffected by metabolic inhibitors and appears to occur primarily by diffusion. When cells incubated with Cu^{2+} are transferred to a copper-free medium, no leakage of Cu^{2+} from the cells occurs. This suggests that once it is inside the cell, copper is rapidly bound in a manner that maintains the Cu^{2+} gradient necessary for diffusion into the cell (Shatzman and Kosman, 1978).

One mechanism for this intracellular sequestering of Cu^{2+} is by binding to proteins. Binding is often achieved via a copper–sulfur linkage to cysteine molecules. Shatzman and Kosman (1979) have described a cysteine-rich copper-binding protein in D. dendroides that may play a role in the chelation, storage, and transport of copper. Several copper-binding proteins are enzymes requiring Cu^{2+} for maximal activity. These include laccase, tyrosinase, ascorbic acid oxidase, uridine nucleosidase, cytochrome oxidase, and superoxide dismutase.

There is much variation among fungi in response to copper. Table 5.6 summarizes the effect of copper on the growth of 11 different species. At pH 7 copper stimulates growth of Penicillium notatum and Monilinia fructicola. For other spe-

Table 5.6 Influence of Copper Sulfate and pH on Growth (mg dry wt) of Various Fungi [Percentage Increase (+) or Decrease (−) from the Copper-Free Control at Each pH is Indicated in Parentheses]

Initial pH	Copper Sulfate (molarity)	Aspergillus flavus	Aspergillus niger	Penicillium notatum	Penicillium simplicissimum	Trichoderma koningi	Aspergillus oryzae	Fusarium culmorum	Fusarium oxysporum	Rhizopus nigricans	Monilinia fructicola	Stemphylium sarcinaeforme
7.0	0	114	115	55	100	163	150	128	113	93	62	165
	0.0004	126(−11)	107(−7)	96(+75)	106(+1)	146(−10)	155(+3)	126(−2)	127(+12)	108(+16)	76(+23)	149(−10)
	0.004	103(−10)	100(−13)	87(+58)	98(−2)	153(−6)	127(−15)	90(−30)	106(−6)	80(−14)	80(+29)	16(−90)
	0.04	97(−15)	28(−76)	91(+65)	56(−44)	121(−26)	65(−57)	0	0	16(−83)	0	0
4.0	0	135	152	94	96	142	150	125	108	119	82	125
	0.0004	112(−17)	157(+2)	92(−2)	107(+11)	134(−6)	141(−6)	135(+8)	88(−19)	106(−11)	104(+27)	163(+30)
	0.004	123(−9)	121(−20)	84(−11)	99(+3)	114(−20)	0	0	31(−71)	56(−53)	80(−2)	0
	0.04	87(−36)	28(−82)	65(−31)	38(−60)	112(−21)	0	0	0	0	0	0
3.0	0	136	173	65	−96	154	146	105	96	128	57	157
	0.0004	134(−2)	177(+2)	85(+31)	99(+3)	117(−24)	119(−18)	59(−44)	90(−6)	110(−14)	110(+93)	0
	0.004	75(−45)	112(−35)	107(+65)	94(−2)	102(−34)	35(−76)	18(−83)	0	67(−48)	50(−12)	0
	0.04	72(−47)	65(−62)	40(−38)	130(+35)	106(−31)	0	0	0	0	0	0
2.0	0	53	153	0	88	134	112	0	70	0	20	0
	0.0004	53(0)	158(+3)	0	83(−6)	132(−1)	106(−5)	0	53(−33)	0	31(+55)	0
	0.004	48(−9)	154(+1)	0	74(−16)	144(+7)	81(−28)	0	0	0	21(+5)	0
	0.04	34(−36)	140(−8)	0	65(−26)	76(−43)	0	0	0	0	0	0

Source: Modified from Starkey (1973).

cies, however, increasing copper results in greater inhibition of growth. It is evident from this table that in many cases the effect of Cu^{2+} is modified by the pH of the culture medium. For many of the species tested, copper inhibition increases with decreasing pH. For example, both *Aspergillus oryzae* and *Fusarium culmorum* will grow in the presence of 0.004 M copper at pH 7.0, but growth is completely inhibited at pH 4.0. For other species, such as *Aspergillus niger*, high pH increases copper toxicity (Starkey, 1973). This sensitivity to pH suggests the importance of binding sites in the mode of action of copper (Ross, 1975).

The presence of other media components also affects the response of fungi to copper. For example, the inhibitory effect of copper on growth of *Aphanomyces cochlioides* can be overcome by the addition of iron to the growth medium (Herr, 1973). In *Candida utilis* the extent of copper inhibition is decreased by either alkaline pH's or the presence of amino acids such as glycine, arginine, ornithine, aspartate, or glutamate. These amino acids form complexes with copper; the greater the proportion of amino acid to copper, the smaller the extent of copper inhibition. Organic acids such as citrate, malate, pyruvate, and oxalate also form complexes with copper; and large quantities of these acids reduce copper toxicity (Avakyan, 1971a,b).

Some fungal species are remarkably tolerant of high environmental copper concentrations. *Penicillium ochro-chloron* can grow in a medium saturated with copper sulfate (40% $CuSO_4$ solution) (Okamoto and Fuwa, 1974). Siegel (1973) described a strain of *Penicillium notatum* that attacks elementary copper and can grow on copper plates covered with a glucose-peptone-yeast extract broth. The growth rate under these conditions is decreased by only 25% even though the intracellular copper level rises to 8% of the dry weight and the mycelium turns blue! The hyphae actually etch the surface of the copper plate, suggesting that this species secretes copper-solubilizing enzymes into the environment.

The mechanisms for copper resistance in these species are not known. Copper-tolerant strains of *Saccharomyces cerevisiae* produce much H_2S (Nakai, 1961a,b), which is thought to be responsible for depositing copper as well as copper sulfide at the cell exterior (Ashida et al., 1963). Within the cell, excess copper may be precipitated as copper oxalate crystals (Levi, 1969) sequestered in copper-sulfur proteins (Kikuchi, 1964), or it may be chelated by fungal metabolites such as citrinin (Ahles et al., 1976). In addition, copper has been shown to have mutagenic properties that induce the formation of yeast strains with increased resistance to copper (Antoine, 1965).

IRON

Iron functions both as an enzyme activator and as a component of hemelike porphyrins, which are important in electron transfer. The role that iron plays in microorganisms, including fungi, has been extensively reviewed by Neilands (1974).

Iron is extremely insoluble, particularly at alkaline pH's; and thus it is not easily accumulated by cells. Ferrous iron is somewhat more soluble than Fe^{3+}. Many fungi have solved the problem of iron uptake by secreting chelators called **siderophores** which solubilize the iron and then transport it across the cell membrane. The most thoroughly studied siderophores are cyclic peptides that are derivatives of hydroxamic acid. Some of the most common siderophores secreted by fungi are illustrated in Figure 5.12. Ferrichrome, for example, is a hexapeptide with three hydroxamic acid derivatives and three molecules of glycine (Emery, 1971). *Aspergillus* species secrete not only ferrichrome but the related compounds ferricrocin, ferrichrysin, ferrirubin, and fusigen as well (Wiebe and Winkelmann, 1975). Coprogen is secreted by *Neurospora crassa* (Winkelmann, 1974). Other fungi secrete organic acids, such as citrate, that chelate iron and transport it into the cell (Winkelmann, 1979).

Siderophores appear to function in two major ways. Extracellular siderophores are excreted in large quantities when the fungus is grown under iron-deficient conditions. Because of their high affinity for iron (the affinity constant of ferrichrome for Fe^{3+} is approximately 10^{30}), the ligands effectively scavenge iron in the medium. The iron–ligand complex then enters the cell by an active transport mechanism, the iron is released, and the ligand is recycled to the outside. Release of iron is achieved either by specific esterases or by reduction of Fe^{3+} to Fe^{2+}, which has a much lower affinity for the siderophore (Emery, 1976; Ernst and Winkelmann, 1974; Straka and Emery, 1979). On the other hand, *Aspergillus nidulans, Penicillium chrysogenum*, and *N. crassa* have an additional set of siderophores, called cellular siderophores, which remain within the cell and are not secreted even under iron deficiency. For example, the cellular siderochrome in *P. chrysogenum* is ferrichrome, but coprogen is secreted when the cells are iron deficient. Cellular siderophores may function as a mechanism for storing iron or for transporting it across organelle membranes (Charlang et al., 1981).

Rhodotorula pilimanae produces a short-chain dihydroxamate called rhodotorulic acid, which chelates iron. Unlike the trihydroxamates discussed above, this compound does not enter the cell but rather functions as an "iron taxi," transporting iron to the cell surface. It then transfers its iron to membrane-bound chelating agents, which transport it into the cell (Carrano and Raymond, 1978).

Pilobolus sphaerosphorus and *P. crystallinus* are two of several fungal species that are unable to produce siderophores of any kind. In order to cultivate these species, one must add an exogenous siderophore to the culture medium. Bourret (1982) found that ascorbic acid, when added with $FeSO_4$, can fully replace ferrichrome and coprogen as a means of keeping iron in solution and thus available to these species under the required alkaline conditions of growth.

Once inside the cell, iron may be stored in various ways. Siderophores appear to be a common storage molecule. In addition, excess iron may be bound to polyphosphate complexes in a manner similar to magnesium. In *Penicillium chrysogenum*, up to 80% of the total iron may be stored in this manner (Okorokov et al., 1974). *Phycomyces blakesleeanus* stores iron in molecules of ferritin, which consist of a massive core of up to 4000 iron atoms surrounded by a protein shell. When examined using electron microscopy, these molecules are clustered in a cystalline pattern

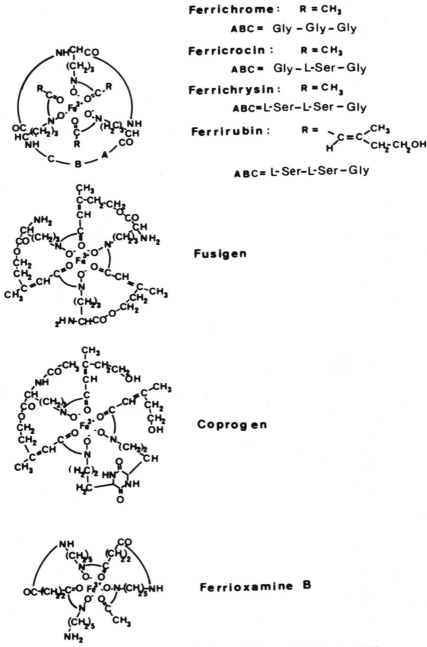

Ferrichrome: R = CH₃
ABC = Gly – Gly – Gly

Ferricrocin: R = CH₃
ABC = Gly – L-Ser – Gly

Ferrichrysin: R = CH₃
ABC = L-Ser – L-Ser – Gly

Ferrirubin: R =

ABC = L-Ser – L-Ser – Gly

Fusigen

Coprogen

Ferrioxamine B

Figure 5.12 Some fungal siderophores. (Wiebe and Winkelmann, 1975)

on the surface of lipid droplets in the cytoplasm (Figure 5.13). Ferritin becomes concentrated in spores of *P. blakesleeanus*, and iron is released during germination for use in biochemical reactions (David, 1974).

For many of its important biochemical roles, iron is transferred to protoporphyrin chelators forming heme and hemelike molecules. Heme is a prosthetic group for cytochromes found in the respiratory electron transport chain and also in enzymes such as catalase and peroxidase. Other enzymes contain iron that is not chelated as heme but rather is bound to sulfhydryl groups. These iron-sulfur proteins include succinic dehydrogenase, NADH dehydrogenase, and aconitase. In most iron-containing enzymes the iron undergoes reversible ferrous-ferric changes and thus participates in electron transfers.

Iron deficiency affects a host of biochemical processes. As would be expected from the involvement of iron in cytochromes, iron-deficient cells have decreased levels of ATP (Thomas and Dawson, 1978). Iron deficiency induced at the proper stage in the cell cycle in *N. crassa* seems to provoke an inhibition of DNA synthesis and causes a concomitant decrease in meiotic recombination frequency (Lander, 1972). Iron has been shown to form complexes with RNA (Pezzano and Coscia, 1970), and iron may interact with DNA in a manner that switches developmental processes on and off (Hall and Axelrod, 1978). In *Colletotrichum musae* iron in conidia inhibits germination, but this inhibition can be overcome by a variety of chelating agents (Harper et al., 1980).

One of the striking growth reponses of fungi to iron is seen with *Colletotrichum lagenarium*. On a purified medium in which iron is omitted, growth is decreased

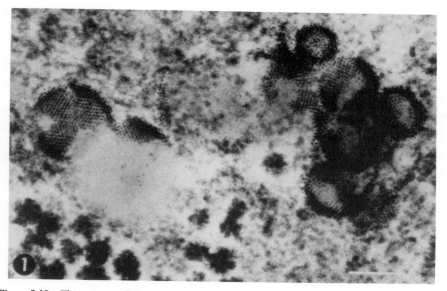

Figure 5.13 Thin section of *Phycomyces* sporangiophore showing lipid droplet covered with crystalline arrays of ferritin molecules. × 83,000. (David, 1974)

Table 5.7   Growth of *Colletotrichum lagenarium* on Purified Media
Deficient in Trace Elements

Deficient Element (−)	Mycelium (mg)	
	In Purified Medium	In Nonpurified Medium
Control[a]	308.0	346.8
− Fe	58.5	212.5
− Zn	100.7	75.3
− Cu	266.8	292.3
− Mn	258.8	360.0
− Mo	309.3	170.5
− Ca	337.5	401.5
− All elements	51.2	72.3

Source: Singh and Garraway (1975).

[a]Control = basal medium added with 1 ppm of Fe, Zn, Cu, Mn, Mo, and Ca.

by 81% compared with the control medium (Table 5.7). In fact, this decrease in growth is comparable to that which occurs when all trace elements are omitted. Mycelial growth increases in a linear manner in response to logarithmic increases in iron from 0.01 to 10 ppm (Figure 5.14). Above 25 ppm growth is inhibited (Singh and Garraway, 1975).

The form in which iron is added to the medium influences the growth rate.

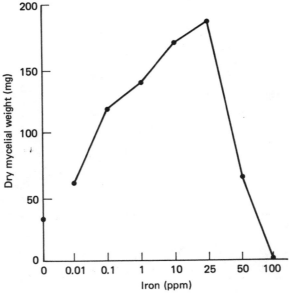

Figure 5.14 Growth of *Colletotrichum lagenarium* at different concentrations of iron. (After Singh and Garraway, 1975)

Table 5.8   Growth of *Colletotrichum lagenarium*
on Glucose–Asparagine Medium
Supplemented with Different Sources
of Iron

Added Compounds	Iron (ppm)	Mycelium (mg)
Control (Zn as basal dose)	—	21.3
FeCl$_3$	1.0	150.0
	0.1	150.7
FeSO$_4$	1.0	149.0
	0.1	143.0
Fe$_2$(SO$_4$)$_3$	1.0	213.0
	0.1	202.8
Fe (OH)$_3$	1.0	236.0
	0.1	48.5
Fe (OH)—(CH$_3$COO)$_2$	1.0	213.0
	0.1	183.0
Fe$_2$O$_3$	1.0	138.0
	0.1	32.8

Source: Singh and Garraway (1975).

Cuppett and Lilly (1973) showed that although 20 isolates of *Phytophthora infestans* can grow on iron supplied as either Fe^{3+} or Fe^{2+}, the lag period is reduced with the latter. For 18 of the isolates, growth on Fe^{3+} is enhanced appreciably by the inclusion of ascorbic acid in the medium. In most of the isolates, the onset of rapid growth on Fe^{3+} corresponds to the appearance of Fe^{2+} in the medium, which suggests that Fe^{2+} is used more rapidly than Fe^{3+}. In the absence of ascorbic acid, any Fe^{2+} in the medium is rapidly oxidized to Fe^{3+}; thus, one mechanism by which ascorbic acid enhances growth may be to convert Fe^{3+} to the reduced form, which is more soluble and thus more readily utilizable by this species.

In *C. lagenarium*, however, more growth is obtained with iron as the trivalent $Fe_2(SO_4)_3$ than with the divalent $FeSO_4$ (Table 5.8). This table also shows that the anion with which iron is supplied is at least as important as the valence. For example, at 1.0 ppm Fe_2O_3 is a relatively poor source of iron, but $Fe(OH)_3$ is an excellent source (Singh and Garraway, 1975).

MANGANESE

It is often difficult to demonstrate that a fungus has a specific requirement for manganese because in many reactions manganese and magnesium function equally well. As we have seen, Mn^{2+} binds to ATP and can function as the required cation in reactions involving the transfer of phosphate groups. We have also seen that although Mn^{2+} and Mg^{2+} may each activate a particular enzyme, the properties of that enzyme may depend on which cation is present. Recall that the manganese

Table 5.9a   Influence of the Addition of Manganese on the Activity of Glucanases and
Proteases in *Aspergillus nidulans*

Culture Conditions	α-1,3, Glucanase[a]	Laminarinase[a]	Amylase[a]	Protease[a]
2.5 μM Mn^{2+}	14.3	16.2	14.3	3.2
0. μM Mn^{2+}	14.1	9.5	5.8	1.0
2.5 μM Mn^{2+} added after 4 days	18.9	9.6	8.5	1.7

Source: Zonneveld (1976).

[a]All enzyme activities were measured on the 7th day and expressed as μM glucose equivalents/min • mg dry weight for the glucanases and μg albumin equivalents/min • mg dry weight for the protease.

form of isocitrate dehydrogenase has a greater affinity for the substrate but is less sensitive to metabolic regulation than the magnesium form. Other enzymes activated by Mn^{2+} include superoxide dismutase, protein kinase, adenyl cyclase, and RNA polymerase.

Manganese may also indirectly affect the activities of other fungal enzymes. For example, when *Aspergillus nidulans* is incubated in a manganese-free medium, the activities of several cell wall-degrading enzymes are low and those of certain enzymes associated with intermediary metabolism are high compared with tissue grown in a medium supplemented with 2.5 μM manganese (Tables 5.9a and 5.9b). In addition, alpha-1,4-glucan, the main source of carbon and energy for reproduction in this species, is absent from the cell wall under manganese deficiency. These changes may be responsible for the inability of A. *nidulans* to form fruiting bodies on a medium lacking manganese (Zonneveld, 1975a,b, 1976).

In *Aspergillus niger*, manganese deficiency results in reduced protein, lipid, and nucleic acid levels (Kubicek et al., 1979; Orthofer et al., 1979). In addition, ammonium and amino acids accumulate intracellularly, and large amounts of amino acids are excreted, perhaps due to an effect of manganese on membrane permeability (Kubicek et al., 1979; Habison et al., 1979).

In most of the species studied, manganese is taken up by the same carrier as magnesium. Transport is active, and in *Saccharomyces cerevisiae* uptake is apparently

Table 5.9b   Influence of the Addition of Manganese on Some Enzymes of Intermediary Metabolism in *Aspergillus nidulans*

Culture Conditions	Glucose-6-*P* DH	Aldolase	Malate DH	Phospho-Glucomutase	Phospho-Glucoisomerase	Glutamate DH
2.5 μM Mn^{2+}	0.7	0.9	44.8	6.1	29.1	0.1
0 μM Mn^{2+}	3.7	3.1	146.8	14.7	49.9	0.8
2.5 μM Mn^{2+} added after 4 days	2.4	1.9	83.7	9.9	35.4	0.4

Source: Zonneveld (1976).

[a]All enzyme activities were measured after 7 days of growth and expressed as nmoles NAD(P)(H)/min • mg dry weight.

associated with the hydrolysis of polyphosphates and ATP and with the efflux of potassium (Okorokov et al., 1979).

The manganese requirements of fungi vary considerably. Barnett and Lilly (1966) examined the effect of manganese on growth of several species (Table 5.10). *Chaetomium globosum* and *C. elatum* exhibit negligible growth when incubated in a manganese-free medium. However, they grow well if the medium is supplemented with as little as 0.01 ppm Mn^{2+}. On the other hand, growth of *C. bostrychodes*, *C. aureum*, and *Alternaria tenuis* is relatively insensitive to manganese concentration. Other species such as *Hypoxylon punctulatum* and *Glomerella cingulata* grow moderately well without added manganese, but growth increases with increasing Mn^{2+} concentration. Those species which grow poorly in the absence of Mn^{2+} have hyphae that are shorter, swollen, and more highly branched than hyphae grown in manganese medium (Figure 5.15). *Aspergillus parasiticus* has a yeastlike morphology when incubated in a manganese-poor medium but grows as a filament when the Mn^{2+} concentration is raised from 0.73 to 7.3 μM (Detroy and Ciegler, 1971).

Metabolite production is greatly influenced by manganese. In *Neurospora crassa* there is an absolute Mn^{2+} requirement for the conversion of mevalonic acid into terpenes (Bobowski et al., 1977). Manganese also promotes the synthesis of squalene by *Rhizopus arrhizus* (Campbell and Weete, 1978) and malformin by *Aspergillus*

Table 5.10   Growth of Selected Fungi on a Basal Glucose-Asparagine Medium, With and Without Added Manganese

Species	Days	Dry Mycelium in Liquid (mg)		
		No Added Mn	0.01 ppm	0.5 ppm
Chaetomium globosum Kunze	7	Trace	47	65
C. bostrychodes Zopf	7	13	50	65
C. bostrychodes	14	50	58	55
C. elatum Kunze	10	Trace	63	61
C. cochliodes Palliser	10	4	82	78
C. aureum Chivers	6	34	31	30
Hypoxylon punctulatum Berk. & Rav.	12	15	27	34
Neocosmospora vasinfecta E. W. Smith	7	54	77	76
Glomerella cingulata (Stonem.) Sp. & V. Schr.	6	62	66	72
Alternaria tenuis Auct.	12	59	63	57
		mm Diam on Agar		
Chaetomium globosum	8	8	45	45
C. bostrychodes	8	6	42	43
Neocosmospora vasinfecta	8	20	20	22

Source: Barnett and Lilly (1966).

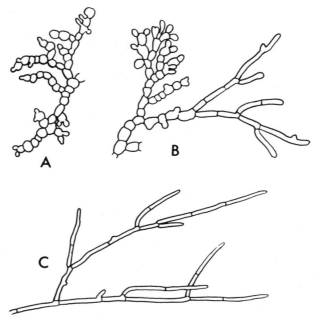

Figure 5.15 Hyphae of *Chaetomium globosum*. (A. and B.) Hyphae growing on basal glucose–aspara-gine agar showing short swollen, globose, or irregularly shaped cells, a characteristic symptom of man-ganese deficiency. (C.) Normal hyphae growing on the basal medium plus 0.5 μg/ml manganese. All ×750. (Barnett and Lilly, 1966)

spp. (Steenbergen and Weinberg, 1968). In the latter case Mn^{2+} is stimulatory only at concentrations between 10^{-5} and 10^{-6} M.

High concentrations of Mn^{2+} can sometimes produce surprising results. In yeast, $MnSO_4$ at 10mM strongly inhibits nuclear replication and partially inhibits the replication of mitochondrial DNA. This concentration also inhibits protein synthesis in both the cytoplasm and the mitochondria. Of the 11 cations tested, only Mn^{2+} had the ability to induce mitochondrial mutations (Putrement et al., 1977).

ZINC

Yeasts such as *Saccharomyces cerevisiae* and *Candida utilis* as well as filamentous fungi such as *Neocosmospora vasinfecta* all accumulate zinc in approximately the same manner. As is the case with many divalent cations, zinc is initially adsorbed to charged sites on the plasma membrane; the cation is taken into the cell via a carrier-mediated active transport mechanism. Very little efflux of zinc occurs (Failla et al., 1976; Paton and Budd, 1972; Ponta and Broda, 1970).

In *N. vasinfecta*, polyphosphate groups located on the cell membrane appear to be the anions to which zinc adsorbs. During uptake a phosphate group is enzymatically cleaved and is transported with Zn^{2+} into the cell. The external polyphosphate is then regenerated using ATP. Accumulation of zinc by this organism is inhibited by both phosphate and glucose. Phosphate indirectly affects uptake at pH 6.5 and above by precipitating zinc from the medium as $Zn_3(PO_4)_2$. The effect of glucose is more direct and appears to be due to a competition between glucose and zinc for polyphosphate at the cell surface (Paton and Budd, 1972).

Zinc is a functional component of a variety of fungal enzymes ranging from those involved in intermediary metabolism to the synthesis of DNA and RNA (Table 5.11). One of the most thoroughly studied zinc enzymes is alcohol dehydrogenase, which contains four zinc atoms; replacing zinc with cobalt results in an 83% decrease in activity (Sytkowski, 1977). In yeast RNA polymerase, zinc aids in attaching the entering nucleotide to the growing RNA chain (Lattke and Weser, 1977).

Because of the large number of enzymes requiring zinc, omission of this cation from a culture medium may have widespread physiological consequences. For example, the RNA content of *Candida albicans* grown for 24 hours in a zinc-deficient medium is one-third that found in cells incubated with zinc (Yamaguchi, 1975). In *Rhodotorula gracilis*, zinc deficiency results not only in decreased RNA and protein synthesis but a decrease in the number of mitochondria and in the organization of the cristae (Cocucci and Rossi, 1972). For *Verticillium albo-atrum* the addition of 5 μg/ml zinc to the growth medium results in an increase in the total cellular nitrogen content and rate of oxygen uptake (Table 5.12). Notice that growth in the presence of zinc is stimulated 326% after only three days! In this organism the response to zinc is greatest when L-alanine is the nitrogen source as compared with nitrate, urea, and ammonium nitrate (Throneberry, 1973).

Table 5.11   Some Zinc-containing Enzymes Found in Fungi

Aldolase
Alcohol dehydrogenase
D-Glyceraldehyde-3-phosphate dehydrogenase
D-Lactate cytochrome c reductase
NADH-dependent lactate dehydrogenase
Nuclease P_1
Phosphomannose isomerase
RNA polymerase
Aminopeptidase
Neutral proteinase I
Pyruvate decarboxylase
Superoxide dismutase
DNA polymerase[a]
Hexokinase[a]
Phosphatase[a]
Phosphoglucomutase[a]

Source: Adapted from Failla (1977).
[a]Probable zinc enzymes, but evidence is incomplete.

Table 5.12   Effect of 72-Hour Exposure to Zinc on Growth and
Metabolism of *Verticillium albo-atrum* Obtained
from Cultures of Varying Ages[a]

Parameter Measured	Culture Age in Days as Inocula Spores			
	3	5	7	10
Total dry wt produced	+ 326[b]	+ 133	+ 42	+ 208
Total N accumulated	+ 435	+ 244	+ 121	+ 396
N content	+ 39	+ 50	+ 60	+ 60
Total O_2 uptake	+ 464	+ 218	+ 92	+ 516
O_2 uptake per unit dry wt	+ 32	+ 37	+ 44	+ 99
O_2 uptake per unit N	+ 2	- 7	- 7	+ 24

Source: Throneberry (1973).

[a]Growth in Czapek-Dox broth; Zn at 5 μg/ml added as $ZnCl_2$.

[b]Data represent percentage changes from controls of equivalent cells grown without
Zn.

In several dimorphic fungi such as *Histoplasma capsulatum* (Pine and Peacock,
1958), *Mucor rouxii* (Bartnicki-Garcia and Nickerson, 1962), and *Candida albicans*
(Widra, 1964), increasing zinc concentrations stimulate the formation of yeastlike
cells. Figure 5.16 illustrates this phenomenon for *C. albicans*. Notice that the
optimum zinc concentration for the formation of yeast cells is greater than that for
growth and that high concentrations are inhibitory to both (Yamaguchi, 1975). This
effect of zinc is specific for yeast and cannot be duplicated by Ca^{2+}, Cu^{2+}, Mn^{2+},
or Co^{2+} (Bedell and Soll, 1979).

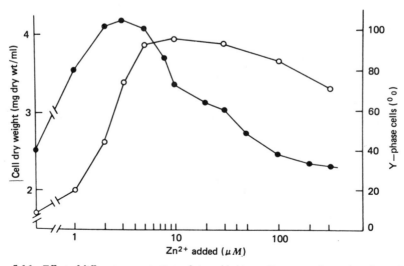

Figure 5.16   Effect of different concentrations of zinc added to medium on total growth and morphology
of *Candida albicans*. ● dry weight of yeast; ○ proportion of yeastlike cells. (Yamaguchi, 1975)

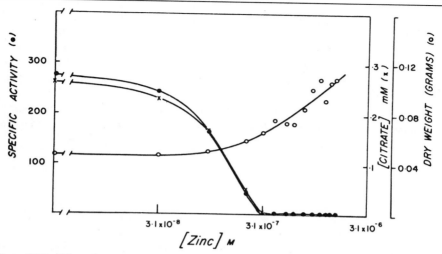

Figure 5.17 Effect of increasing amounts of zinc on growth and citrate accumulation by *Aspergillus niger*. (Wold and Suzuki, 1976)

Lastly, zinc affects the production of a variety of metabolites. In *Aspergillus niger*, maximum citrate production occurs with low concentrations of zinc in the medium. As zinc concentration increases, citrate production drops off but growth increases (Figure 5.17). (Wold and Suzuki, 1976). Zinc produces a similar effect on the production of the pigment cynodontin by *Helminthosporium cynodontis* (White and Johnson, 1971). These studies provide support for the widely reported observation that metabolites are excreted under conditions that are suboptimal for growth.

MOLYBDENUM

Molybdenum has been most thoroughly studied in relation to its role as a cofactor for nitrate reductase (Chapter 4). Radioactively labeled Mo^{2+} is incorporated into nitrate reductase of *Neurospora crassa*, and Mo^{2+} enhances the *in vitro* assembly of this enzyme (Lee et al., 1974b). Similarly, Garrett and Cove (1976) demonstrated that a molybdenum-containing component is necessary for nitrate reductase activity in *Aspergillus nidulans*.

Because of the specific molybdenum requirement for this enzyme, all fungi incubated in a synthetic medium containing nitrate as the sole nitrogen source would be expected to show a requirement for molybdenum. This is not always seen, however, probably due to the presence of trace amounts of Mo^{2+} in the medium. Thind and Rawla (1967) found that when two species of *Helminthosporium* are incubated in a nitrate medium, only *H. oryzae* is substantially affected by the addition of Mo^{2+}; in *H. sacchari* growth is only slightly stimulated.

These differing responses to Mo^{2+} among species may be due to variations in the extent of involvement of this cation in other cellular reactions. Molybdenum is a required cofactor for xanthine dehydrogenase (Lewis and Scazzochio, 1977), and molybdate (MoO_4^{2-}) can substitute for phosphate in certain reactions (Byers et al., 1979). Molybdate at high concentrations inhibits the growth of *Saccharomyces cerevisiae* primarily by binding to sugars such as glucose in the growth medium. The addition of nonfermentable sugars removes this inhibition because these sugars compete with glucose for molybdate (Zemek et al., 1975).

The uptake of Mo^{2+} is assumed to be the same as for other divalent cations (Kotyk and Janacek, 1975), but in *Penicillium chrysogenum* and *A. nidulans*, molybdenum and molybdate are transported by the sulfate carrier (Tweedie and Segel, 1970).

SODIUM

Marine fungi are the most thoroughly studied species known to require sodium for growth. Several species of the Oomycete *Thraustochytrium* will not develop without Na^+ in the growth medium. The optimal concentration (as NaCl) is 2.5–3.0%, and development is inhibited above 5% (Goldstein, 1963).

Because Na^+ can function as an osmoticum and as a nutrient, experiments are often necessary to distinguish the physical from the chemical effects of this element. Figure 5.18 indicates that a marine species of *Labyrinthula* grows best at a NaCl concentration of 2.4% (Sykes and Porter, 1973). To determine if sodium is functioning primarily as an osmoticum, growth was measured using a basal medium containing decreasing amounts of NaCl and sufficient mannitol or KCl to bring the final osmotic potential of the medium equal to that attained with 2.4% NaCl. As evident from Figure 5.18, neither mannitol nor KCl can substitute for suboptimal NaCl. In this species, therefore, the effect of Na^+ is not due to any osmotic effect nor can K^+ substitute for Na^+.

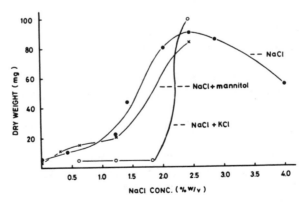

Figure 5.18   Effect of NaCl concentration on growth of *Labyrinthula*. ● NaCl alone, osmotic concentration increasing. ×NaCl with mannitol, osmotic concentration brought up to the equivalent of 2.4% (w/v) NaCl with mannitol. ○ NaCl + KCl. Osmotic concentration brought up to the equivalent of 2.4% (w/v) NaCl with KCl. (Sykes and Porter, 1973)

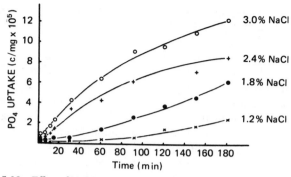

Figure 5.19   Effect of NaCl concentration on phosphate uptake by *Labyrinthula*.

The uptake of Na⁺ has already been discussed in relation to the uptake of other elements. Sodium is taken up both by co-transport with phosphate and by an active $Na^+/K^+/H^+$ exchange system. One reason for the specific Na^+ requirement by *Thraustochytrium* and *Labyrinthula* mentioned above has been shown to be its stimulatory effect on phosphate uptake (Figure 5.19) (Siegenthaler et al., 1967; Sykes and Porter, 1973).

Most fungi, however, are inhibited by sodium at concentrations above a certain low level. There have been many studies examining the adverse effects of Na^+ as well as the mechanisms by which fungi are able to survive in saline environments. Tresner and Hayes (1971) tested 975 species for their tolerance to NaCl (Table 5.13). Basidiomycetes are the most sensitive to NaCl; more than half the species in this group are unable to withstand NaCl concentrations greater than 2%. Members of the genera *Penicillium* and *Aspergillus* are among the most resistant; most of the species in this group are able to grow in media containing NaCl at 20% or more.

The inability of most fungi to grow in environments containing high concentrations of NaCl may be due to several factors. Salinity places a severe osmotic strain on the cells by causing the efflux of water. When *Debaryomyces hansenii*, a salt-tolerant species, is grown in a medium containing increasing salt concentrations, the

Table 5.13   NaCl Tolerance of Major Fungal Classes

Class	Number of Strains	Species	Genera	<5%	5	10	15	20
Basidiomycetes	162	104	47	93.3	8.6	1.1	—	—
Ascomycetes	196	160	87	17.5	33.1	35.6	11.9	1.9
Deuteromycetes	428	358	182	8.4	22.9	41.1	21.5	6.1
Phycomycetes	330	148	33	12.8	9.5	40.5	34.5	2.7

The header spans: *Number of* (Strains, Species, Genera); *Number of Species Tolerating the Indicated Percentage of NaCl* (<5%, 5, 10, 15, 20).

Source: Adapted from Tresner and Hayes (1971).

fungus synthesizes increasing amounts of glycerol. The accumulation of glycerol minimizes the effects of increased salt by contributing to the osmotic balance of the cell (Gustafsson and Norkrans, 1976).

High sodium concentrations have been shown to affect fungal growth and development in a variety of ways. For both *Aspergillus niger* and *Paecilomyces lilacinum*, the greatest number of conidia are produced at 1% NaCl; salinity values higher and lower than this are inhibitory. NaCl also inhibits germination of sporangia and the release of zoospores by *Phytophthora cactorum* (Gisi et al., 1977). High intracellular sodium concentrations have been shown to inhibit several biochemical processes including respiration and fermentation (Norkrans, 1968), the synthesis of aflatoxins (Uriah and Chipley, 1976), and cellulolytic activity (Malik et al., 1980). Marine fungi thrive because they have evolved mechanisms for preventing the intracellular accumulation of sodium. These include membranes with a low passive permeability to Na^+ as well as a specific ion pump for transporting K^+ in and Na^+ out (Jennings, 1972.)

Indeed, Jennings (1972) has suggested that a major cause of reduced growth of fungi in media containing high concentrations of Na^+ is its effect on membrane permeability. The addition of 50 mM sodium to the growth medium causes a doubling of the efflux of K^+ from the mycelium of *Dendryphiella salina* (Jennings and Aynsley, 1971). Sodium also increases the permeability of the membrane to sugars and sugar alcohols (Allaway and Jennings, 1970; Jennings and Austin, 1973). Thus sodium may inhibit growth by causing the efflux of critical cellular constituents (Jones and Jennings, 1965).

COBALT

Although cobalt participates in many cellular reactions, it is uncertain how essential this element is for fungi. Cobalt functions primarily as an enzyme activator, but the enzymes activated by Co^{2+} generally are activated equally well by other divalent cations. For example, cobalt can substitute for zinc in aldolase (Egorov and Kulaev, 1976) and glutamine synthetase (Kretovich et al., 1978).

In animal cells, cobalt is a required component of vitamin B_{12} (cobalamin). Until recently plants and fungi have generally been assumed not to synthesize or require this vitamin. In 1979, Poston and Hemmings demonstrated that cobalamin is present in *Candida utilis* and that this species incorporates ^{57}Co from the medium into this vitamin. Cobalt as cobalamin was found to be involved in the synthesis of methionine in this organism and to stimulate the enzyme leucine-2,3-aminomutase. Thus, vitamin B_{12} may likewise be synthesized by other fungi from trace contaminants of Co^{2+} in the medium, and the requirement for cobalt by fungi may be more widespread than it once appeared.

In *Saccharomyces cerevisiae* and *Neurospora crassa*, cobalt is transported via the general cation uptake system (Fuhrmann and Rothstein, 1968; Venkateswerlu and Sastry, 1970). Uptake consists of an initial binding to the cell surface followed by active transport across the membrane. Two K^+ are released for each Co^{2+} taken up (Norris and Kelly, 1977).

Yeast mitochondria accumulate Co^{2+} in proportion to the concentration of Co^{2+} in the medium. The Co^{2+} accumulation causes a small increase in the density of mitochondria and a change in their morphology. Cobalt at concentrations of 10^{-5} to 10^{-4} M reversibly stimulates the rate of respiration in intact mitochondria in the presence of α-ketoglutarate. With succinate as the substrate, respiratory stimulation of Co^{2+} is irreversible. These and other data indicate that Co^{2+} may act on at least two sites in yeast mitochondria (Tuppy and Siegart, 1973).

OTHER INORGANIC ELEMENTS

A variety of other inorganic elements are used in fungal nutrition studies, although little is known about their essentiality or modes of action. We have mentioned that inorganic elements such as cadmium, nickel, selenium, and strontium have been used primarily to test for possible inhibitory effects on the uptake of other elements. Studies with uranium have shown that Saccharomyces cerevisiae can adsorb heavy metals at the cell surface, and thus fungi may play an increasingly important role in the industrial recovery and/or decontamination of these metals (Strandberg et al., 1981).

Boron stimulates the growth of several fungi, such as Cercospora dolichi (Rawla et al., 1977), Myrothecium verrucaria, Stachybotrys atra, and Alternaria tenuis (Naplekova and Anikina, 1970).

Scandium can partially replace the requirement for calcium in Alternaria solani (Steinberg, 1950). More recently, Chahal and Rawla (1977) demonstrated that scandium is one of five trace elements required for the proper growth of Cercospora granati.

Vanadium is present in appreciable quantities in yeast extract and is responsible for the stimulatory effect of yeast extract on the synthesis of the glycoprotein mating factor in Hansenula wingei (Crandall and Caulton, 1973). Some of the vanadium in yeast cells appears to be complexed with RNA (Pezzano and Coscia, 1970). It also accumulates in high concentrations in Amanita muscaria (Meisch et al., 1977, 1978) and is required for growth of the thermophilic yeast Candida slooffii (Roitman et al., 1969). In Glomerella cingulata vanadium depresses lipid biosynthesis (Anekwe, 1976), and in Neurospora it inhibits growth on nitrate medium by acting as a competitive inhibitor of nitrate reductase (Lee et al., 1974a).

SUMMARY

Inorganic nutrients are critically important for the growth and development—even the survival—of fungi. This chapter has demonstrated the varied roles that each element plays both within the individual and among different species. We should emphasize that even though much elegant work has been done clarifying the many effects of these elements, very little is known about the details of their uptake, storage, and utilization in most fungi.

Because they are present in relatively high concentrations and thus are more easily studied, macronutrients and their modes of action are more thoroughly understood than are micronutrients.

Magnesium stabilizes ATP, RNA, and enzymes and may also affect spindle fibers. Phosphorus is not only a structural component of nucleic acids, phospholipids, and vitamins but as polyphosphate acts as a mechanism for storing cations such as magnesium and iron. Potassium, by shunting back and forth across the membrane, maintains the electrical and osmotic balances of the cell. Sulfur is a component of certain amino acids, vitamins, and metabolites. Calcium produces many effects by binding to the cell membrane, and it also influences the functioning of microtubules and microfilaments.

The roles of the micronutrients are poorly understood and for some fungi it is difficult to determine their essentiality. All function as enzyme cofactors: copper for various oxidases, iron for heme and nonheme iron proteins, zinc for many enzymes including alcohol dehydrogenase, molybdenum for nitrate reductase, and manganese for many enzymes including some of those affected by magnesium. Sodium is also required by certain marine species. Other elements such as cobalt, boron, scandium, and vanadium may also be required by fungi but in quantities too minute to detect using present methods.

Problems of precipitation and contaminants as well as complications caused by interference among elements all present challenges to fungal physiologists and others who seek to understand the roles of inorganic nutrients in the growth and development of fungi.

REFERENCES

Ahles, L. A., Q. Fernando, O. R. Rodig, and J. M. Quante. 1976. Interaction of copper (II) with the fungal metabolite citrinin. *Bioinorg. Chem.* **5**: 361–366.

Ahluwalia, B., J. H. Duffus, L. J. Paterson, and G. M. Walker. 1978. Synchronization of cell division in the fission yeast *Schizosaccharomyces pombe* by ethylenediaminetetra-acetic acid. *J. Gen. Microbiol.* **106**: 261–264.

Aiking, H., A. Sterkenburg, and D. W. Tempest. 1977. The occurrence of polyphosphates in *Candida utilis* NCYC 321, grown in chemostat culture under conditions of potassium- and glucose-limitation. *FEMS Microbiol. Lett.* **1**: 251–254.

Allaway, A. E., and D. H. Jennings. 1970. The influence of cations on glucose transport and metabolism by, and the loss of sugar alcohols from, the fungus *Dendryphiella salina*. *New Phytol.* **69**: 581–593.

Anekwe, G. E. 1976. Effect of vanadium pentoxide on the incorporation of (2-^{14}C)-acetate into fungal lipids. *Physiol. Plant.* **38**: 305–306.

Antoine, A. 1965. The resistance of yeast to copper ions: III. *Saccharomyces cerevisiae*, yeast foam, nature of two resistant forms. *Exp. Cell. Res.* **40**: 570–584.

Ashida, J., N. Higashi, and T. Kikuchi. 1963. An electron microscopic study on copper precipitations by copper resistant yeast cells. *Protoplasma* **57**: 27–32.

Avakyan, Z. A. 1971a. Comparative toxicity of free ions and complexes of copper with organic acids for *Candida utilis*. *Mikrobiologiya* **40**: 305–310.

Avakyan, Z. A. 1971b. Comparative toxicity of free ions and complexes of copper with amino acids for *Candida utilis*. *Mikrobiologiya* **40**: 417–423.

Barnett, H. L., and V. G. Lilly. 1966. Manganese requirements and deficiency symptoms of some fungi. *Mycologia* **58**: 585–591.

Barratt, D. W., and R. A. Cook. 1978. Role of metal cofactors in enzyme regulation. Differences in the regulatory properties of the *Neurospora crassa* nicotinamide adenine dinucleotide specific isocitrate dehydrogenase depending on whether Mg^{2+} or Mn^{2+} serves as divalent cation. *Biochemistry* **17**: 1561–1566.

Bartnicki-Garcia, S., and W. J. Nickerson. 1962. Nutrition, growth and morphogenesis of *Mucor rouxii*. *J. Bacteriol.* **84**: 841–858.

Bedell, G. W., and D. R. Soll. 1979. Effects of low concentrations of zinc on the growth and dimorphism of *Candida albicans*: Evidence for zinc-resistant and -sensitive pathways for mycelium formation. *Infect. and Immun.* **26**: 348–354.

Bobowski, G. S., W. G. Barker, and R. E. Subden. 1977. The conversion of (2-^{14}C)-mevalonic acid into triterpenes by cell-free extracts of a *Neurospora crassa* albino mutant. *Can. J. Bot.* **55**: 2137–2141.

Borisy, G. E., J. B. Olmstead, J. H. Marcum, and C. Allen. 1974. Microtubule assembly in vitro. *Fed. Proc.* **33**: 167–174.

Borst-Pauwels, G. W. F. H. 1967. A study of the factors causing a decrease in the rate of phosphate uptake during phosphate accumulation. *Acta Bot. Neerl.* **16**: 115–131.

Bourret, J. A. 1982. Iron nutrition in *Pilobolus*: ascorbic acid as a replacement for siderphores. *Exp. Mycol.* **6**: 210–215.

Boutry, M., F. Foury, and A. Goffeau. 1977. Energy-dependent uptake of calcium by the yeast *Schizosaccharomyces pombe*. *Biochim. Biophys. Acta* **464**: 602–612.

Breton, A., and Y. Surdin-Kerjan. 1977. Sulfate uptake in *Saccharomyces cerevisiae*: Biochemical and genetics study. *J. Bacteriol.* **132**: 224–232.

Budd, K. 1979. Magnesium uptake in *Neocosmospora vasinfecta*. *Can. J. Bot.* **57**: 491–496.

Byers, L. D., H. S. She, and A. Alayoff. 1979. Interaction of phosphate analogues with glyceraldehyde-3-phosphate dehydrogenase. *Biochemistry* **18**:2471–2480.

Campbell, O. A., and J. D. Weete. 1978. Squalene biosynthesis by a cell-free extract of *Rhizopus arrhizus*. *Phytochemistry* **17**: 431–434.

Carrano, C. J., and K. N. Raymond. 1978. Coordination chemistry of microbial iron transport compounds: rhodotorulic acid and iron uptake in *Rhodotorula pilimanae*. *J. Bacteriol.* **136**: 69–74.

Chahal, S. S., and G. S. Rawla. 1977. Trace element requirement for growth of *Cercospora granati*. *Indian Phytopathol.* **30**: 47–50.

Chalutz, E., A. K. Matoo, J. D. Anderson, and M. Lieberman. 1978. Regulation of ethylene production by phosphate in *Penicillium digitatum*. *Plant Cell Physiol.* **19**: 189–196.

Charlang, G., B. Ng, N. H. Horowitz, and R. M. Horowitz. 1981. Cellular and extracellular siderophores of *Aspergillus nidulans* and *Penicillium chrysogenum*. *Mol. Cell. Biol.* **1**: 94–100.

Cocucci, M. C., and G. Rossi. 1972. Biochemical and morphological aspects of zinc deficiency in *Rhodotorula gracilis*. *Arch. Microbiol.* **85**: 267–279.

Crandall, M., and J. H. Caulton. 1973. Induction of glycoprotein mating factors in diploid yeast of *Hansenula wingei* by vanadium salts or chelating agents. *Exp. Cell. Res.* **82**: 159–167.

Cuppett, V. M., and V. G. Lilly. 1973. Ferrous iron and the growth of twenty isolates of *Phytophthora infestans* in synthetic media. *Mycologia* **65**: 67–77.

David, C. N. 1974. Ferritin and iron metabolism in *Phycomyces*. In J. B. Neilands (Ed.), *Microbial Iron Metabolism: A Comprehensive Treatise*, pp. 149–158. New York: Academic.

Dawson, P. S. S., and L. P. Steinhauer. 1977. Radiorespirometry of *Candida utilis* in phased culture under nitrogen-, carbon-, and phosphorus-limited growth. *Can. J. Microbiol.* **23**: 1689–1694.

Detroy, R. W., and A. Ciegler. 1971. Induction of yeast-like development in *Aspergillus parasiticus*. *J. Gen. Microbiol.* **65**: 259–264.

Duffus, J. H., and L. J. Paterson. 1974a. Control of cell division in yeast using the ionophore A23187 with calcium and magnesium. *Nature* **251**: 626–627.

Duffus, J. H., and L. J. Paterson. 1974b. The cell cycle in the fission yeast *Schizosaccharomyces pombe*: Changes in activity of magnesium dependent ATPase and in the total internal magnesium in relation to cell division. *Z. Mikrobiol.* **14**: 727–729.

Durr, M., K. Urech, T. Boller, A. Wiemken, J. Schwencke, and M. Nagy. 1979. Sequestration of arginine by polyphosphate in vacuoles of yeast (*Saccharomyces cerevisiae*). *Arch. Microbiol.* **121**: 169–175.

Eddy, A. A., K. J. Indge, K. Backen, and J. A. Nowacki. 1970. Interactions between potassium ions and glycine transport in the yeast *Saccharomyces carlsbergensis*. *Biochem. J.* **120**: 845–852.

Eddy, A. A., and J. A. Nowacki. 1971. Stoichiometrical proton and potassium ion movements accompanying the absorption of amino acids by the yeast *Saccharomyces carlsbergensis*. *Biochem. J.* **122**: 701–711.

Egorov, S. N., and I. S. Kulaev. 1976. Isolation and properties of tripolyphosphates from the fungus *Neurospora crassa*. *Biokhimiya* **41**: 1958–1976.

Elliott, C. G. 1972. Calcium chloride and growth and reproduction of *Phytophthora cactorum*. *Trans. Brit. Mycol. Soc.* **58**: 169–172.

Emery, T. 1971. Role of ferrichrome is a ferric ionophore in *Ustilago sphaerogena*. *Biochemistry* **10**: 1483–1488.

Emery, T. 1976. Fungal ornithine esterases: Relationship to iron transport. *Biochemistry* **15**: 2723–2728.

Ernst, J., and G. Winkelmann. 1974. Metabolic products of microorganisms: 135. Uptake of iron by *Neurospora crassa*: IV. Iron transport properties of semisynthetic coprogen derivatives. *Arch. Microbiol.* **100**: 271–282.

Ezhov, V. A. 1978. Influence of phosphate and glucose on the kinetics of growth and phosphohydrolase activity of *Penicillium brevicompactum*. *Mikrobiologiya* **47**: 997–1003.

Ezhov, V. A., S. I. Bezborodova, and N. I. Santsevich. 1978. Regulation of the biosynthesis of extracellular phosphohydrolases in *Penicillium brevicompactum*. *Mikrobiologiya* **47**: 665–671.

Failla, M. L. 1977. Zinc: Functions and transport in microorganisms. In E. D. Weinberg (Ed.), *Microorganisms and Minerals*, pp. 151–214. New York: Dekker.

Failla, M. L., C. D. Benedict, and E. D. Weinberg. 1976. Accumulation and storage of Zn^{2+} by *Candida utilis*. *J. Gen. Microbiol.* **94**: 23–36.

Faller, L. D., B. M. Baroudy, A. M. Johnson, and R. X. Ewall. 1977. Magnesium ion requirements for yeast enolase activity. *Biochemistry* **16**: 3864–3869.

Fletcher, J. 1979. An ultrastructural investigation into the role of calcium in oospore-initial development in *Saprolegnia diclina*. *J. Gen. Microbiol.* **113**: 315–326.

Fuhrmann, G. F., and A. Rothstein. 1968. The transport of Zn^{2+}, Co^{2+}, and Ni^{2+} into yeast cells. *Biochim. Biophys. Acta* **163**: 32–330.

Gale, E. F. 1974. The release of potassium ions from *Candida albicans* in the presence of polyene antiobiotics. *J. Gen. Microbiol.* **80**: 451–465.

Garrett, R. H., and D. J. Cove. 1976. Formation of NADPH-nitrate reductase activity in vitro from *Aspergillus nidulans* MiaD and cnx mutants. *Mol. Gen. Genet.* **149**: 179–186.

Gisi, U., J. J. Oertli, and F. J. Schwinn. 1977. Wasser- und Salsbeziehungen der Sporangien von *Phytophthora cactorum* (Leb. et Cohn). Schroet. in vitro. *Phytopathol. Z.* **89**: 261–284.

Gold, M. H., and H. J. Hahn. 1979. Effect of divalent metal ions on the synthesis of oligosaccharide side chains of *Neurospora crassa* glycoproteins. *Phytochemistry* **18**: 1269–1272.

Goldstein, S. 1963. Development and nutrition of new species of *Thraustochytrium*. *Amer. J. Bot.* **50**: 271–279.

Griffin, D. H. 1966. Effect of electrolytes on differentiation in *Achlya* spp. *Plant Physiol.* **41**: 1254–1256.

Gustafsson, L., and B. Norkrans. 1976. On the mechanism of salt tolerance. Production of glycerol and heat during growth of *Debaryomyces hansenii*. *Arch. Microbiol.* **110**: 177–183.

Habison, A., C. P. Kubicek, and M. Rohr. 1979. Phosphofructokinase as a regulatory enzyme in citric acid producing *Aspergillus niger*. *FEMS Microbiol. Lett.* **5**: 39–42.

Hall, N. E. L., and D. E. Axelrod. 1978. Sporulation competence in *Aspergillus nidulans*: A role for iron in development. *Cell Diff.* **7**: 73–82.

Hammond, S. M., and B. N. Kliger. 1976. Differential effects of monovalent and divalent ions upon the mode of action of the polyene antibiotic candicidin. *J. Appl. Bacteriol.* **41**: 59–68.

Hammond, S. M., P. A. Lambert, and B. Kliger. 1974. The mode of action of polyene antibiotics: induced potassium leakage in *Candida albicans*. *J. Gen. Microbiol.* **81**: 325–330.

Hankin, L., and S. L. Anagnostakis. 1975. The use of solid media for detection of enzyme production by fungi. *Mycologia* **67**: 597–607.

Harada, T., and B. Spencer. 1960. Choline sulfate in fungi. *J. Gen. Microbiol.* **22**: 520–527.

Harper, D. B., T. R. Swinburne, S. Moore, A. E. Brown,, and H. Graham. 1980. A role for iron in germination of conidia of *Colletotrichum musae*. *J. Gen. Microbiol.* **121**: 169–174.

Hemmings, B. A. 1978. Phosphorylation of NAD-dependent glutamate dehydrogenase from yeast. *J. Biol. Chem.* **253**: 5255–5258.

Hendrix, J. W., and S. M. Guttman. 1970. Sterol or calcium requirement by *Phytophthora parasitica* var. *nicotianae* for growth on nitrate nitrogen. *Mycologia* **62**: 195–198.

Herr, L. 1973. Growth of *Aphanomyces cochlioides* in synthetic media as affected by carbon, nitrogen, methionine and trace elements. *Can. J. Bot.* **51**: 2495–2503.

Holmes, R. P., and P. R. Stewart. 1977. Calcium uptake during mitosis of the myxomycete *Physarum polycephalum*. *Nature* **269**: 592–594.

Holmes, R. P., and P. R. Stewart. 1979. The response of *Physarum polycephalum* to extracellular Ca^{2+}: Studies on Ca^{2+} nutrition, Ca^{2+} influxes and Ca^{2+} compartmentalization. *J. Gen. Microbiol.* **113**: 275–285.

Howes, N. K.,, and K. J. Scott. 1972. Sulphur nutrition of *Puccinia graminis* f. sp. *tritici* in axenic culture. *Can. J. Bot.* **50**: 1165–1170.

Howes, N. K., and K. J. Scott. 1973. Sulphur metabolism of *Puccinia graminis* f. sp. *tritici* in axenic culture. *J. Gen. Microbiol.* **76**: 345–354.

Hutchinson, T. E., M. E. Cantino, and E. C. Cantino. 1977. Calcium is a prominent constituent of the gamma particle in the zoospore of *Blastocladiella emersonii* as revealed by X-ray microanalysis. *Biochem. Biophys. Res. Commun.* **74**: 336–342.

Jennings, D. H. 1972. Cations and filamentous fungi. In W. P. Anderson (Ed.), Ion Transport in Plants, pp. 323–335. New York: Academic.

Jennings, D. H., and S. Austin. 1973. The stimulatory effect of the non-metabolized sugar 3-O-methyl glucose on the conversion of mannitol and arabitol to polysaccharide and other insoluble compounds in the fungus *Dendryphiella salina*. *J. Gen. Microbiol.* **75**: 287–294.

Jennings, D. H., and J. S. Aynsley. 1971. Compartmentation and low temperature fluxes of potassium of mycelium of *Dendryphiella salina*. *New Phytol.* **70**: 713–723.

Jones, D., W. J. McHardy, and M. J. Wilson. 1976. Ultrastructure and chemical composition of spines in Mucorales. *Trans. Brit. Mycol. Soc.* **66**: 153–157.

Jones, E. B. G., and D. H. Jennings. 1965. The effect of cations on the growth of fungi. *New Phytol.* **64**: 86–100.

Juliani, M. H., and C. Klein. 1977. Calcium ion effects on cyclic adenosine 3':5'-monophosphate bindings to the plasma membrane of *Dictyostelium discoideum*. *Biochim. Biophys. Acta* **497**: 369–376.

Kato, T., and Y. Tonomura. 1975a. Ca^{2+}-sensitivity of actinomysin ATPase purified from *Physarum polycephalum*. *J. Biochem.* **77**: 1127–1134.

Kato, T., and Y. Tonomura. 1975b. *Physarum* tropomyosin-troponin complex. Isolation and properties. *J. Biochem.* **78**: 583–588.

Khovrychev, M. P. 1973. Adsorption of copper ions by cells of *Candida utilis*. *Mikrobiologiya* **42**: 745–749.

Kikuchi, T. 1964. Comparison of original and secondarily developed copper resistance of yeast strains. *Bot. Mag.* (Tokyo) **77**: 395–402.

Killham, K., N. D. Lindley, and M. Wainwright. 1981. Inorganic sulfur oxidation by *Aureobasidium pullulans*. *Appl. Environ. Microbiol.* **42**: 629–631.

Kotyk, A., and K. Janacek. 1975. Cell membrane transport: Principles and techniques. New York: Plenum.

Kretovich, V. L., T. L. Auerman, and T. I. Uralets. 1978. Thermal stability of glutamine synthestase in the feed yeast *Candida tropicalis* in Mg-, Mn-, and Co-activated systems. *Mikrobiologiya* **47**: 217–219.

Kubicek, C. P., W. Hampel, and M. Rohr. 1979. Manganese deficiency leads to elevated amino acid pools in citric acid accumulating *Aspergillus niger*. *Arch. Microbiol.* **123**: 73–79.

Kybal, J., J. Majer, I. Komersova, and W. D. Wani. 1968. Phosphorus content during development of *Claviceps purpurea*. *Phytopathology* **58**: 647–650.

Landner, L. 1972. A possible causal relationship between iron deficiency, inhibition of DNA synthesis and reduction of meiotic recombination frequency in *Neurospora crassa*. *Mol. Gen. Genet.* **119**:103–117.

Lattke, H. and U. Weser. 1977. Functional aspects of zinc in yeast RNA polymerase B. *FEBS Lett.* **83**: 297–300.

Lee, K., R. Erickson, S. Pan, G. Jones, F. May, and A. Nason. 1974a. Effect of tungsten and vanadium on the in vitro assembly of assimilatory nitrate reductase utilizing *Neurospora* mutant nit-1. *J. Biol. Chem.* **249**: 3953–3959.

Lee, K. Y., S. S. Pan, R. Erickson, and A. Nason. 1974b. Involvement of molybdenum and iron in the in vitro assembly of assimilatory nitrate reductase utilizing *Neurospora* mutant nit-1. *J. Biol. Chem.* **249**: 3941–3952.

Lehninger, A. L. 1975. *Biochemistry*, 2nd ed. New York: Worth.

LeJohn, H. B., L. E. Cameron, R. M. Stevenson, and R. U. Meuser. 1974. Influence of cytokinins and sulfhydryl group reacting agents on calcium transport in fungi. *J. Biol. Chem.* **249**: 4016–4020.

Lenny, J. F., and H. W. Klemmer. 1966. Factors controlling sexual reproduction and growth in *Pythium graminicola*. *Nature* **209**: 1365–1366.

Levi, M. P. 1969. The mechanism of action of copper-chromium-arsenate preservatives against wood destroying fungi. British Wood Producers Association Annual Convention.

Lewis, M. J., and P. C. Patel. 1978. Isolation and identification of the cytoplasmic membrane from *Saccharomyces carlsbergensis* by radioactive labeling. *Appl. Environ. Microbiol.* **36**: 851–856.

Lewis, M. J., and C. Scazzocchio. 1977. The genetic control of molybdoflavoproteins in *Aspergillus nidulans*: A xanthine dehydrogenase. I. Half-molecule in cnx⁻ mutant strains of *Aspergillus nidulans*. *Eur. J. Biochem.* **76**: 441–446.

Lotan, R., I. Berdicevsky, D. Merzback, and N. Grossowicz. 1976. Effect of calcium ions on growth and metabolism of *Saccharomyces carlsbergensis*. *J. Gen. Microbiol.* **92**: 76–80.

Lowendorf, H. S., G. P. Bazinet, and C. W. Slayman. 1975. Phosphate transport in *Neurospora*. Derepression of a high-affinity transport system during phosphate starvation. *Biochim. Biophys. Acta* **389**: 541–549.

Lukasziewicz, Z., and J. J. Pieniazek. 1972. Mutations increasing the specificity of the sulphate permease in *Aspergillus nidulans*. *Bull. Acad. Pol. Sci.* **20**: 833–836.

Maas, J. L. 1976. Stimulation of sporulation of *Phytophthora fragariae*. *Mycologia* **68**: 511–522.

Mahler, H. R., and E. H. Cordes. 1971. *Biological Chemistry*, 2nd ed. New York: Harper & Row.

Malik, K. A., F. Kauser, and F. Azam. 1980. Effect of sodium chloride on the cellulolytic ability of some *Aspergilli*. *Mycologia* **72**: 229–243.

Marzluf, G. A. 1973. Regulation of sulfate transport in *Neurospora* by transinhibition and by inositol depletion. *Arch. Biochem. Biophys.* **156**: 244–254.

Mason, J. W., H. Rasmussen, and F. Dibella. 1971. 3'5'AMP and Ca²⁺ in slime mold aggregation. *Exp. Cell Res.* **67**: 156–160.

Matoo, A. K., E. Chalutz, J. D. Anderson, and M. Lieberman. 1979. Characterization of the phosphate-mediated control of ethylene production by *Penicillium digitatum*. *Plant Physiol.* **64**: 55–60.

McGuire, W. G., and G. A. Marzluf. 1974. Sulfur storage in *Neurospora*: Soluble sulfur pools of several developmental stages. *Arch. Biochem. Biophys.* **161**: 570–580.

McReady, R. G. L., and G. A. Din. 1974. Active sulfate transport in *Saccharomyces cerevisiae*. *FEBS Lett.* **38**: 361–363.

Meisch, H., J. A. Schmitt, and W. Reinle. 1977. Heavy metals in higher fungi: Cadmium, zinc and copper. *Z. Naturforsch., Sect. C.* **32**: 172–181.

Meisch, H., J. A. Schmitt, and W. Reinle. 1978. Heavy metals in higher fungi: III. Vanadium and molybdenum. Z. Naturforsch., Sect C. 33: 1–6.

Morzycha, E., and A. Paszewski. 1979. Regulation of S-amino acid biosynthesis in Saccharomyces lipolytica. Mol. Gen. Genet. 174: 33–38.

Nakai, N. 1961a. Studies on the adaptation of yeast to copper: XX. Production of hydrogen sulphide by a copper-resistant strain. Science Reports (Faculty of Liberal Arts and Education, Gifu University [Natural Sciences]) 2: 498–508.

Nakai, N. 1961b. Effect of nutritional conditions on hydrogen sulphide generation by yeast. Science Reports (Faculty of Liberal Arts and Education, Gifu University [Natural Sciences]) 2: 509–519.

Naplekova, N. N., and A. P. Anikina. 1970. Assimilation of boron by cellulose-decomposing microorganisms. Mikrobiologiya 39: 547.

Neilands, J. B. (Ed.). 1974. Microbial Iron Metabolism: A Comprehensive Treatise. New York: Academic.

Norkrans, B. 1968. Studies on marine occurring yeasts: Respiration, fermentation and salt tolerance. Arch. Microbiol. 62: 358–372.

Norkrans, B., and A. Kylin. 1969. Regulation of the potassium to sodium ratio and of the osmotic potential in relation to salt tolerance in yeasts. J. Bacteriol. 100: 836–845.

Norris, P. R., and D. P. Kelly. 1977. Accumulation of cadmium and cobalt by Saccharomyces cerevisiae. J. Gen. Microbiol. 99: 317–324.

Okamoto, K., and K. Fuwa. 1974. Copper tolerance of a new strain of Penicillium ochro-chloron. Agric. Biol. Chem. 38: 1405–1406.

Okorokov, L. A., V. M. Kadomtseva, and B. I. Titovskii. 1979. Transport of manganese into Saccharomyces cerevisiae. Folia Microbiol. 24: 240–246.

Okorokov, L. A., L. P. Lichko, V. M. Kadomtseva, V. P. Kholodenko, and I. S. Kulaev. 1974. Metabolism and physiochemical state of Mg^{2+} ions in fungi. Mikrobiologiya 43: 410–416.

Okorokov, L. A., L. P. Lichko, V. P. Kholodenko, V. M. Kadomtseva, S. B. Petrikevich, E. I. Zaichkin, and A. M. Karimiva. 1975. Free and bound magnesium in fungi and yeasts. Folia Microbiol. 20: 460–466.

Orthofer, R., C. P. Kubicek, and M. Rohr. 1979. Lipid levels and manganese deficiency in various citric acid producing strains of Aspergillus niger. FEMS Lett. 5: 403–406.

Paton, W. H. N., and K. Budd. 1972. Zinc uptake by Neocosmospora vasinfecta. J. Gen. Microbiol. 72: 173–184.

Pezet, R., and V. Pont. 1975. Elemental sulfur responsible for self-inhibition of spores of Phomopsis viticola. Experientia 31: 439–449.

Pezet, R., and V. Pont. 1977. Elemental sulfur: Accumulation in different species of fungi. Science 196: 428–429.

Pezzano, H., and L. Coscia. 1970. The presence of iron and vanadium in ribonucleic acid from yeast. Acta Physiol. Latino Amer. 20: 402–420.

Pine, L., and C. L. Peacock. 1958. Studies on the growth of Histoplasma capsulatum: IV. Factors influencing conversion of the mycelial phase to the yeast phase. J. Bacteriol. 75: 167–174.

Ponta, H., and E. Broda. 1970. (The mechanism of uptake of zinc by baker's yeast). Planta 95: 18–26.

Poston, J. M., and B. A. Hemmings. 1979. Cobalamins and cobalamin-dependent enzymes in Candida utilis. J. Bacteriol. 140: 1013–1016.

Putrement, A., H. Baranowska, A. Ejchart, and W. Jachymczyk. 1977. Manganese mutagenesis in yeast: VI. Mn^{2+} uptake, mitDNA replication and ER induction—Comparison with other divalent cations. Mol. Gen. Genet. 151: 69–76.

Ramsey, A. M. and L. J. Douglas. 1979. Effects of phosphate limitation on the cell wall and lipid composition of Saccharomyces cerevisiae. J. Gen. Microbiol. 110: 185–191.

Rawla, G. S., H. S. Rewal, and S. S. Chahal. 1977. Organic factors and trace elements for Cercospora dolichi and their gross effects on growth. Indian Phytopathol. 30: 189–194.

Reichert, U., R. Schmidt, and M. Foret. 1975. A possible mechanism for energy coupling in purine transport in Saccharomyces cerevisiae. FEBS Lett. 52: 100–102.

Reinert, W. G., and G. A. Marzluf. 1974. Regulation of sulfate metabolism in Neurospora crassa: transport and accumulation of glucose-6-sulfate. Biochem. Genet. 12: 97–108.

Ridgway, E. B., and A. C. H. Durham. 1976. Oscillations of calcium ion concentrations in *Physarum polycephalum*. *J. Cell Biol.* **69**: 223–226.

Roberts, K. R., and G. A. Marzluf. 1971. Specific interaction of chromate with the dual sulfate permease systems of *Neurospora crassa*. *Arch. Biochem. Biophys.* **142**: 651–659.

Rodriguez-Navarro, A., and E. D. Sancho. 1979. Cation exchanges of yeast in the absence of magnesium. *Biochim. Biophys. Acta* **552**: 322–330.

Roitman, I., L. R. Travassos, H. P. Azevedo, and A. Cury. 1969. Choline, trace elements and amino acids as factors for growth of an enteric yeast, *Candida slooffii*. *Sabouraudia* **7**: 15–19.

Roomans, G. M., F. Blasco, and G. W. F. H. Borst-Pauwels. 1977. Cotransport of phosphate and sodium by yeast. *Biochim. Biophys. Acta* **467**: 65–71.

Roomans, G. M., and G. W. F. H. Borst-Pauwels. 1977. Interaction of phosphate with monovalent cation uptake in yeast. *Biochim. Biophys. Acta* **470**: 84–91.

Roomans, G. M., and G. W. F. H. Borst-Pauwels. 1979. Interaction of cations with phosphate uptake by *Saccharomyces cerevisiae*. *Biochem. J.* **178**: 521–527.

Roomans, G. M., A. P. R. Theuvenet, T. P. R. van den Berg, and G. W. F. H. Borst-Pauwels. 1979. Kinetics of Ca^{2+} and Sr^{2+} uptake by yeast. Effects of pH, cations, and phosphate. *Biochim. Biophys. Acta* **551**: 187–196.

Ross, I. S. 1975. Some effects of heavy metals on fungal cells. *Trans. Brit. Mycol. Soc.* **64**: 175–193.

Rothstein, A., A. Hayes, D. Jennings, and D. Hooper. 1958. The active transport of Mg^{++} and Mn^{++} into the yeast cell. *J. Gen. Physiol.* **41**: 585–594.

Scopes, R. K. 1978. The steady-state kinetics of yeast phosphoglycerate kinase; anomalous kinetic-plots and the effects of salts on activity. *Eur. J. Biochem.* **85**: 503–516.

Seaston, A., G. Carr, and A. A. Eddy. 1976. The concentration of glycine by preparations of the yeast *Saccharomyces carlsbergensis* depleted of adenosine triphosphate. *Biochem. J.* **154**: 669–676.

Shamis, D. L., N. S. Gelikova, and L. N. Sarsenova. 1968. Effect of different sources of phosphorus nutrition on the biomass of fodder yeasts and the nitrogen and phosphorous content of them. *Trans. Inst. Mikrobiol. Virusol. Akad. Nauk. Kay. SSR* **11**: 18–24.

Shatzman, A. R., and D. J. Kosman. 1978. The utilization of copper and its role in the biosynthesis of copper-containing proteins in the fungus *Dactylium dendroides*. *Biochim. Biophys. Acta* **544**: 163–179.

Shatzman, A. R., and D. J. Kosman. 1979. Characterization of two copper-binding components of the fungus *Dactylium dendroides*. *Arch. Biochem. Biophys.* **194**: 226–235.

Siegel, S. M. 1973. Solubilization and accumulation of copper from elementary surfaces by *Penicillium notatum*. *Environ. Biol. Med.* **2**: 19–22.

Siegenthaler, P. A., M. Belesky, and S. Goldstein. 1967. Phosphate uptake in an obligately marine fungus: A specific requirement of sodium. *Science* **155**: 93–94.

Singh, B. P., and R. N. Tandon. 1967a. Sulfur and phosphorus requirements of *Alternaria tenuis* Auct. isolated from papaya (*Carica papaya* L.) leaf. *Proc. Natl. Acad. Sci. India, Sect. B* **37**: 199–203.

Singh, B. P., and R. N. Tandon. 1967b. Phosphorus requirements of certain isolates of *Alternaria tenuis* Auct. *Proc. Natl. Acad. Sci. India, Sect B* **37**: 131–134.

Singh, R. D., and M. O. Garraway. 1975. Role of trace elements in the growth and sporulation of *Colletotrichum lagenarium*. *Indian Phytopathol.* **28**: 468–475.

Slayman, C. L., and C. W. Slayman. 1968. Net uptake of potassium in *Neurospora*: Exchange for sodium and hydrogen ions. *J. Gen. Physiol.* **52**: 424–443.

Slayman, C. W., and C. L. Slayman. 1970. Potassium transport in *Neurospora crassa*. Evidence for a multisite carrier at high pH. *J. Gen. Physiol.* **55**: 758–791.

Slayman, C. W., and E. L. Tatum. 1964. Potassium transport in *Neurospora*: I. Intracellular sodium and potassium concentrations and cation requirements for growth. *Biochim. Biophys. Acta* **88**: 578–592.

Slayman, C. W., and E. L. Tatum. 1965. Potassium transport in *Neurospora*: II. Measurement of steady-state potassium fluxes. *Biochim. Biophys. Acta* **102**: 149–160.

Starkey, R. L. 1973. Effect of pH on toxicity of copper to *Scytalidium* sp., a copper-tolerant fungus, and some other fungi. *J. Gen. Microbiol.* **78**: 217–225.

Steenbergen, S. T., and E. D. Weinberg. 1968. Trace metal requirements for malformin biosynthesis. *Growth* **32**: 125–134.

Steinberg, R. A. 1950. Growth on synthetic nutrient solutions of some fungi pathogenic to tobacco. *Am. J. Bot.* **37**: 711–714.

Straka, J. G., and T. Emery. 1979. The role of ferrichrome reductase in iron metabolism and *Ustilago sphaerogena*. *Biochim. Biophys. Acta* **569**: 277–286.

Strandberg, G. W., S. E. Shumate, and J. R. Parrott. 1981. Microbial cells as biosorbents for heavy metals: Accumulation of uranium by *Saccharomyces cerevisiae* and *Pseudomonas aeruginosa*. *Appl. Environ. Microbiol.* **41**: 237–245.

Sussman, M., and C. Boschwitz. 1975. Adhesive properties of cell ghosts derived from *Dictyostelium discoideum*. *Dev. Biol.* **44**: 362–368.

Suzuki, T., S. Akiyama, S. Fujimoto, M. Ishikawa, Y. Nakao, and H. Fukuda. 1976. The aconitase of yeast: IV. Studies of iron and sulfur in yeast aconitase. *J. Biochem. (Tokyo)* **80**: 199–804.

Sykes, E. E., and D. Porter. 1973. Nutritional studies of *Labyrinthula* sp. *Mycologia* **65**: 1302–1311.

Sytkowski, A. J. 1977. Metal stoichiometry, coenzyme binding, and zinc and cobalt exchange in highly purified yeast alcohol dehydrogenase. *Arch. Biochem. Biophys.* **184**: 505–517.

Thind, K. S., and G. S. Rawla. 1967. Trace element studies on six species of *Helminthosporium*. *Proc. Indian Acad. Sci., Sect. B.* **66**: 259–272.

Thomas, K. C., and P. S. S. Dawson. 1978. Relationship between iron-limited growth and energy limitation during phased cultivation of *Candida utilis*. *Can. J. Microbiol.* **24**: 440–447.

Throneberry, G. O. 1973. Some physiological responses of *Verticillium albo-atrum* to zinc. *Can. J. Bot.* **51**: 57–59.

Tresner, H. D., and J. A. Hayes. 1971. Sodium chloride tolerance of terrestrial fungi. *Appl. Microbiol.* **22**: 210–213.

Tuppy, H., and Siegart. 1973. (Effect of Co^{2+} on yeast mitochondria). *Monatsch. Chem.* **104**: 1433–1443.

Tweedie, J. W., and I. H. Segel. 1970. Specificity of transport processes for sulfur, selenium and molybdenum anions by filamentous fungi. *Biochim. Biophys. Acta* **196**: 95–106.

Upadhyay, R. K., and R. S. Dwivedi. 1977. Sulphur and phosphorus requirements of *Pestalotiopsis funerea* causing leaf spot of *Eucalyptus globulus*. *Proc. Indian Natl. Sci. Acad., Part B.* **43**: 125–129.

Urbanus, J. F. L. M., H. van den Ende, and B. Koch. 1978. Calcium oxalate crystals in the wall of *Mucor mucedo*. *Mycologia* **70**: 829–842.

Uriah, N., and J. R. Chipley. 1976. Effects of various acids and salts on growth and aflatoxin production by *Aspergillus flavus* NRRL 3145. *Microbios* **17**: 51–59.

Urech, K., M. Durr, T. Boller, and A. Wiemken. 1978. Localization of polyphosphate in vacuoles of *Saccharomyces cerevisiae*. *Arch. Microbiol.* **116**: 275–278.

Vallee, M., and I. H. Segel. 1974. Sulfate transport by protoplasts of *Neurospora crassa*. *Microbios* **4**: 21–31.

Venkateswerlu, G., and K. S. Sastry. 1970. The mechanism of uptake of cobalt ions by *Neurospora crassa*. *Biochem. J.* **118**: 497–503.

Weimberg, R. 1975. Polyphosphate levels in nongrowing cells of *Saccharomyces mellis* as determined by magnesium ion and the phenomenon of "UberKompensation". *J. Bacteriol.* **121**: 1122–1130.

Weinberg, E. D. 1974. Secondary metabolism: control by temperature and inorganic phosphate. *Dev. Ind. Microbiol.* **15**: 70–81.

Weisenberg, R. C. 1972. Microtubule formation in vitro in solutions containing low calcium concentrations. *Science* **177**: 1104–1105.

White, J. P., and G. T. Johnson. 1971. Zinc effect on growth and cynodontin production of *Helminthosporium cynodontis*. *Mycologia* **63**: 548–561.

Widra, A. 1964. Phosphate directed Y-M variation in *Candida albicans*. *Mycopathol. Mycol. Appl.* **23**: 197–202.

Wiebe, C., and G. Winkelmann. 1975. Kinetic studies on the specificity of chelate-iron uptake in *Aspergillus*. *J. Bacteriol.* **123**: 837–842.

Winkelmann, G. 1974. Metabolic products of microorganisms: 132. Uptake of iron by *Neurospora crassa*. III. Iron transport studies with ferrichrome-type compounds. *Arch. Microbiol.* **98**: 39–50.

Winkelmann, G. 1979. Surface iron polymers and hydroxyacids: A model of iron supply in sideramine-free fungi. *Arch. Microbiol.* **121**: 43–51.

Wold, W. S. M., and I. Suzuki. 1976. The citric acid fermentation of *Aspergillus niger*: Regulation by zinc of growth and acidogenesis. *Can. J. Microbiol.* **22**: 1083–1092.

Yamaguchi, H. 1975. Control of dimorphism in *Candida albicans* by zinc: Effect on cell morphology and composition. *J. Gen. Microbiol.* **86**: 370–372.

Yamamoto, L. A., and I. H. Segel. 1966. The inorganic sulfate transport system of *Penicillium chrysogenum*. *Arch. Biochem. Biophys.* **114**: 523–538.

Zemek, J., V. Bilik, and L. Zakutna. 1975. Effect of some aldolases on growth of *Saccharomyces cerevisiae* inhibited with molybdenum. *Folia Microbiol.* **20**: 467–469.

Zonneveld, B. J. M. 1975a. Sexual differentiation in *Aspergillus nidulans*: The requirement for manganese and the correlation between phosphoglucomutase and the synthesis of reserve material. *Arch. Microbiol.* **105**: 105–108.

Zonneveld, B. J. M. 1975b. Sexual differentiation in *Aspergillus nidulans*: The requirement for manganese and its effect on beta-1,3-glucan synthesis and degradation. *Arch. Microbiol.* **105**: 101–104.

Zonneveld, B. J. M. 1976. The effect of glucose and manganese on adenosine-3'5'-monophosphate levels during growth and differentiation of *Aspergillus nidulans*. *Arch. Microbiol.* **108**: 41–44.

6
VITAMINS AND GROWTH FACTORS

Fungi require small amounts of various organic compounds for their growth. These compounds are not nutrients in the sense of the carbon, nitrogen, and inorganic compounds discussed previously but function in more subtle ways. Vitamins produce a growth or developmental response at 0.01 to 1.0 ppm and typically have a catalytic function in the cell as coenzymes or constituents of coenzymes. A number of organic compounds not categorized as vitamins are also active at low concentrations, usually above 10 ppm. These compounds, which we will arbitrarily place under the rubric "organic growth factors," include inositol, fatty acids, and the growth hormones of higher plants.

VITAMINS

All fungi need vitamins. Organisms that can synthesize required vitamins from simple precursors are called **auxoautotrophs. Auxoheterotrophs,** on the other hand, must have one or more vitamins supplied exogenously because the organism cannot synthesize them from simple precursors.

Fungi differ dramatically in their vitamin requirements; auxoautotrophs and auxoheterotrophs for various vitamins are often found even among different strains of the same species. There are different degrees of auxoheterotrophy: An isolate may have the capacity to synthesize all but one of the vitamin's subunits or may synthesize the entire vitamin but in suboptimal amounts. In this latter case the organism may survive without exogenous supplies of the vitamin, but the addition of the vitamin to the growth medium enhances growth. On the other hand, exogenous applications of a vitamin for which an individual is already auxoautotrophic may raise the endogenous concentration to inhibitory levels.

Our discussion will focus on the vitamins thiamine, biotin, riboflavin, pyridoxine, niacin, pantothenate, and folic acid.

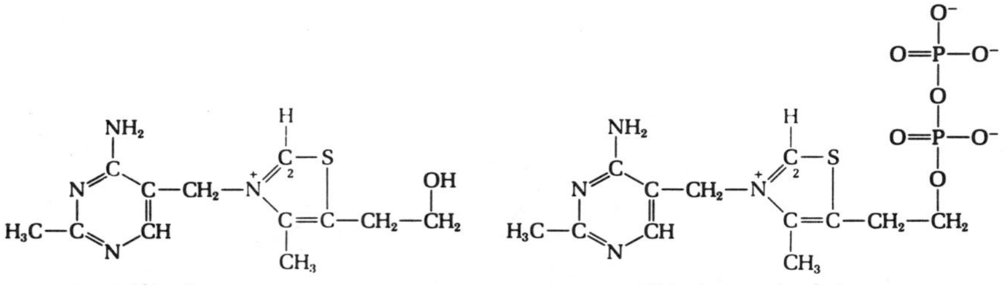

Figure 6.1   Thiamine and thiamine pyrophosphate. (Lehninger, 1975)

Thiamine

Fungi are more often auxoheterotrophic for thiamine (vitamin B_1) than for any other vitamin. The thiamine molecule is composed of two subunits: a pyrimidine moiety and a thiazole group (Figure 6.1). Fungi that are auxoheterotrophic for thiamine may require either the complete molecule or merely one of the subunits. Ridings and co-workers (1969) used this requirement as a basis for distinguishing isolates of *Pythium*. Some species, such as *P. deliense*, do not require exogenous sources of thiamine or its subunits (Table 6.1). Others, such as *P. anandrum*, require only the pyrimidine moiety, and all isolates of *P. vexans* require the intact thiamine molecule. Notice in this last species that even the combination of thiazole plus pyrimidine cannot satisfy the thiamine requirement. This suggests that the fungus may lack the enzyme necessary to link the subunits.

Typical responses to thiamine are illustrated for two species of *Ramulispora* (Figure 6.2a). *R. sorghi* is auxoautotrophic and grows well regardless of the thiamine concentration in the medium. However, *R. sacchari* is auxoheterotrophic and exhibits the greatest growth when supplied with 2 ppm thiamine. Similar types of responses are observed for other vitamins and growth factors (Figure 6.2b–d) (Rawla and Chahal, 1975).

High endogenous thiamine levels may explain why several species of fungi are adversely affected by exogenously applied thiamine. Examples of fungi in which

Table 6.1   Growth of Species of *Pythium* in a Glucose-L-Asparagine Basal Medium (BM), and the BM Supplemented with the Thiazole (Tz) and Pyrimidine (Py) Moieties as well as with Thiamine

Species	Mycelial Dry Weight (mg) After 10 Days in the Various Media				
	BM	BM + Tz	BM + Py	BM + Tz + Py	BM + Thiamine
Pythium deliense	63	60	60	60	61
P. anandrum	4	4	62	65	64
P. vexans	3	3	3	3	100

Source: Adapted from Ridings et al. (1969).

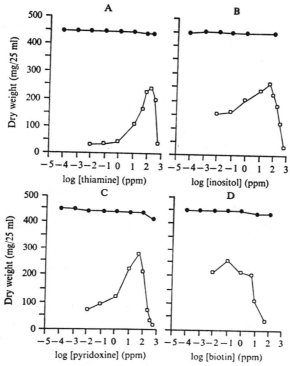

Figure 6.2 The relationship between concentrations of organic growth factors and growth of *Ramulispora sacchari* (O—O) and *R. sorghi* (●—●). (Rawla and Chahal, 1975)

thiamine inhibits growth and/or reproduction include *Rhizoctonia solani* (Khan and Azam, 1975), *Suillus variegatus* (Langkramer, 1969), and *Helminthosporium nodulosum* (Hegde and Ranganathaiah, 1971). In *Bipolaris maydis*, sporulation is inhibited by more than 70% when the basal medium is supplemented with 1 μg/ml thiamine (Figure 6.3). This inhibition is observed at all glucose concentrations greater than 4 g/liter and when either L-asparagine or ammonium citrate is the nitrogen source (Garraway, 1973).

The requirements for thiamine, as well as other vitamins, are frequently influenced by environmental conditions. In the dimorphic fungus *Mucor rouxii*, nutritional requirements become more complex in the absence of oxygen. Abundant growth is obtained aerobically in a medium lacking vitamins. However, under anaerobic conditions both thiamine and nicotinic acid are required by both the yeast and filamentous forms (Bartnicki-Garcia and Nickerson, 1961). In *Sordaria fimicola*, growth and perithecia production is inhibited at pH 3.4–3.8. However, this inhibition can be overcome by the addition of thiamine or its pyrimidine moiety to the medium (Lilly and Barnett, 1947). *Pythium butleri* requires thiamine as soon as the salt content of the medium exceeds a certain level (Robbins and Kavanagh, 1938).

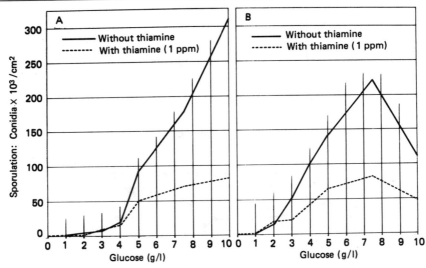

Figure 6.3 Sporulation response of *Bipolaris maydis* race T to various glucose concentrations on media containing (A) L-asparagine or (B) ammonium citrate as the nitrogen source with (-----) or without (———) 1 ppm thiamine. (Garraway, 1973)

Studies with species that are auxoheterotrophic for thiamine have demonstrated the existence of specific thiamine transport systems. In *Saccharomyces cerevisiae* thiamine uptake is carrier-mediated and energy dependent with a pH optimum of 4.5. Transport is inhibited by short-chain fatty acids, which cause an efflux of thiamine from the cell, and by thiamine analogs such as pyrithiamine which act as competitive inhibitors (Iwashima et al., 1973, 1975). Two thiamine-binding proteins have been isolated: one from the plasma membrane, which appears to be directly involved in transport, and another from the soluble fraction whose function is unknown. Both the extent of thiamine binding and the rate of thiamine uptake are higher in cells that have been incubated in a thiamine-deficient medium rather than a medium containing thiamine. This suggests that an intracellular accumulation of thiamine represses the synthesis of the carrier protein or proteins (Iwashima et al., 1979; Iwashima and Nose, 1976; Iwashima and Nishimura, 1979).

Thiamine accumulates intracellularly in *S. cerevisiae* and only later is it converted to thiamine-pyrophosphate (TPP), its biologically active form (Iwashima et al., 1973). This phosphorylation is catalyzed by thiamine pyrophosphokinase, a regulatory enzyme that is inhibited by several thiamine derivatives such as pyrithiamine and oxythiamine (Sicho and Kralova, 1968).

Thiamine-pyrophosphate serves as the required coenzyme for several enzymes of intermediary metabolism that catalyze the removal or transfer of aldehyde groups. These include pyruvate decarboxylase, transketolase, pyruvate dehydrogenase, and α-ketoglutarate dehydrogenase. In pyruvate decarboxylase and transketolase, TPP is an obligatory cofactor for reconstituting the enzyme from its subunits. Analogs of TPP interfere with this reconstitution and thus are inhibitory (Gounaris et al., 1975;

Tomita et al., 1973, 1974). In *S. cerevisiae* thiamine can induce the synthesis of the pyruvate decarboxylase and transketolase apoenzymes (Witt and Neufang, 1970).

Because of its importance as a cofactor, thiamine (as TPP) has a wide variety of biochemical effects. For example, the addition of 1 μg/ml thiamine to a basal glucose-asparagine medium significantly stimulates the production of ethanol by *Bipolaris maydis* (Figure 6.4a). This increase is accompanied by a drop in the levels of both intracellular and extracellular pyruvate (Figure 6.4b,c). These observations focus attention on two enzymes associated with alcoholic fermentation: pyruvate decarboxylase (PD) and alcohol dehydrogenase (ADH) (Figure 6.5). *In vitro* assays of *B. maydis* tissue incubated with various concentrations of thiamine revealed that the activity of PD increases with increasing thiamine concentration in the medium (Table 6.2). The activity of ADH, on the other hand, is low and unaffected by thiamine. These data suggest that the increase in PD activity is not due merely to activation of pre-existing enzyme but rather to increased synthesis. Thus, when thiamine is present, PD is induced, resulting in an increase in the rate of conversion of pyruvate to acetaldehyde. Acetaldehyde is then reduced to ethanol in the presence of ADH, and ethanol accumulates in the medium. In tissue grown without thiamine, pyruvate accumulates due to low PD activity, and little ethanol is produced. Notice in Figure 6.5 that instead of reacting with PD, pyruvate can also be converted to acetyl CoA by pyruvate dehydrogenase (PDH), an enzyme also requiring TPP as a cofactor. It is not known how the activity of PDH affects the thiamine-mediated flow of carbon from pyruvate to ethanol (Evans and Garraway, 1976).

Thiamine has been shown to affect the intracellular concentration of a host of

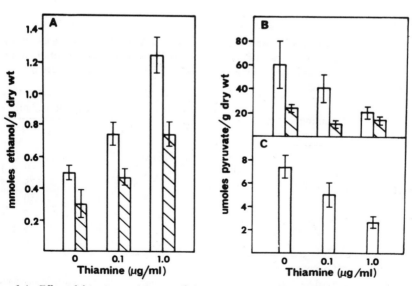

Figure 6.4 Effect of thiamine on (A) extracellular ethanol and (B) extracellular pyruvate produced by *Bipolaris maydis* after 3 days (□) and 6 hours (◣) incubation, and (C) on the level of intracellular pyruvate assayed after 3 days. (Evans and Garraway, 1976)

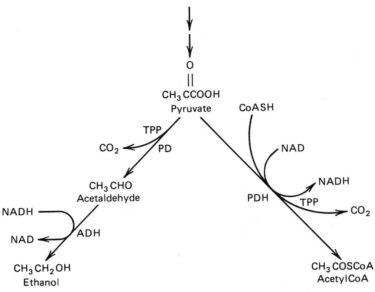

Figure 6.5 Two pathways for pyruvate metabolism involving thiamine.

Table 6.2 Activities of Pyruvate Decarboxylase and Alcohol Dehydrogenase from *Bipolaris maydis* in Relation to Both Thiamine-HCl Content of Culture Media and the Form of Thiamine in the Enzyme Assay Reaction Mixture[a]

Enzyme	Thiamine Derivative	Thiamine-HCl Content of Medium (μg/ml)		
		0	0.1	1.0
		nmoles NADH oxidized/min/mg protein $\pm$ SE		
PD	None	31 $\pm$ 2	27 $\pm$ 5	50 $\pm$ 11
PD	Thiamine-PP	53 $\pm$ 18	83 $\pm$ 10	125 $\pm$ 22
PD	Thiamine-HCl	28 $\pm$ 5	25 $\pm$ 6	36 $\pm$ 9
ADH	None	4 $\pm$ 2	8 $\pm$ 3	7 $\pm$ 2
ADH	Thiamine-PP	6 $\pm$ 1	6 $\pm$ 1	5 $\pm$ 0
ADH	Thiamine-HCl	8 $\pm$ 3	5 $\pm$ 0	6 $\pm$ 0

Source: Evans and Garraway (1976).

[a]The tissue was incubated for 3 days in basal medium containing different concentrations of thiamine-HCl. Enzyme preparations of both pyruvate decarboxylase (PD) and alcohol dehydrogenase (ADH) were made, and the activities of each were determined in the presence of no thiamine derivative, 7.5 mM thiamine-pyrophosphate (thiamine-PP), or 7.5 mM thiamine-HCl.

other metabolites including glycerol (Liebert, 1970), lipids (Nishikawa et al., 1977), amino acids (Rozenfeld and Disler, 1971), organic acids (Tachibana and Siode, 1971), nicotinic acid (Kawasaki et al., 1970), and aflatoxins (Basappa et al., 1967).

Cells of *Saccharomyces carlsbergensis* grown on a basal medium containing 1 μg/ml thiamine have a significantly lower respiration rate when compared with a thiamine-free control (Figure 6.6a). The activity of cytochrome oxidase and the total cytochrome contents of the cells are also low. These observations may explain the

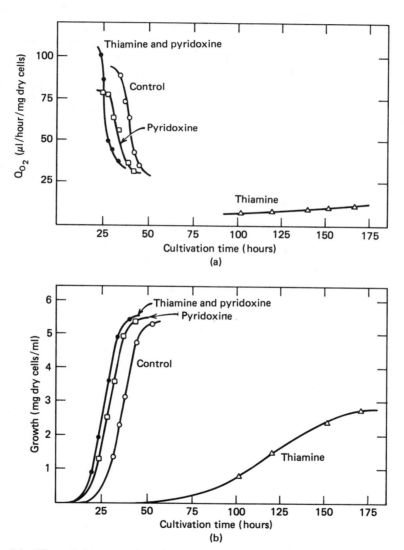

Figure 6.6 Effects of thiamine and pyridoxine on (a) respiration and (b) growth of *Saccharomyces carlsbergensis*. (Nakamura et al., 1974)

low growth of this species on a thiamine medium (Figure 6.6b). These inhibitory effects of thiamine can be nullified by the addition of pyridoxine (vitamin B_6) to the growth medium. Among its many functions pyridoxine is a required cofactor for the synthesis of heme and consequently for the synthesis of cytochromes (Nakamura et al., 1974). Because the pyridoxine content of S. *carlsbergensis* is decreased when the cells are incubated on a thiamine medium (Kishi, 1969), the inhibitory effect of thiamine on respiration may be due primarily to its effect on endogenous pyridoxine; the thiamine-induced decrease in pyridoxine levels results in decreased heme synthesis, low cytochrome content, and thus low respiration rates (Nakamura et al., 1981). This indirect effect of thiamine on heme synthesis is supported by the observation that thiamine also causes a decrease in levels of unsaturated fatty acids, particular palmitoleic and oleic acids, and a concomitant increase in the levels of saturated fatty acids in this species. The desaturation process is known to occur by a reaction requiring cytochrome b_5 and thus indirectly requiring pyridoxine (Nishikawa et al., 1974).

Biotin

Biotin (Figure 6.7a) is probably second only to thiamine in the frequency with which it is required for the growth of fungi. In its biologically active form, biotin is cova-

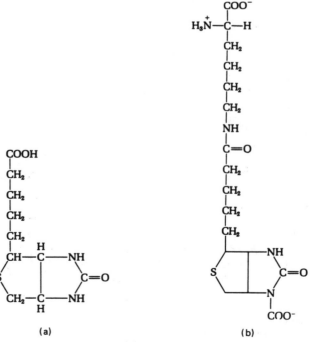

Figure 6.7 (a) Biotin. (b) Biocytin with CO_2 attached to biotin moiety. (Lehninger, 1975)

lently linked to a lysine residue at the active site of certain carboxylase enzymes. This biotin–lysine complex is called biocytin (Figure 6.7b) and functions as a coenzyme in the transfer of either CO_2 or carboxyl groups. Enzymes requiring biotin include pyruvate carboxylase, acetyl-CoA carboxylase, and urea carboxylase. The role of biotin is apparently the same in each case: biotin first binds to the CO_2 or carboxyl group and then transfers it to the substrate. For example, in pyruvate carboxylase the following reaction occurs:

$$CO_2 + \text{biotin-enzyme} \longrightarrow CO_2\text{-biotin-enzyme}$$
$$CO_2\text{-biotin-enzyme} + \text{pyruvate} \longrightarrow \text{oxaloacetate} + \text{biotin-enzyme}$$

(Moss and Lane, 1971). Understanding the involvement of biotin in these and other reactions has been facilitated by the use of avidin, a glycoprotein that forms bioinactive complexes with biotin (Rogers and Lichstein, 1969b).

In *S. cerevisiae*, biotin is taken up by a carrier-mediated active transport system that is repressed by high extracellular biotin concentrations (Rogers and Lichstein, 1969a,b). Excess accumulation is avoided by the leakage of biotin from cells by diffusion (Becker and Lichstein, 1972). The carrier system of repressed cells is kinetically similar to that of derepressed cells except that repressed cells have a lower V_{max}. This indicates that reduced uptake in the presence of high extracellular biotin concentrations is due merely to a decreased number of carrier molecules instead of the formation of a carrier with a decreased affinity. The biotin carrier is tightly bound to the cell membrane and cannot be released by cold osmotic shock nor removal of the cell wall (Cicmanec and Lichstein, 1974). There are approximately 4000–8000 transport sites for biotin per yeast cell (Bayer et al., 1978).

The biosynthesis of biotin by auxoautotrophs is influenced by nutrients in the culture medium. For example, in *Phycomyces blakesleeanus* biotin synthesis is stimulated by hexoses and maltose as carbon sources and by L-asparagine and urea as nitrogen sources (Chenouda and El-Awamry, 1971b). More than 92% of the total biotin and biotin derivatives synthesized by this species is excreted into the culture medium (Chenouda and El-Awamry, 1971a). If the medium contains high concentrations of selenium, a biotin analog containing selenium instead of sulfur is synthesized and excreted. This selenobiotin can replace biotin in a number of its biological properties (Lindblow-Kull et al., 1980).

The importance of biotin in cellular processes can be illustrated by its role in lipid synthesis. As mentioned above, biotin is a cofactor for acetyl-CoA carboxylase, an enzyme important in the synthesis of fatty acids. In several fungi including *Aspergillus nidulans* (Rao and Modi, 1968), *Candida tropicalis* (Kacchy et al., 1972), and *Hypomyces chlorinus* (Gontier and Touze-Soulet, 1971) which are auxoheterotrophic for biotin, the lack of this vitamin results in decreased cellular lipid content (Figure 6.8). This in turn causes an abnormal increase in membrane permeability (Nomdedu et al., 1974) and a variety of ultrastructural changes including decreased numbers of organelles, hypertrophy of cells, and thickened cell walls (Figure 6.9) (Dargent and Touze-Soulet, 1976). Often the addition of fatty acids to the growth medium can partially compensate for biotin deficiency (Shimada et al., 1978).

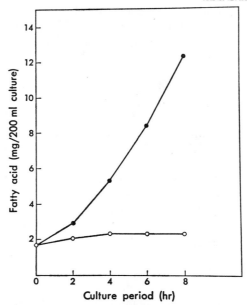

Figure 6.8   Change in the total amount of fatty acids of yeast cells per culture in (○) biotin-free or (●) biotin supplemented aspartate media. (Mizunaga et al., 1975)

   Biotin-induced changes in the levels of fatty acids are also responsible for the phenomenon of "cell death" in *S. cerevisiae*. Strains of this species that require biotin for growth can survive in a biotin-deficient medium containing most nitrogen sources; however, the levels of aspartate and asparagine are particularly important. In these strains aspartate is formed by the transamination of oxaloacetate, which in turn arises by the carboxylation of pyruvate via the biotin-mediated pyruvate carboxylase. Under conditions in which biotin and aspartate are lacking, little growth occurs owing to the absence of this important amino acid in proteins. If aspartate is present but biotin is lacking, the proper proteins can be formed but fatty acid synthesis cannot proceed due to low acetyl-CoA carboxylase activity. In this latter case, death is thought to result from defective cell membranes due to lowered cellular fatty acid levels. The addition of oleic, palmitoleic, or linolenic acid can partially overcome this biotin deficiency (Figure 6.10) (Mizunaga et al., 1975).
   Biotin also affects the levels of other cellular constituents, but the mechanisms for this are not known. For example, the addition of biotin to the culture medium has been shown to decrease the activities of enzymes such as isocitrate lyase (Nabeshima et al., 1977), glucose-6-phosphate dehydrogenase, and glutamine synthetase (Desai and Modi, 1977); to increase the content of DNA, RNA, and protein (Aurich et al., 1967); and to decrease the concentration of organic acids (Kacchy et al., 1972). Biotin has also been shown to affect the composition and ultrastructure of cell walls. In *Candida albicans*, 10 ng/ml biotin promotes the formation of yeastlike cells whereas 0.1 ng/ml biotin results in filamentous growth. These latter cells have walls with greater amounts of glucan-protein but smaller amounts of

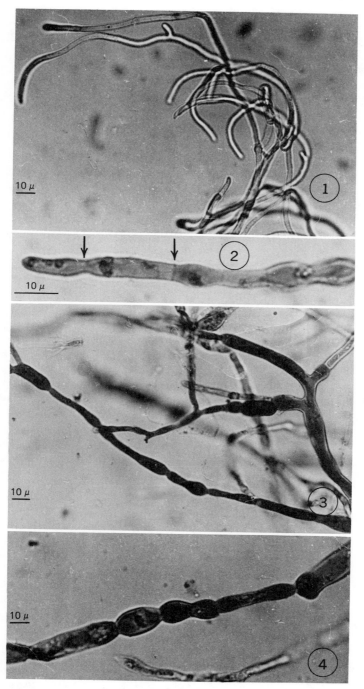

Figure 6.9 Effect of biotin on the morphology of *Hypomyces chlorinus*. (1) Colonies incubated in the presence of biotin. The cells are regular, rectangular, and have a relatively constant diameter. ×625. (2) Apical cell of a colony grown in a biotin-free medium. Constrictions (*arrows*) are present at the location of septa. ×1500. (3) Colony grown on a biotin-free medium. Numerous deformations give the hyphae an irregular appearance. ×625. (4) Irregular aspect of the mycelium. X1000. (Dargent and Touze-Soulet, 1976)

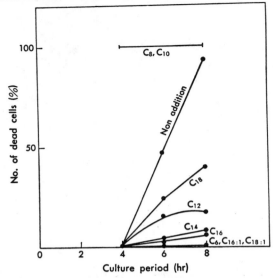

Figure 6.10　Effect of various fatty acids on the death rate of yeast cells cultured in biotin-free aspartate medium. (Mizunaga et al., 1975)

glucan-mannan compared with yeast-phase cells. Thus, dimorphism in this species may be triggered by an alteration in the metabolism of cell wall constituents brought about by biotin deficiency (Yamaguchi, 1974).

The requirements of certain fungi for biotin can often be strikingly different. Figure 6.11 illustrates that isolate V_1 of *Verticillium dahliae* is auxoheterotrophic for biotin whereas isolate V_2 is auxoautotrophic (Milton and Isaac, 1967). Notice also

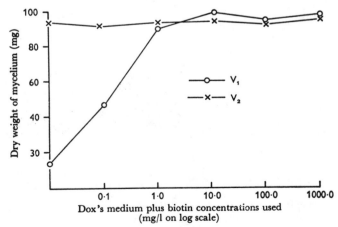

Figure 6.11　The effect of biotin concentration on the growth of isolates V_1 and V_2 of *Verticillium dahliae*. (Milton and Isaac, 1967)

that raising the biotin concentration above 1 mg/liter has little effect on the growth of either isolate.

Biotin also affects the reproduction of many species. Growth of three pathogenic strains of *Phialophora asteris* is slow on a defined medium lacking biotin. However, the addition of 0.5 μg/ml biotin results in a marked stimulation of growth, pigmentation, and sporulation (Burge and Isaac, 1977). In *Sordaria macrospora* the development of fruiting bodies is blocked at the stage of small protoperithecia in the absence of biotin. The addition of as little as 3 μg/liter biotin is sufficient for development to continue (Molowitz et al., 1976). *Neurospora crassa* has a circadian rhythm of conidiation in a medium containing biotin at 0.025 μg/liter but not at 5 μg/liter (Figure 6.12). Because the rhythm can also be inhibited by high CO_2 levels, the role of biotin in this case may be to fix CO_2 via carboxylase enzymes into intermediates that are inhibitory to the maintenance of the rhythm (West, 1975).

The presence of other nutrients in the medium may also affect the response of the organism to biotin. For example, the biotin content of *Fusarium moniliforme* ranges from 2.0 to 17.5 μg/g dry weight depending on the nitrogen source used; ammonium nitrogen results in the greatest vitamin content (Shcherbina, 1973). *Candida tropicalis* requires 3–5 times as much biotin on a medium containing glucose as on a medium containing hexadecane as the carbon source. The requirement for biotin is also greater at 39°C than at 29°C (Todosuchuk et al., 1976).

The pH of the medium can also be a crucially important factor in biotin utilization. Exogenous biotin is required for growth of the yeast *Lipomyces starkeyi* when incubated at pH 5.5–6.5 but not at pH values below this level. Apparently, enzymes

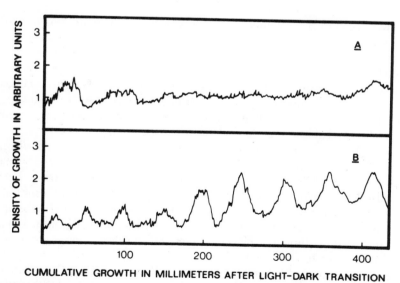

Figure 6.12 Densitometer tracings of wild-type cultures of *Neurospora crassa* grown on a basal medium with (a) 5.0 or (b) 0.025 μg biotin/liter. (West, 1975)

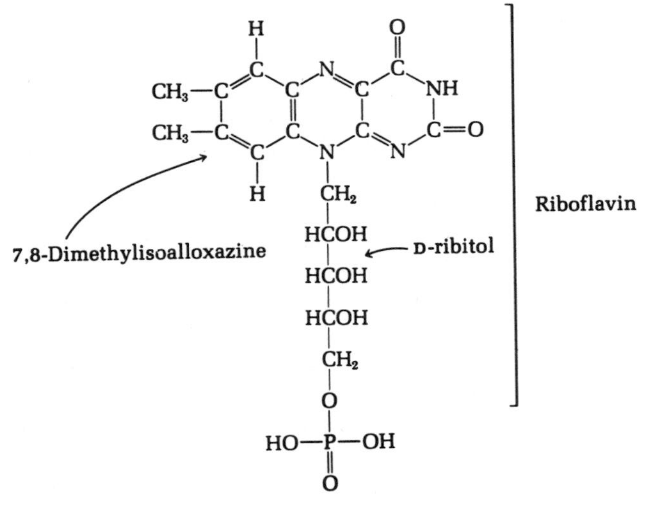

Figure 6.13 Riboflavin and its derivatives. (a) Flavin mononucleotide (FMN). (b) Flavin adenine dinucleotide (FAD). (Lehninger, 1975)

involved in biotin synthesis are inhibited by low pH in this species (Uzuka et al., 1974).

Riboflavin

Riboflavin (vitamin B_2) is composed of a substituted isoalloxazine ring to which a ribitol side chain is attached (Figure 6.13a). Riboflavin is a component of the coenzymes flavin mononucleotide (FMN, also called riboflavin phosphate) and flavin adenine dinucleotide (FAD)—the so-called flavin coenzymes (Figure 6.13b). These coenzymes participate in the transfer of electrons and hydrogens catalyzed by a variety of enzymes, including pyruvate dehydrogenase, D-amino acid oxidase, and succinic dehydrogenase. In addition, flavin coenzymes are important components of the mitochondrial electron transport chain.

Riboflavin is readily transported into cells of S. cerevisiae and Pichia guilliermondii that are auxoheterotrophic for this vitamin, but it is not transported in auxoautotrophs (Perl et al., 1976; Sibirnyi et al., 1977a,b). Figure 6.14 illustrates that wild type cells of S. cerevisiae have only a slight ability to take up riboflavin; but two mutant strains that have blocks at specific steps in riboflavin synthesis take up this vitamin readily, particularly during the early to mid-log phase of growth. Uptake is carrier mediated, is stimulated by certain monovalent cations, and is greater under anaerobic than aerobic conditions. The carrier is specific for flavins, and riboflavin analogs are good competitive inhibitors. In both S. cerevisiae and P. guilliermondii there is also good evidence for a second carrier that is responsible for the efflux of riboflavin. This carrier, present in both wild type and mutants, is different from the

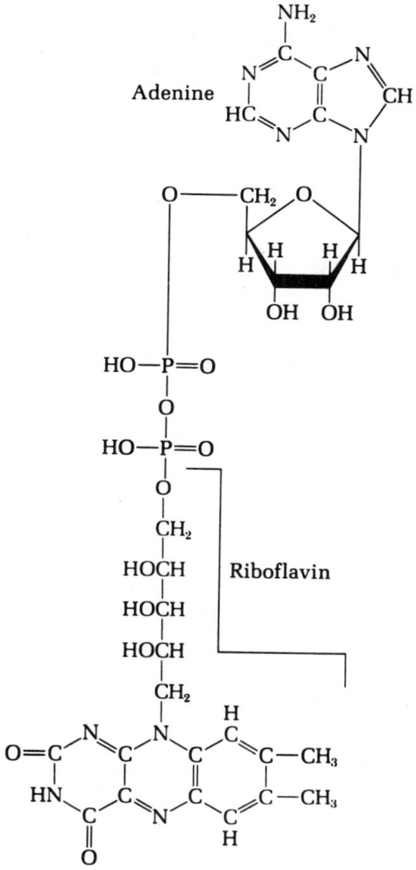

Adenine

NH$_2$

O—CH$_2$ O

H H H

OH OH

HO—P=O

O

HO—P=O

O

CH$_2$

HOCH Riboflavin

HOCH

HOCH

CH$_2$

O=C N C N C C—CH$_3$

HN C C N C C—CH$_3$

O

(b)

Figure 6.13 *(Continued)*

uptake carrier in being stimulated by sugars such as glucose, mannose, and fructose, and it does not seem to vary with the phase of growth. The relative rates of uptake and efflux are dependent on pH, temperature, and extracellular riboflavin concentration. It is not known whether uptake in *S. cerevisiae* occurs by active transport or facilitated diffusion; although uptake appears to proceed against a concentration gradient it is not certain if metabolic energy is required. However, in *P. guilliermondii* and *Schwanniomyces occidentalis*, metabolic uncouplers inhibit uptake. There is an inverse relationship between the intracellular concentration or flavin coenzymes and the activity of the uptake system. These flavins may thus act as repressors of the carrier protein (Perl et al., 1976).

Work with riboflavin auxoautotrophs of *P. guilliermondii* indicates that guanosine triphosphate (GTP) is a precursor for riboflavin biosynthesis (Figure 6.15). The enzyme GTP-cyclohydrolase, which catalyzes the first step in the pathway, plays an

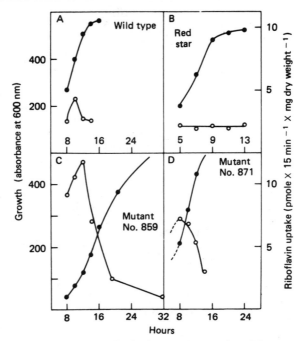

Figure 6.14   Variations of riboflavin uptake during anaerobic growth in different types of yeast. (Perl et al., 1976)

important regulatory role. For example, in conditions of iron deficiency. *P. guilliermondii* produces large quantities of riboflavin. These cells have activities of GTP-cyclohydrolase 20–40 times those of iron-sufficient cells; other enzymes are not affected (Shavlovskii et al., 1980). Iron appears to cause a repression of GTP-cyclohydrolase synthesis. This enzyme can also be regulated by feedback inhibition of FAD but not FMN or riboflavin (Shavlovskii et al., 1977).

Nutritional factors also affect riboflavin biosynthesis. In *C. tropicalis* (Kvasnikov et al., 1977), *Erythrothecium ashbyii* (Sato et al., 1966), and several strains of *Aspergillus* (Redchyts, 1972), transferring the tissue from a carbohydrate to a hydrocarbon medium results in a large increase in riboflavin production. In *P. guilliermondii*, riboflavin synthesis is enhanced by the addition of either EDTA (Qudeer and Igbal, 1971) or cobalt (Schlee, 1977). Both compounds act by reducing the intracellular concentration of iron—the former by chelation and the latter by inhibiting iron uptake.

As we have seen with strains of *S. cerevisiae* and *P. guilliermondii* (Figure 6.14), exogenously applied riboflavin causes increased growth in fungi that are auxoheterotrophic for this vitamin. Riboflavin is also required by isolates of *Fusarium caerulum* (Vir and Grewal, 1973) and *Helminthosporium nodulosum* (Hegde and Ranganathaiah, 1971). The adverse effects of high temperature on growth of *Penicillium vermiculatum* may be alleviated somewhat by riboflavin. At 30° C and above, mu-

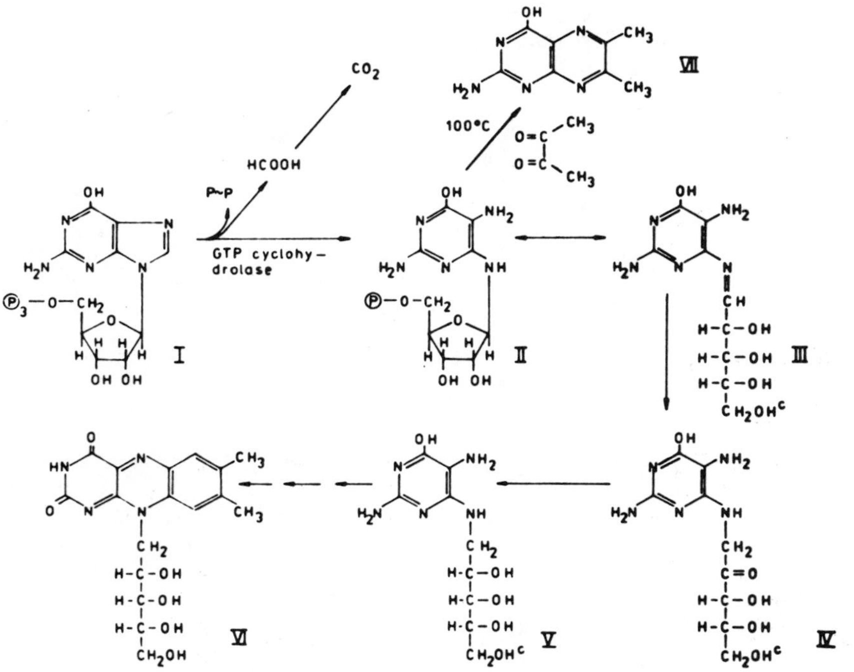

Figure 6.15 Early steps in riboflavin biosynthesis in yeast. (I) GTP. (VI) Riboflavin. (After Shavlovskii et al., 1980)

tants of this species grow maximally only when supplied with riboflavin (Mitra and Chaudhuri, 1974).

Riboflavin absorbs light in the near-UV and blue regions of the spectrum and has been implicated as the principal photoreceptor in several light-induced developmental processes. These processes have been recently reviewed by Tan (1978) and include conidiation, spore germination, and fruiting body initiation. Tschabold (1967) has demonstrated that light-induced perithecial production by *Hyphomyces solani* f. *cucurbitae* is stimulated by increasing concentrations of riboflavin. Production is inhibited by mepacrine, a compound that interferes with several flavoprotein electron transfer reactions. The addition of riboflavin to the growth medium also significantly increases light-induced sporulation in *Cercospora personata* (Bhama and Swamy, 1976). In addition, the key photoreceptor involved in phototropism in *Phycomyces blakesleeanus* appears to be riboflavin (Otto et al., 1981).

Pyridoxine

Pyridoxine (also called pyridoxol or vitamin B_6) as well as the related compound pyridoxal are precursors for the coenzyme pyridoxal phosphate (Figure 6.16). This coenzyme is involved in reactions where amino groups are transferred from one molecule to another, thus it is a particularly important cofactor in transamination

Vitamin B$_6$
Pyridoxine
(pyridoxol)

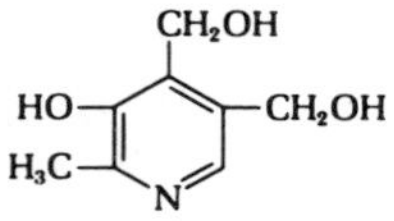

Pyridoxal

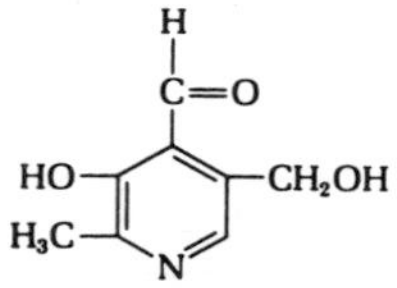

Pyridoxamine

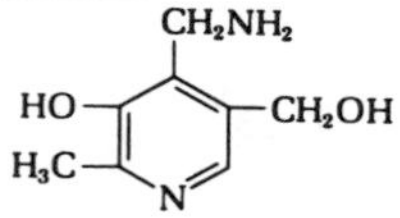

Pyridoxal phosphate

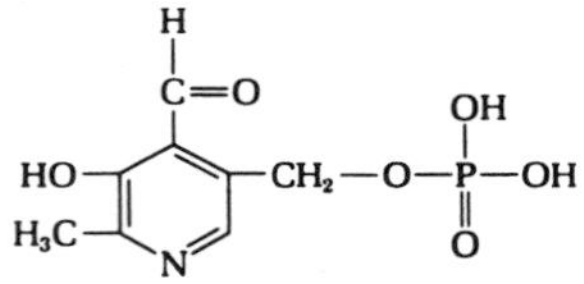

Figure 6.16 Pyridoxine and some of its derivatives. (Lehninger, 1975)

reactions. In addition to transaminases, specific enzymes that require pyridoxal phosphate include tryptophan synthase (Hilgenberg and Hofman, 1977), kynureninase (Soda et al., 1975), and serine transhydroxymethylase. The latter enzyme is protected from heat, urea, and trypsin digestion by pyridoxal phosphate, which suggests that the binding of this cofactor confers stability to the conformation of the protein (Yamagata and Takeshima, 1976).

Many cellular processes are affected by pyridoxine. Recall that in *Saccharomyces carlsbergensis* the thiamine-induced inhibition of respiration and growth can be overcome by the addition of pyridoxine (Figure 6.6a,b). Similarly, nicotinic acid synthesis, which involves two B$_6$ enzymes, is also low in tissues grown in thiamine media but increases upon the addition of B$_6$ (Kawasaki et al., 1970). Thiamine apparently interferes with the synthesis of B$_6$, which in turn is required for the synthesis of heme, cytochromes, and nicotinic acid. The inhibitory effects of thiamine on respiration and growth can thus be explained by B$_6$ deficiency (Kishi, 1969). In addition, studies using pyridoxineless mutants of *Aspergillus nidulans* have shown that B$_6$ deficiency results in increased levels of total lipids, sterols, phospholipids, and triacylglycerols but decreased levels of fatty acids (Mohana and Shanmugasundarum, 1978).

As with other vitamins, the response of fungal cells to vitamin B$_6$ is dependent on whether they are auxoautotrophic or auxoheterotrophic. As indicated in Figure

6.2, *Ramulispora sorghi* is auxoautotrophic and is not affected by exogenous pyridoxine; *R. sacchari*, on the other hand, exhibits a growth peak at 1.5 ppm pyridoxine (Rawla and Chahal, 1975).

Barnett (1968) reports an interesting interaction between the fungal parasite *Gonatobotryum fuscum* and its fungal host *Graphium* sp. Both the host and the parasite require vitamin B_6 for growth, although the parasite requires a higher concentration. The requirement for B_6 can be satisfied equally well by pyridoxine phosphate, pyridoxal phosphate, pyridoxal-HCl, and pyridoxamine-2HCl. However, growth of the parasite on a B_6-containing medium is strongly inhibited by light. Evidence suggests that this inhibition is due to the photodestruction of B_6: The parasite grows well when incubated on a B_6 medium in the dark, but growth in the dark is poor if this same medium is exposed to light for four days before seeding with tissue. This growth inhibition can be overcome by the addition of fresh B_6 to the medium.

Biotin can partially replace the B_6 requirement of the parasite. As seen in Table 6.3 the presence of 5 µg/liter biotin in the medium results in good growth of the parasite even when the B_6 concentration is suboptimal; growth of the host is unaffected by biotin. Evidently, under conditions of suboptimal B_6, biotin stimulates the synthesis by the host of a nutrient that is required by the parasite. Using *Sordaria fimicola*, a species totally auxoheterotrophic for biotin, Barnett (1968) demonstrated that the amount of biotin synthesized by the host is correlated with the amount of B_6 added to the medium. Thus, biotin may substitute for B_6 in critical cellular reactions, or the parasite may synthesize B_6 from biotin present in the host.

Table 6.3   The Effects of B_6 Concentration and Presence of Biotin on
Development of *Gonatobotryum fuscum* on *Graphium* sp.
Cultures 6 Days Old, Incubated in Darkness

Vitamins Present in Medium		Average Diameter of Colony (mm)	
µg B_6/liter	µg biotin/liter	Host	Parasite
50	0	30	64
50	5	28	70
25	0	26	57
25	5	26	65
12.5	0	26	50
12.5	5	26	70
6	0	27	[a]trace
6	5	26	60
3	0	27	[a]trace
3	5	27	65
0	0	22 (sparse)	0
0	5	22 (sparse)	35 (variable)

[a]A few hyphae confined to colony of host.
Source: Barnett (1968).

Niacin

Niacin, also called nicotinic acid, is converted intracellularly to nicotinamide and as such is a key component of the coenzymes nicotinamide adenine dinucleotide (NAD) and nicotinamide adenine dinucleotide phosphate (NADP) (Figure 6.17). The nicotinamide moiety is able to undergo reversible oxidation and reduction, as illustrated in this figure, and these coenzymes are thus of critical importance in

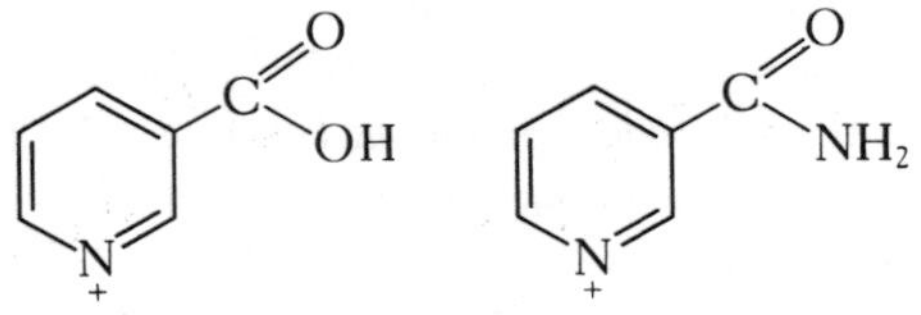

Niacin (nicotinic acid)   Nicotinamide

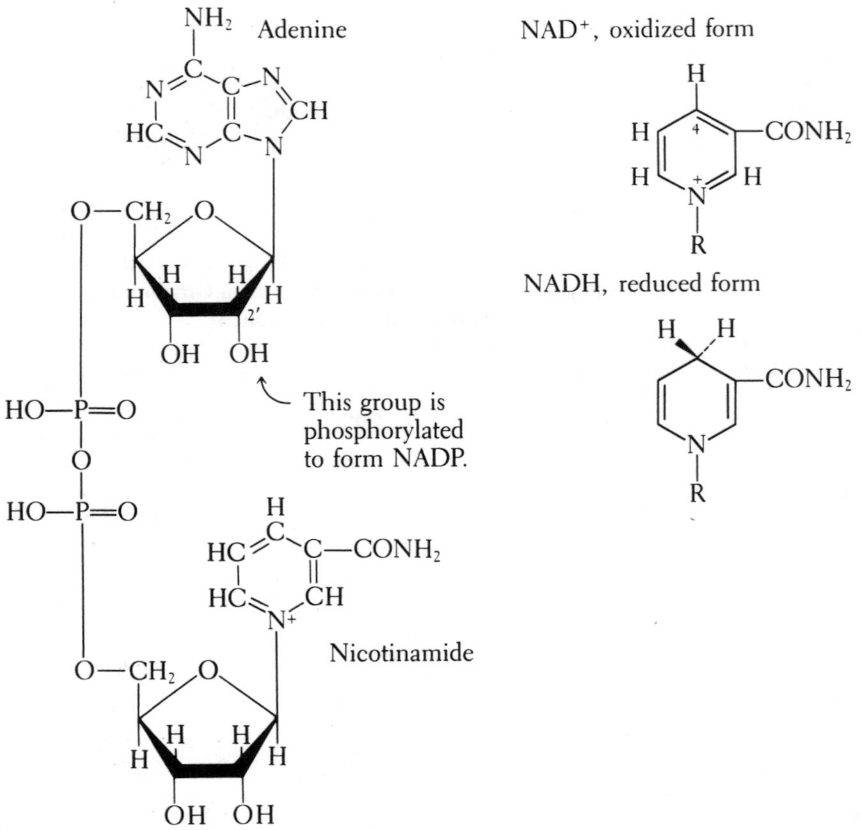

Figure 6.17   Niacin and its derivatives. (After Lehninger, 1975)

electron transfer reactions. Some of the many enzymes requiring NAD/NADH are glucose-6-phosphate dehydrogenase, glutamate dehydrogenase, glyceraldehyde-3-phosphate dehydrogenase, isocitrate dehydrogenase, nitrate reductase, and the NAD dehydrogenase of the respiratory electron transport chain.

Both niacin and nicotinamide are taken up by *Aspergillus niger* and then incorporated into nicotinamide ribose diphosphate ribose, a metabolite of the NAD biosynthetic pathway (Kuwahara, 1976). Fungi that are auxoautotrophic for niacin synthesize it primarily from tryptophan, although alternative pathways have been reported. For example, *S. cerevisiae* synthesizes niacin from tryptophan under aerobic conditions; under anaerobiosis aspartate is the precursor (Yanofsky, 1955; Heilman and Lingens, 1968).

As with other vitamins, the synthesis of niacin is affected by the presence of other nutrients. In *Fusarium moniliforme*, the greatest amount of niacin is obtained using a medium with a carbon-to-nitrogen ratio of 2:1 (Bilai et al., 1974). In *Candida tropicalis*, more niacin is produced when the fungus is grown with hydrocarbons than when glucose is the carbon source (Kvasnikov et al., 1967; 1977). As mentioned previously, in *S. carlsbergensis* thiamine inhibits niacin biosynthesis via an effect on pyridoxine levels.

Exogenous niacin has been shown to stimulate growth in many fungal species, including isolates of *Suillus variegatus* (Langkramer, 1969), *Helminthosporium nodulosum* (Hedge and Ranganathiah, 1971), and *Rhodotorula glutinis* (Samsonova et al., 1974). Niacin also promotes growth of a pyridoxineless strain of *Aspergillus nidulans; in vitro* studies show that niacin may be converted to pyridoxine (Vatsala et al., 1976).

Niacin and its derivatives have been implicated in several developmental processes in fungi. For example, *Physarum polycephalum* undergoes sporulation if exposed to light following four days incubation in darkness on a medium containing niacin, nicotinamide, tryptophan, or certain other niacin precursors. Analogs of niacin inhibit sporulation when added with niacin at the beginning of the dark period (Daniel and Rusch, 1962). One or more 10,000 molecular weight molecules, which stimulate sporulation, are secreted, but it is not known whether niacin stimulates the synthesis of these large molecules or whether two independent pathways are involved (Wilkins and Reynolds, 1979).

Because nicotinamide coenzymes are involved in so many reactions, it is not surprising that changes in the ratio of oxidized to reduced coenzymes are correlated with certain developmental processes. For example, a mutant of *Neurospora crassa* exhibits a rhythmic cycle of conidiation in which the levels of NADH, NADPH, and NADP are low in the conidiating band of tissue while the NAD level is high (Figure 6.18). In the interband region the ratio of NAD to NADH is approximately 4:3 whereas in the conidiating band it is 9:3. On the other hand, the ratio of NADP to NADPH is 2:3 in the interband area and 1:3 in the conidiating band. Although it is not known what role these nicotinamide nucleotides play in the conidiation process, changes in the levels of these key coenzymes could have far-reaching effects on many aspects of metabolism (Brody and Harris, 1973).

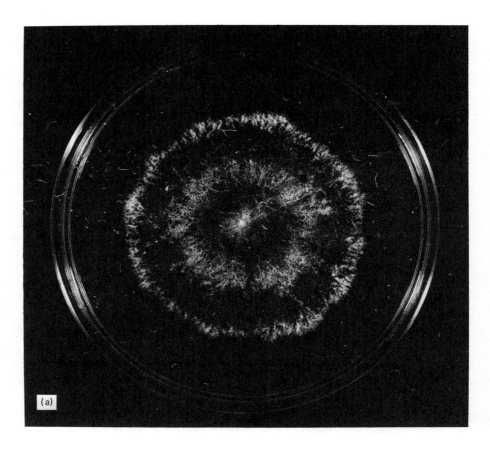

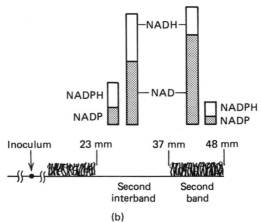

(b)

Figure 6.18 Rhythmic conidiation in *Neurospora crassa*. (a) Concentric areas of spore formation. (b) Schematic illustration of rhythmic spore formation and regional pyridine nucleotide composition 53 hours after inoculation. (Brody and Harris, 1973)

Pantothenic Acid

Pantothenic acid is a component of coenzyme A (CoA) (Figure 6.19) and as such is important in the transfer of acyl (CH₃CO-) groups. One of the most important CoA derivatives is acetyl-CoA (CH₃CO-CoA). As we shall see in Chapter 10, acetyl-CoA plays a central role in the synthesis and degradation of carbohydrates, fatty acids, steroids, and amino acids.

Fungi deficient in pantothenate have a greatly altered metabolism that can be traced to low intracellular levels of CoA. For example, cells of *Saccharomyces cerevisiae* deficient in pantothenate have levels of CoA that are one-fifth those of pantothenate-sufficient cells. As a result of pantothenate deficiency these cells have a low cytochrome content and therefore low respiration rates, low contents of fatty acids and lipids and thus increased membrane permeability, and high rates of H₂S production. The addition of pantothenate to the medium restores these parameters to normal (Hosono et al., 1972; Hosono and Aida, 1974; McCready et al., 1979).

To illustrate the widespread role of pantothenate in fungal metabolism, consider the hypothesis put forth to explain high H₂S production in pantothenate-deficient yeasts. Cysteine accumulates in these cells, and it is this compound which appears to be the source of H₂S. The addition of either methionine or pantothenate lowers the cysteine levels and consequently the level of H₂S. The relationship involving pantothenate, cysteine, methionine, and H₂S is diagrammed in Figure 6.20. As indicated, acetyl-CoA is required for the synthesis of methionine from cysteine; cysteine is also required for the synthesis of CoA. Under conditions of pantothenate deficiency, cysteine accumulates because it cannot be converted to either CoA or methionine. Tissue in which this accumulation occurs has a high activity of cysteine desulfhydrase, an enzyme that converts cysteine to H₂S and pyruvate. This enzyme is inhibited by S-adenosylmethionine, ATP, and GTP. When pantothenate is present, the resulting CoA can participate in respiration and methionine synthesis, and the endproducts of each reaction inhibit the desulfhydrase and the concomitant production of H₂S (Wainwright, 1970; Tokuyama et al., 1973).

The principal precursor of pantothenate in many fungi is β-alanine, and this amino acid can substitute for pantothenate in strains of *S. cerevisiae* and *S. carlsbergensis* (Arst, 1978; Smashevskii, 1967). As with other vitamins, the nutrients in the medium affect the synthesis of pantothenic acid. Transferring *Candida lipolytica* from a glucose to a paraffin medium, for example, results in a dramatic decrease in the cellular content of this vitamin (Borukaeva, 1967).

There are fewer reports in the literature of fungi that require exogenous pantothenate compared with those that require other vitamins. However, Ng (1976) found that all strains of yeast isolated from sour dough have an absolute requirement for pantothenate and a partial requirement for biotin (Table 6.4). Inositol stimulates growth in all strains tested except *S. cerevisiae*, and all strains of *S. exiguus* require niacin and thiamine. In addition, *S. inusitatus* requires folic acid. Pantothenate is also required for optimal growth of *Marsonnina brunea* (Simpson and Hayes, 1978).

The addition of various substances to the growth medium may affect the ability of a species to use pantothenate. Small amounts of casein hydrolysate alter the response of *S. carlsbergensis* to pantothenate. Casein hydrolysate is inhibitory to growth at pantothenate concentrations less than 0.1 μg but is stimulatory at higher

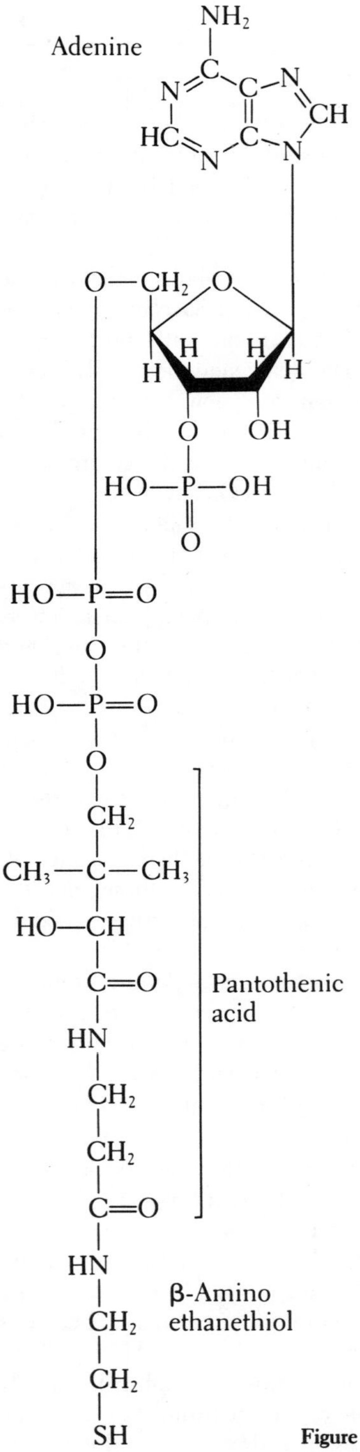

Figure 6.19 The structure of coenzyme A. (After Lehninger, 1975)

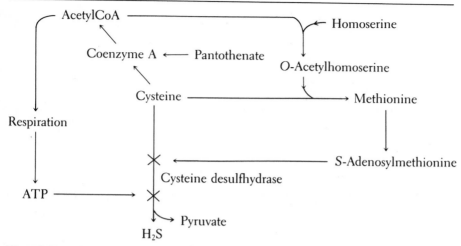

Figure 6.20 Relationship between pantothenate and H₂S production in yeast. (Modified from Tokuyama et al., 1973)

concentrations. Since this organism is routinely used in a bioassay for pantothenate, care must be taken in adjusting the concentrations of media constituents (Waller, 1970).

Folic Acid

Derivatives of folic acid are important coenzymes involved in the transfer of one-carbon compounds. As indicated in Figure 6.21, folic acid is composed of three

Table 6.4   Vitamin Requirements of Various Species of *Saccharomyces*

Medium	Turbidity at 675 nm After 48 Hours at 30° C						
	PBY	CY2	M95	98	M58	M86	M116
Complete	100[a]	100	100	100	100	100	100
	(0.90)	(0.95)	(0.84)	(0.85)	(0.90)	(0.94)	(0.80)
No biotin	57	65	67	23	48	47	35
No pantothenic acid	9	9	21	2	3	4	12
No folic acid	106	100	10	95	101	101	102
No inositol	106	81	90	88	97	90	90
No niacin	100	93	98	18	8	13	19
No *p*-aminobenzoic acid	103	100	108	94	100	98	81
No pyridoxine	106	100	99	92	93	82	85
No riboflavin	103	100	98	85	92	94	75
No thiamine	100	87	106	2	1	1	1

[a]Turbidity in complete medium is taken as 100%, and the actual value is shown in parentheses.
PBY = *S. cerevisiae*, CY2 = *S. uvarum*, M95 = *S. inusitatus*; 98, M58, M86, and M116 are different strains of *S. exiguus*.
Source: Ng (1976).

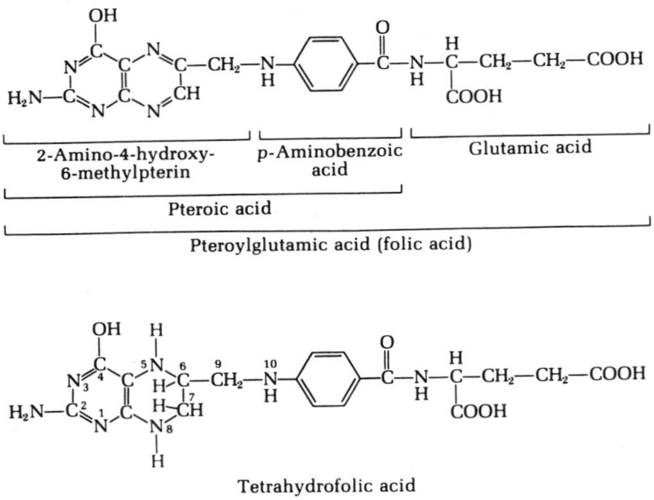

Figure 6.21 Structure of folic acid and tetrahydrofolic acid. The N^5 and N^{10} nitrogen atoms participate in the transfer of one-carbon groups. (Lehninger, 1975)

subunits: *p*-aminobenzoic acid, a substituted pterin, and glutamic acid. In the cell this molecule may have several glutamyl residues; *Saccharomyces cerevisiae*, for example, has hexa-, hepta-, and octo-glutamyl forms of folic acid present (Bassett et al., 1976). Such folylpolyglutamates may function as storage forms of folic acid and may aid in its intracellular retention. On the other hand, they substitute for folic acid in some reactions but are inhibitory in others. Thus, folate metabolism may be regulated by the relative concentrations of the various folylglutamate molecules (McGuire and Bertino, 1981).

Folic acid becomes biologically active after a two-step reduction to tetrahydrofolic acid (THFA). Dihydrofolate reductase, which catalyzes the second step, is a key enzyme (Benkovic, 1980). Its absence from cells of *Pediococcus cerevisiae* may be the reason that this species is auxoheterotrophic for folic acid (Iwai et al., 1977). In addition, this enzyme can be competitively inhibited by analogs of folic acid such as aminopterin and amethopterin (methotrexate), which are clinically important in the treatment of leukemias.

The role of THFA in one-carbon transfers can be illustrated by the conversion of serine to glycine via serine-hydroxymethyltransferase (Figure 6.22). In this reaction the β-carbon of serine becomes bound between the N^5 and N^{10} nitrogen atoms of THFA, forming N^5N^{10}-methylene tetrahydrofolate and releasing glycine. While attached to THFA the one-carbon unit can be modified in various ways (Figure 6.23), resulting in a family of interconvertible THFA coenzymes. McKenzie and Jones (1977) isolated 13 mutants of *S. cerevisiae* that lack one or more of the major enzymes for these interconversions. Mutants that had reduced activity of one enzyme generally had reduced activities of the others. This suggests that these interconversion enzymes may exist as a complex at one genetic locus.

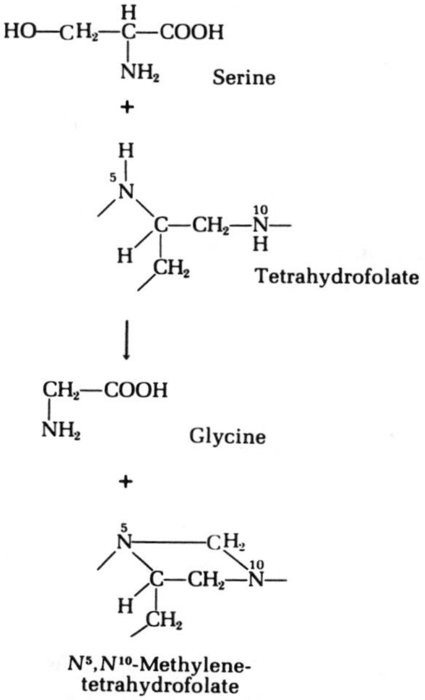

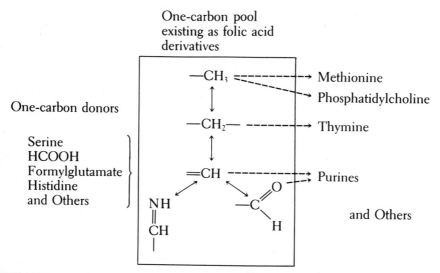

Figure 6.22 Formation of glycine and N^5,N^{10}-methylenetetrahydrofolic acid from serine. (Lehninger, 1975)

Figure 6.23 General interconversion and metabolism of one-carbon units mediated by tetrahydrofolic acid (THFA).

This pool of one-carbon units bound to THFA is extremely important for the synthesis of amino acids, lipids, and nucleic acids. In *Candida guilliermondii* incubated on a medium containing ^{14}C-labeled formaldehyde, the label enters this pool and rapidly accumulates in the amino acids serine and histidine and the purines adenine and guanine (Schlee et al., 1970). Cells of *Rhodotorula flava* incubated on a folate-deficient medium have reduced levels of RNA and DNA; the addition of folic acid or its precursor *p*-aminobenzoic acid restores normal levels (Odintsova and Zhukova, 1968; Zhukova, 1972). In addition, mutants with low activities of interconversion enzymes have a requirement for adenine (Lam and Jones, 1973; Jones and Lam, 1973).

Folic acid derivatives play a particularly important role in aspects of protein synthesis. In prokaryotic cells and the mitochondria of eukaryotic cells, the first amino acid of a polypeptide chain to be specified by the codons of mRNA is methionine. However, this "initiator amino acid" is modified by the addition of a formyl group transferred by THFA while the amino acid is attached to its tRNA but before the tRNA has attached to the ribosome (Figure 6.24). The codon specifying this N-formylmethionine tRNA is an initiator codon that enables the ribosome to recognize where to begin translation (Bianchetti et al., 1977). Inhibitors of dihydrofolate reductase block protein synthesis in yeast mitochondria, presumably by creating a shortage of formylmethionine (Stone and Wilkie, 1974).

A number of species such as *P. cerevisiae* (Iwai et al., 1977) and *R. flava* (Zhukova, 1972) require exogenous folic acid or its derivatives for growth. Some strains of *S. cerevisiae* show a growth response when supplied with the folate precursor *p*-aminobenzoic acid (Bassett et al., 1976). In addition, this precursor is required for maximal growth of two *Rhizoctonia* endophytes of orchid (Stephen and Fung, 1971).

Recently, folic acid has been identified as a chemo-attractant for cells of *Dictyostelium discoideum*. When supplied in pulses to cultures of this organism, folic acid increases the levels of cyclic AMP and cyclic GMP and stimulates aggregation and

$$N^{10}\text{-formyl-THFA} + \text{Methionine-tRNA} \longrightarrow$$

$$\text{THFA} + \text{N-formylmethionine-tRNA}$$

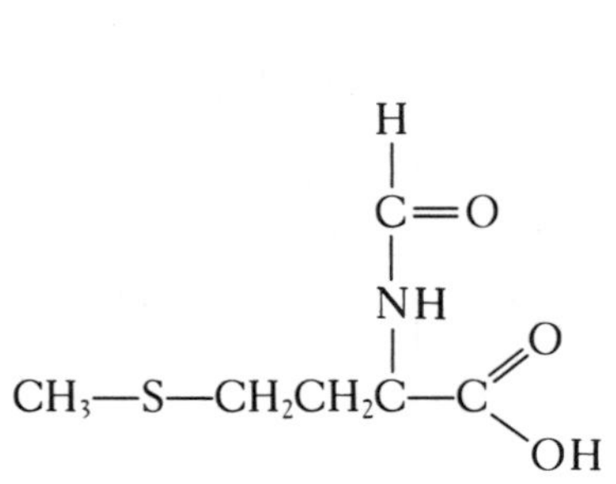

N-formylmethionine

Figure 6.24 Role of tetrahydrofolic acid (THFA) in the formation of N-formylmethionine during protein synthesis.

fruiting body formation. When it is supplied continuously, folic acid has no effect (Wurster and Schubiger, 1977; Wurster et al., 1979). Folic acid binds to specific sites on the plasma membrane, but it is uncertain what role this vitamin plays in the aggregation process (Van Driel, 1981).

GROWTH FACTORS

A wide variety of other organic molecules have been shown to affect fungal growth and development in small concentrations. These compounds, variously called growth factors or growth regulators, differ from vitamins in that they do not appear to function as coenzymes and are usually effective in slightly higher concentrations. Examples of growth factors include such diverse compounds as certain amino acids (Weigl et al., 1975), ascorbic acid (Shukla and Sarkar, 1972), ethyl alcohol, acetaldehyde, acetic acid, lactic acid, pyrogallol (Kratka, 1973), succinic acid, adenine (Kitamoto et al., 1974), and choline (Juretic, 1977). In most cases the mode of action of these compounds is not known, and they exert different effects in different species. We discuss in detail one of these factors, inositol, because its cellular role is relatively well known; and we mention more briefly three other widely studied factors: fatty acids, the hormones of higher plants, and volatile substances.

Inositol

Inositol is the fully hydroxylated derivative of cyclohexane (Figure 6.25). One of its stereoisomers, myoinositol, is often considered a vitamin even though it functions primarily as a structural component of certain phospholipids. In *Saccharomyces cerevisiae*, 90% of the cellular inositol occurs in phosphoinositol-containing sphingolipids and their precursor phosphatidylinositol (Smith and Lester, 1974; Becker and Lester, 1980). In this species, myoinositol is taken up by a single, specific active transport system (Nikawa et al., 1982).

Understanding the important biochemical roles of inositol has been greatly facilitated by the use of mutants that are auxoheterotrophic for this vitamin. For example, in inositol-requiring mutants of S. *cerevisiae* the addition of labeled inositol to the medium results in the initiation of growth and the appearance of label first in phosphatidylinositol and later in more polar lipids. High resolution autoradiography

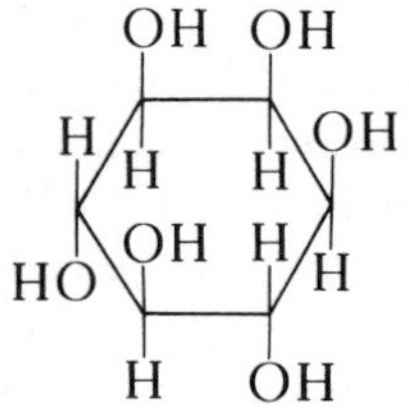

Figure 6.25   Myoinositol.

indicates that inositol is first incorporated into internal membranes and is later transferred to the plasma membrane. When this mutant is incubated with suboptimal concentrations of inositol, alterations are observed in the activities of several membrane-bound enzymes, suggesting that the membrane may be structurally modified due to differences in the phospholipid composition (Dominguez et al., 1978). Enzymes sensitive to inositol deficiencies include monoamine oxidase and cytochrome c oxidase of the inner and outer mitochondrial membranes (Bednarz-Prashad and Mize, 1974), phosphofructokinase, ATP-citrate lyase, aldolase and citrate synthase (Hayashi et al., 1978), acid phosphatase (Yamamoto et al., 1973), and adenylcyclase (Scott et al., 1978).

In addition, inositol has been shown to alter other membrane-associated phenomena such as temperature tolerance (Hayashi et al., 1978; Sullivan and Debusk, 1973) and osmotic relations (Atkinson et al., 1977). The latter workers noted that in the absence of inositol-containing lipids, cell surface expansion of S. cerevisiae terminates after one doubling of whole cells. In spheroplasts, this cessation of membrane expansion is followed by the rapid development of an osmotic imbalance and then lysis.

Henry and co-workers (1977) studied the sequence of events accompanying inositol starvation in mutants of S. cerevisiae. Upon transfer of these cells to an inositol-deficient medium, the synthesis of phosphatidylinositol ceases within 30 minutes. Cell division and expansion continue for another 1.5 hours, and the synthesis of DNA, RNA, proteins, and various lipids does not stop until an hour after that. Cellular K^+ and ATP levels drop after a total of 4.5 hours, and death rapidly ensues. These workers suggest that this "inositol-less death" is due to the imbalance arising between a functioning, expanding cytoplasm and a cell surface that has ceased to grow.

Obviously, the localization of inositol in membrane phospholipids suggests a mechanism by which inositol could affect the synthesis of cell wall polysaccharides. When mutants of S. cerevisiae are transferred to an inositol-deficient medium the synthesis of glycans (primarily beta-1,3-glucans) declines rapidly, even before there is any detectable slowing in the growth rate or in the synthesis of macromolecules. Thus, inositol may be required for certain steps in glycan synthesis, perhaps by (1) functioning as a cofactor, (2) charging and orienting a portion of the membrane for a particular synthetic step, or (3) functioning in the packaging or secreting of glycans in membrane vesicles (Hanson and Lester, 1980).

The involvement of inositol in cell wall metabolism has also been observed in mutants of Neurospora crassa. In this species, considerable amounts of inositol are associated with the cell wall, and mutations affecting inositol biosynthesis result in altered cell wall composition and a highly branched mycelium. Notice in Table 6.5 that in one of these mutants increasing the concentration of inositol in the medium from 3.3 to 55.4 μM results in increased levels of inositol in the cell wall but decreased wall weight. Also, the glucosamine content of the wall increases by approximately 50%, but the galactosamine content decreases by about the same amount. On the other hand, the inositol content of the medium has little effect on the amounts of amino acids or hexoses present in the wall. Thus, the highly branched

Table 6.5 Levels of Inositol and the Amino Sugars in the Cell Walls of the Mycelium of the Inl Mutant of *Neurospora crassa* Grown on Varying Concentrations of Inositol

Inositol in Medium (μM)	Cell Wall Content (mg of cell wall/g [dry wt] mycelia)	Concentration (μmol/100 mg of cell wall)		
		Glucosamine	Galactosamine	Inositol
3.3	22.4	14.9	31.0	0.15
8.3	15.5	28.2	16.6	0.18
27.7	15.8	24.6	17.0	0.25
55.4	12.8	29.5	19.0	0.22

Source: Hanson and Brody (1979).

mycelium observed under inositol deficiency may be a consequence of a weakened wall brought about by low glucosamine and thus low chitin levels. As in *S. cerevisiae,* it is uncertain how inositol affects either the levels of amino sugars or the morphology of the hyphae. The plasma membrane and consequently the cell wall may be physically disrupted by a deficiency of membrane phospholipids. This may be particularly important at the hyphal tips where membrane lipid may be so altered that polysaccharides are not properly aligned on the membrane surface (Hanson and Brody, 1979).

Another morphological effect of inositol is illustrated by its involvement in yeast–mold dimorphism. In *Candida tropicalis* the conversion of the yeastlike form to the filamentous form is stimulated by low concentrations of ethanol. However, as shown in Table 6.6, inositol at 5 μg/ml can nullify the effect of ethanol and result in yeastlike cells (Tani et al., 1979).

Table 6.6 Effect of Myoinositol on the Morphological Changes in *Candida tropicalis* Caused by Ethanol

Ethanol (%)	Myoinositol (μg/ml)	Cell Form
0	—	Yeastlike
1.5	0	Filamentous
	1.3	Filamentous (some yeasts)
	2.5	Yeastlike (some filaments)
	5.0	Yeastlike
2.5	0	Filamentous
	1.3	Filamentous
	2.5	Filamentous (some yeasts)
	5.0	Yeastlike

Source: Adapted from Tani et al. (1979).

Fatty Acids

Fatty acids at low concentrations have been shown to stimulate growth and development in certain fungi. Mutants of *Neurospora crassa* require an exogenous supply of fatty acids for growth because they have lost the capacity to synthesize these compounds (Barber and Lands, 1973; Scott, 1977). In other cases, fatty acids are not required for survival but rather modulate patterns of development. For example, oleic, linoleic, and linolenic acids promote the inititation and growth of rhizomorphs in *Armillaria mellea*. These compounds are effective only when added as supplements to a growth medium and not as the sole carbon sources, indicating that they are functioning more as growth factors than merely as carbon skeletons (Moody and Weinholds, 1972).

Growth factors that are stimulatory to some fungi are inhibitory to others. Teh (1974) observed that the same short-chain fatty acids that promote rhizomorph formation in *A. mellea* are toxic to growth of *Cladosporium resinae* grown on a glucose medium.

As we have seen with other nutrients, environmental conditions often modify the effects of growth factors such as fatty acids. For example, *S. cerevisiae* has a requirement for a sterol and an unsaturated fatty acid as growth factors when grown under anaerobic conditions but has no such requirement under aerobicity (Alterthum and Rose, 1973).

Hormones of Higher Plants

Because of the important role of fungi in plant disease, there have been numerous studies testing the effect of naturally occurring plant growth hormones and their synthetic analogs on fungal growth and development. In many cases these hormones have stimulatory effects. Gibberellic acid (GA) and the auxins indole acetic acid (IAA), naphthaleneacetic acid (NAA), and 2,4-dichlorophenoxyacetic acid (2,4-D) stimulate conidial germination of *Neurospora crassa* (Nakamura et al., 1978) and mycelial growth of *Rhizoctonia solani* (Khan and Azam, 1975). However, plant hormones have varied effects on fruiting body formatin in *Lentinus tigrinus* (Table 6.7). IAA causes a significant decrease in the number of fruiting bodies at 100 ppm but a significant increase at 300 ppm. IAA also stimulates fruiting body fresh weight, cap diameter, and stalk length. GA affects only the number of fruiting bodies and stalk length; kinetin affects only cap diameter and stalk length. Notice that for both these hormones there are instances where they are stimulatory at one concentration, inhibitory at another, and have no effect at still another (Sladky and Tichy, 1974). Such varied effects of plant hormones are commonly observed with other species.

Volatile and Other Miscellaneous Compounds

Numerous reports have appeared on the effects of volatile compounds on fungal growth. Volatiles believed to be precursors of trisporic acids (Chapter 9) are produced

Table 6.7 Stimulation of the Formation of Fruiting Bodies of *Lentinus tigrinus* by Growth Regulators[a]

		Number of Fruiting Bodies on Cultivation Cylinder	Fruiting Body Fresh Weight Production from One Cultivation Cylinder (g)	Diameter of the Cap (mm)	Length of the Stalk (mm)
Control		6.3 ± 0.3	6.5 ± 0.3	22.6 ± 1.5	28.8 ± 1.4
IAA	50	6.5 ± 0.5	7.3 ± 0.3	28.9 ± 3.1	29.6 ± 2.0
(ppm)	100	5.0 ± 0.1[++]	6.5 ± 2.8	31.3 ± 3.6[+]	43.8 ± 2.6[++]
	200	6.5 ± 0.5	6.8 ± 0.9	23.1 ± 3.7	27.6 ± 1.7
	300	9.0 ± 0.1[++]	8.1 ± 0.3[++]	20.6 ± 3.1	27.8 ± 1.7
GA₃	100	6.5 ± 0.5	5.4 ± 1.4	18.6 ± 4.9	39.9 ± 2.3[++]
(ppm)	200	9.0 ± 1.0[+]	6.5 ± 1.8	24.6 ± 2.6	40.3 ± 4.0[++]
	300	9.0 ± 0.1[++]	7.7 ± 0.8	20.8 ± 3.7	32.6 ± 1.9
	400	5.0 ± 0.1[++]	6.0 ± 1.2	22.5 ± 3.3	32.0 ± 3.0
KIN	100	7.5 ± 1.5	7.2 ± 4.2	27.3 ± 3.4	32.9 ± 3.0
(ppm)	200	6.5 ± 0.5	5.1 ± 2.5	35.2 ± 1.8[++]	37.5 ± 1.9[++]
	300	7.5 ± 0.5	5.1 ± 1.9	35.0 ± 3.8[++]	33.0 ± 9.9
	400	9.5 ± 1.5	6.1 ± 0.4	13.6 ± 2.9[++]	34.6 ± 4.7

Source: Sladky and Tichy (1974).

[a] [+]Statistically significant, $P < 0.05$.
[++]Statistically significant, $P < 0.01$.

by *Mucor mucedo* (Mesland et al., 1974) and *M. racemosus* (Mooney and Sypherd, 1976) and induce the morphogenesis of sexual structures. In a number of instances volatiles are stimulatory to spore germination and mycelial growth (Afifi, 1975); however, many are toxic (Nandi, 1977; Pavlica et al., 1978). Flodin and Fries (1978) studied volatiles emanating from pine wood on the growth of five wood-rotting species (Table 6.8). Notice that volatiles from fresh wood are generally inhibitory, but after the wood is heat-treated the amount of inhibition decreases, and in some cases the volatiles are quite stimulatory. When *Stereum sanguinolentum* is exposed to monoterpenes known to be produced from fresh pine wood, growth is generally inhibited. However, growth of these species is stimulated by hexanal and hexanal plus α-pinene, which are produced by heat-dried wood.

Five species of mycoparasites including *Gonatrobotryum simplex* have an absolute requirement for an extract from the host or other fungi. This growth factor has been partially purified and given the name mycotrophein, but its chemical identity has not been completely determined (Whaley and Barnett, 1963; Joron and Barnett, 1978).

Table 6.8 Effects of Volatiles from Fresh and Heat-Dried Wood on Some Wood-Rotting Fungi[a]

Fungus	Amount of Wood Chips (g/6l)	Fresh Wood Dry Weight Rel. Control	Fresh Wood Significance Level	Heated Wood Dry Weight Rel. Control	Heated Wood Significance Level
Stereum sanguinolentum	0	1.00		1.00	
	1	0.50	− − −	0.55	− − −
	3	0.43	− − −	0.51	− − −
	10	0.44	− − −	0.58	− − −
	30	0.28	− − −	0.71	− − −
	100	0.29	− − −	1.06	0
Fomes annosus	0	1.00		1.00	
	1	0.53	− − −	0.72	−
	3	0.66	− −	1.27	0
	10	0.68	−	1.13	0
	30	0.84	0	1.26	0
	100	0.81	0	1.31	0
Trametes pini	0	1.00		1.00	
	1	1.01	0	0.94	0
	3	1.11	0	0.96	0
	10	1.25	+	1.03	0
	30	1.38	+ +	0.78	−
	100	0.83	−	1.05	0
Coniophora puteana	0	1.00		1.00	
	1	0.74	−	1.24	0
	3	1.19	0	1.11	0
	10	0.78	0	0.76	0
	30	0.90	0	0.70	0
	100	0.97	0	0.79	0
Lentinus lepideus	0	1.00		1.00	
	1	1.15	0	1.22	+
	3	1.14	0	1.22	0
	10	1.06	0	1.46	+ + +
	30	0.95	0	1.74	+ + +
	100	0.92	0	2.00	+ + +

Source: Flodin and Fries (1978).

[a] + = stimulation, − = inhibition; the number of plus or minus signs corresponds to the significance level as expressed by: * = 5%, ** = 1%, *** = 0.1%.

SUMMARY

Vitamins are organic compounds needed for growth, development, and metabolite production. Typically they function as coenzymes or constituents of coenzymes and consequently are needed in minute amounts by the fungal cell. Many fungi can synthesize their own vitamins from simple precursors and are called auxoautotrophs. Others, called auxoheterotrophs, must have one or more vitamins supplied exogenously. Fungi are most often auxoheterotrophic for thiamine and biotin.

The major biochemical roles that the most common vitamins have as coenzymes have been determined. Thiamine is involved in reactions in which aldehyde groups are removed or transferred. Biotin functions in carboxylation reactions. Riboflavin is involved in both electron transfer and photoreception. Pyridoxine functions in transamination reactions. Niacin is a component of NAD(H) and NADP(H) and thus is important in electron transfer reactions. Panthothenic acid is a constituent of coenzyme A and thus is involved in the transfer of acyl groups. Folic acid plays an important role in one-carbon transfers.

A variety of organic compounds that are effective promoters of fungal growth and development function neither as coenzymes nor as nutrients in the sense of carbon and nitrogen sources. These "organic growth factors" produce a response at concentrations substantially above that produced by typical vitamins and include myoinositol, fatty acids, plant growth hormones, and a host of other compounds. Several compounds in this group are volatile and some may be in inhibitory at high concentrations. The mechanism of action of these compounds either as promotors or inhibitors of growth and development is usually not known.

REFERENCES

Afifi, A. 1975. Effect of volatile substances from species of Labiatae on rhizospheric and phyllospheric fungi of *Phaseolus vulgaris*. *Phytopathol. Z.* **83**: 296–302.

Alterthum, F., and A. H. Rose. 1973. Osmotic lysis of spheroplasts from *Saccharomyces cerevisiae* grown anaerobically in media containing different unsaturated fatty acids. *J. Gen. Microbiol.* **77**: 371–382.

Arst, H. N. 1978. GABA transaminase provides an alternate route of beta-alanine synthesis in *Aspergillus nidulans*. *Mol. Gen. Genet.* **163**: 23–27.

Atkinson, K. D., A. I. Kolat, and S. A. Henry. 1977. Osmotic imbalance in inositol-starved sphaeroplasts of *Saccharomyces cerevisiae*. *J. Bacteriol.* **132**: 806–817.

Aurich, H., W. Neumann, and H. P. Kleber. 1967. Nucleic acids and proteins in *Neurospora* in biotin and pyridoxine deficiency. *Acta Biol. Med. Ger.* **19**: 221–229.

Barber, E. D., and W. E. M. Lands. 1973. Quantitative measurement of the effectiveness of unsaturated fatty acids required for the growth of *Saccharomyces cerevisiae*. *J. Bacteriol.* **115**: 543–551.

Barnett, H. L. 1968. The effects of light, pyridoxine and biotin on the development of the mycoparasite, *Gonatobotryum fuscum*. *Mycologia* **60**: 244–251.

Bartnicki-Garcia, S., and W. J. Nickerson. 1961. Thiamine and nicotinic acid: Anaerobic growth factors for *Mucor rouxii*. *J. Bacteriol.* **82**: 142–148.

Basappa, S. C., A. Jayraraman, V. Sreenivasmurthy, and H. A. B. Parpia. 1967. Effect of B-group vitamins and ethyl alcohol on aflatoxin production by *Aspergillus oryzae*. *Indian J. Exp. Biol.* **5**: 262–263.

Bassett, R., D. G. Weir, and J. M. Scott. 1976. The identification of hexa-, hepta- and octoglutamates as the polyglutamyl forms of folate found throughout the growth cycle of yeast. *J. Gen. Microbiol.* **93**: 169–172.

Bayer, E. A., E. Skutelsky, T. Viswanatha, and M. Wilchek. 1978. Specific localization and quantification of biotin transport components in yeast by use of a biotin-conjugated, impermeant, electron-dense label. *Mol. Cell. Biochem.* **19**: 23–30.

Becker, G. W., and R. L. Lester. 1980. Biosynthesis of phosphoinositol containing sphingolipids from phosphatidylinositol by a membrane preparation of *Saccharomyces cerevisiae. J. Bacteriol.* **142**: 747–754.

Becker, J. M., and H. C. Lichstein. 1972. Transport overshoot during biotin uptake by *Saccharomyces cerevisiae. Biochim. Biophys. Acta* **282**: 409–420.

Bednarz-Prashad, A. J., and C. E. Mize. 1974. Mitochondrial membranes of inositol-requiring *Saccharomyces carlsbergensis.* Covalent binding of a radioactive marker to the outer membrane. *Biochemistry* **13**: 4237–4262.

Benkovic, S. J. 1980. On the mechanism of action of folate- and biopterin-requiring enzymes. *Ann. Rev. Biochem.* **49**: 227–251.

Bhama, K. S., and R. N. Swamy. 1976. Sporulation of *Cercospora personata* in culture: III. Effect of riboflavin in the medium. *Kavaka* **4**: 65–68.

Bianchetti, R., G. Lucchini, P. Crosti, and P. Tortora. 1977. Dependence of mitochondrial protein synthesis on formulation of the initiator methionyl-tRNAf. *J. Biol. Chem.* **252**: 2519–2523.

Bilai, V. I., S. M. Shcherbina, L. O. Bohomolova, and N. S. Proskuryakova. 1974. Effect of different ratios of carbon and nitrogen on fusaric and nicotinic acids biosynthesis and respiration in *Fusarium. Mikrobiol. Zh.* **36**: 293–299.

Borukaeva, M. R. 1967. Comparative study of biotin, inositol, and nicotinic acid biosynthesis by *Candida* cultures on different carbon sources. *Mikrobiologiya* **36**: 376–379.

Brody, S., and S. Harris. 1973. Circadian rhythms in *Neurospora:* Spatial differences in pyridine nucleotide levels. *Science* **180**: 498–500.

Burge, M. N., and I. Isaac. 1977. Growth responses to biotin by *Phialophora asteris. Trans. Brit. Mycol. Soc.* **68**: 451–453.

Chenouda, M. S., and Z. A. El-Awamry. 1971a. Biosynthesis of biotin by micro-organisms: I. Biotin vitamers accumulated in the culture medium of *Phycomyces blakesleeanus* and the time course of synthesis and in relation to metabolic activities. *J. Gen. Appl. Microbiol.* **17**: 335–343.

Chenouda, M. S., and Z. A. El-Awamry. 1971b. Biosynthesis of biotin by micro-organisms: II. Biochemical factors affecting the synthesis of biotin vitamers in *Phycomyces blakesleeanus. J. Gen. Appl. Microbiol.* **17**: 345–352.

Cicmanec, J. F., and H. C. Lichstein. 1974. Biotin uptake by cold-shocked cells, sphaeroplasts, and repressed cells of *Saccharomyces cerevisiae:* Lack of feed-back control. *J. Bacteriol.* **119**: 718–725.

Daniel, J. W., and H. P. Rusch. 1962. Niacin requirement for sporulation in *Physarum polycephalum. J. Bacteriol.* **83**: 1244–1250.

Dargent, R., and J. Touze-Soulet. 1976. Ultrastructure of the hyphae of *Hypomyces chlorinus* Tul. cultivated in the presence or in the absence of biotin. *Protoplasma* **89**: 49–71.

Desai, J. D., and V. V. Modi. 1977. Growth, glucose metabolism and melanin formation in biotin deficient *Aspergillus nidulans. Folia Microbiol.* **22**: 55–60.

Dominguez, A., J. R. Villaneuva, and R. Sentandreu. 1978. Inositol deficiency in *Saccharomyces cerevisiae* NCYC 86. *Antonie van Leeuwenhoek J. Microbiol. Serol.* **44**: 25–34.

Evans, R. C., and M. O. Garraway. 1976. Effect of thiamine on ethanol and pyruvate production in *Helminthosporium maydis. Plant Physiol.* **57**: 812–816.

Flodin, K., and N. Fries. 1978. Studies on volatile compounds from *Pinus silvestris* and their effect on wood-decomposing fungi: II. Effects of some volatile compounds on fungal growth. *Eur. J. For. Pathol.* **8**: 300–310.

Garraway, M. O. 1973. Sporulation in *Helminthosporium maydis:* inhibition by thiamine. *Phytopathology* **63**: 900–902.

Gontier, C., and J. Touze-Soulet. 1971. The influence of optimum and below doses of biotin on the metabolism of *Hypomyces chorinus* Tul: Evolution of the quantities of free and combined lipids in the mycelia during development. *C. R. Hebd. Seances Acad. Sci., Ser. D. Natur. (Paris)* **272**: 1354–1356.

Gounaris, A. D., I. Turkenkopf, L. L. Averchia, and J. Greenlie. 1975. Pyruvate decarboxylase: 3. Specificity restrictions for thiamine pyrophosphate in the protein association step—subunit structure. *Biochim. Biophys. Acta* **405**: 492–499.

Hanson, B. A., and S. Brody. 1979. Lipid and cell wall changes in an inositol-requiring mutant of *Neurospora crassa. J. Bacteriol.* **138**: 461–466.

Hanson, B. A., and R. L. Lester. 1980. Effects of inositol starvation on phospholipid and glycan synthesis in *Saccharomyces cerevisiae. J. Bacteriol.* **142**: 79–89.

Hayashi, E., R. Hasegawa, and T. Tomita. 1978. The fluctuation of various enzyme activities due to myo-inositol deficiency in *Saccharomyces carlsbergensis. Biochim. Biophys. Acta* **540**: 231–237.

Hegde, R. K., and K. G. Ranganathaiah. 1971. Vitamin requirements of two isolates of *Helminthosporium nodulosum* of Ragi. *Mysore J. Agric. Sci.* **5**: 101–104.

Heilman, H. D., and F. Lingens. 1968. On the regulation of nicotinic acid biosynthesis in *Saccharomyces cerevisiae. Hoppe-Seyler's Z. Physiol. Chem.* **349**: 231–236.

Henry, S. A., K. D. Atkinson, A. L. Kolat, and M. R. Culbertson. 1977. Growth and metabolism of inositol-starved *Saccharomyces cerevisiae. J. Bacteriol.* **130**: 472–484.

Hilgenberg, W., F. Hofmann. 1977. Tryptophan synthase in *Phycomyces blakesleeanus:* I. Characterization of the enzyme. *Physiol. Plant.* **40**: 181–185.

Hosono, K., and K. Aida. 1974. Lipid composition of *Saccharomyces cerevisiae* defective in mitochondria due to pantothenic acid deficiency. *J. Gen. Microbiol.* **20**: 47–58.

Hosono, K., K. Aida, and T. Uemura. 1972. Effect of pantothenic acid on respiratory activity of aerobically grown *Saccharomyces cerevisiae. J. Gen. Appl. Microbiol.* **18**: 189–199.

Iwai, K., M. Ikeda, and S. Fujino. 1977. Nutritional requirements for folate compounds and some enzyme activities involved in folate biosynthesis. *J. Nutr. Sci. Vitaminol.* **23**: 95–100.

Iwashima, A., and H. Nishimura. 1979. Isolation of a thiamine-binding protein from *Saccharomyces cerevisiae. Biochim. Biophys. Acta* **577**: 217–220.

Iwashima, A., H. Nishimura, and Y. Nose. 1979. Soluble and membrane-bound thiamine-binding proteins from *Saccharomyces cerevisiae. Biochim. Biophys. Acta* **557**: 460–468.

Iwashima, A., H. Nishino, and Y. Nose. 1973. Carrier-mediated transport of thiamine in baker's yeast. *Biochim. Biophys. Acta* **330**: 222–234.

Iwashima, A., and Y. Nose. 1976. Regulation of thiamine transport in *Saccharomyces cerevisiae. J. Bacteriol.* **128**: 855–857.

Iwashima, A., Y. Wakabayashi, and Y. Nose. 1975. Thiamine transport mutants of *Saccharomyces cerevisiae. Biochim. Biophys. Acta* **413**: 243–247.

Jones, E. W., and K. Lam. 1973. Mutations affecting levels of tetrahydrofolate interconversion enzymes in *Saccharomyces cerevisiae:* II. Mafo positions on chromosome VII of ade 3-31 and ade 15. *Mol. Gen. Genet.* **123**: 209–218.

Joron, E. G., and H. L. Barnett. 1978. Nutrition and parasitism of *Melanospora zamiae. Mycologia* **70**: 300–312.

Juretic, D. 1977. Lecithin requirement for the sporulation process in *Neurospora crassa. J. Bacteriol.* **130**: 524–525.

Kacchy, A. N., A. M. Madia, V. V. Modi, and N. Parekh. 1972. Octadecanol metabolism in *Candida tropicalis. Indian J. Exp. Biol.* **10**: 246–248.

Kawasaki, C., T. Kishi, and T. Nishihara. 1970. The relation between thiamine and vitamin B_6 on *Saccharomyces carlsbergensis* 4228: X. Indirect inhibition of biosynthesis of nicotinic acid by thiamine. *Vitamins (Kyoto)* **41**: 107–112.

Khan, M. W., and F. F. Azam. 1975. In vitro response of cauliflower isolate of *Rhizoctonia solani* to certain growth factors. *Indian Phytopathol.* **28**: 202–205.

Kishi, T. 1969. The relation between thiamine and vitamins B_6 on *Saccharomyces carlsbergensis* 4228: IX. Relations between growth inhibition by thiamine and vitamin B_6 content in the cells. *Vitamins (Kyoto)* **40**: 385–391.

Kitamoto, Y., N. Yamane, N. Hosoi, and Y. Ichikawa. 1974. Nutritional aspects of fruit-body formation of *Favolus arcularius* in replacement culture. *Trans. Brit. Mycol. Soc. Jpn.* **15**: 60–71.

Kratka, K. 1973. The effect of some organic substances on the germinability of the clamydospores of *Ustilago nuda* (Jens.) *Sb. Uvti. Ochr. Rostl.* **9**: 73–98.

Kuwahara, M. 1976. Formation of nicotinamide ribose diphosphate ribose, a new metabolite of the NAD pathway, by growing mycelium of *Aspergillus niger. Agric. Biol. Chem.* **40**: 1573–1580.

Kvasnikov, E. I., D. M. Isakova, G. S. Eliseeva, and Z. I. Loike. 1977. Patterns of growth and metabolism of thermotolerant microorganisms on media containing carbohydrates and hydrocarbons. *Prikl. Biokhim. Mikrobiol.* **13**: 881–892.

Kvasnikov, E. I., D. M. Isakova, and V. T. Vaskivnyuk. 1967. Accumulation dynamics of B-group vitamins in *Candida tropicalis* during the utilization of petroleum hydrocarbons. *Mikrobiologiya* **36**: 932–937.

Lam, K. B., and E. W. Jones. 1973. Mutations affecting levels of tetrahydrofolate interconversion enzymes in *Saccharomyces cerevisiae*: I. Enzyme levels in ade 3-41 and ade 15, a dominant adenine auxotroph. *Mol. Gen. Genet.* **123**: 199–208.

Langkramer, O. 1969. Effect of some vitamins on the growth of *Suillus variegatus* (Sev. ex Fr.) O. Kuntze. *Ceska Mykol.* **23**: 53–60.

Lehninger, A. I. 1975. *Biochemistry*, 2nd ed. New York: Worth.

Leibert, H. P. 1970. Influence of malic and tartaric acids and of vitamin B_1 on glycerol production during alcoholic fermentation. *Zentralbl. Bakteriol. Parasetenkd. Infektionskr. Hyg. Abt. 2 Naturwiss.* **126**: 162–170.

Lilly, V. G., and H. L. Barnett. 1947. The influence of pH and certain growth factors on mycelial growth and perithecial formation by *Sordaria fimicola. Am. J. Bot.* **34**: 131–138.

Lindblow-Kull, C., F. J. Kull, and A. Shrift. 1980. Evidence for the biosynthesis of selenobiotin. *Biochem. Biophys. Res. Commun.* **93**: 572–576.

McCready, R. G. L., G. A. Din, and H. R. Krouse. 1979. The effect of pantothenate on sulfate metabolism and sulfur isotope fractionation by *Saccharomyces cerevisiae. Can. J. Microbiol.* **25**: 1139–1144.

McGuire, J. J., and J. R. Bertino. 1981. Enzymatic synthesis and function of folypolyglutamates. *Mol. Cell. Biochem.* **38**: 20–48.

McKenzie, K. Q., and E. W. Jones. 1977. Mutants of the tetrahydrofolate interconversion pathway of *Saccharomyces cerevisiae. Genetics* **86**: 85–102.

Mesland, D. A. M., J. G. Huisman, and H. van den Ende. 1974. Volatile sexual hormones in *Mucor mucedo. J. Gen. Microbiol.* **80**: 111–117.

Milton, J. M., and I. Isaac. 1967. Studies on a biotin requiring strain of *Verticillium dahliae. Trans. Brit. Mycol. Soc.* **50**: 539–547.

Mitra, J., and K. L. Chaudhuri. 1974. Temperature sensitive riboflavin mutants of *Penicillium vermiculatrens* Dangeard. *Nucleus (Calcutta)* **17**: 194–198.

Mizunaga, T., H. Kuraishi, and K. Aida. 1975. Mechanism of unbalanced growth and cell death of biotin-deficient yeast cells in the presence of aspartic acid. *J. Gen. Appl. Microbiol.* **21**: 305–316.

Mohana, K., and E. R. B. Shanmugasundarum. 1978. Pyridoxine and its relation to lipids: studies with pyridoxineless mutants of *Aspergillus nidulans. J. Nutr. Sci. Vitaminol.* **24**: 397–404.

Molowitz, R., M. Bahn, and B. Hock. 1976. The control of fruiting body formation in the ascomycete *Sordaria macrospora* Auersw. by arginine and biotin: A two-factor analysis. *Planta* **128**: 14–148.

Moody, A. R., and A. R. Weinhold. 1972. Fatty acids and naturally occurring plant lipids as stimulants of rhizomorph formation in *Armillaria mellea. Phytopathology* **62**: 264–267.

Mooney, D. T., and P. S. Sypherd. 1976. Volatile factor involved in dimorphism of *Mucor racemosus. J. Bacteriol.* **126**: 1266–1270.

Moss, J., and M. D. Lane. 1971. The biotin-dependent enzymes. *Adv. Enzymol.* **35**: 321–442.

Nabeshima, S., A. Tanaka, and S. Fukui. 1977. Effect of biotin on the level of isocitrate lyase in *Candida tropicalis. Agr. Biol. Chem.* **41**: 281–285.

Nakamura, I., N. Isobe, N. Nakamura, T. Kamihara, and S. Fukui. 1981. Mechanism of thiamine-induced respiratory deficiency in *Saccharomyces carlsbergensis. J. Bacteriol.* **147**: 954–961.

Nakamura, T., Y. Kawanabe, E. Takiyama, N. Takahashi, and T. Murayama. 1978. Effects of auxin and gibberellin on conidial germination in *Neurospora crassa. Plant Cell Physiol.* **19**: 705–709.

Nakamura, I., Y. Kishikawa, T. Kamihara, and S. Fukui. 1974. Respiratory deficiency in *Saccharomyces carlsbergensis* 4228 caused by thiamine and its prevention by pyridoxine. *Biochem. Biophys. Res. Commun.* **59**: 771–776.

Nandi, B. 1977. Effect of some volatile aldehydes, ketones, esters and terpenoids on growth and development of fungi associated with wheat grains in the field and in storage. *Z. Pflanzenkr. Pflanzenschutz.* **84**: 114–128.

Ng, H. 1976. Growth requirements of San Francisco sour dough yeast and Baker's yeast. *Appl. Environ. Microbiol.* **31**: 395–398.

Nikawa J., T. Nagumo, and S. Yamashita. 1982. Myo-inositol transport in *Saccharomyces cerevisiae. J. Bacteriol.* **150**: 441–446.

Nishikawa, Y., I. Kanamura, T. Kamihara, and S. Fukui. Effects of thiamine and pyridoxine on the composition of fatty acids in *Saccharomyces carlsbergensis* 4228. *Biochem. Biophys. Res. Commun.* **59**: 777–780.

Nishikawa, Y., I. Kanamura, T. Kamihari, and S. Fukui. 1977. Effects of thiamine and pyridoxine on the lipid composition of *Saccharomyces carlsbergensis* 4228. *Biochim. Biophys. Acta* **486**: 483–489.

Nomdedu, L., M. Sancholle, and J. Touze-Soulet. 1974. (Biotin deficiency and cellular permeability of *Hypomyces chlorinus*.) *Can. J. Bot.* **52**: 1867–1874.

Odintsova, E. N., and S. V. Zhukova. 1968. Effects of para-aminobenzoic acid and folic acid on biosynthesis of nucleic acids. *Mikrobiologiya* **37**: 788–791.

Ott, M. K., M. Jayaram, R. M. Hamilton, and M. Delbruck. 1981. Replacement of riboflavin by an analogue in the blue-light photoreceptor of *Phycomyces. Proc. Natl. Acad. Sci. USA* **78**: 266–269.

Pavlica, D. A., T. S. Hora, J. J. Bradshaw, R. K. Skogerboe, and R. Baker. 1978. Volatiles from soil influencing activities of soil fungi. *Phytopathology* **68**: 758–765.

Perl, M., E. B. Kearney, and T. P. Singer. 1976. Transport of riboflavin into yeast cells. *J. Biol. Chem.* **251**: 3221–3228.

Qudeer, M. A., and M. P. Igbal. 1971. Effect of EDTA on the production of riboflavin by *Candida guilliermondii. Pak. J. Sci. Ind. Res.* **14**: 129–133.

Rao, K. K., and V. V. Modi. 1968. Metabolic change in biotin deficient *Aspergillus nidulans. Can. J. Microbiol.* **14**: 813–815.

Rawla, G. S., and S. S. Chahal. 1975. Comparative trace element and organic growth factor requirements of *Ramulispora sacchari* and *R. sorghi. Trans. Brit. Mycol. Soc.* **64**: 532–535.

Redchyts, T. I. 1972. Riboflavin formation by fungi of the genus *Aspergillus* Mich. during cultivation on hydrocarbon media. *Mikrobiol. Zh.* **34**: 729–753.

Ridings, W. H., M. E. Gallegly, and V. G. Lilly. 1969. Thiamine requirements helpful in distinguishing isolates of *Pythium* from those of *Phytophthora. Phytopathology* **59**: 737–742.

Robbins, W. J., and F. Kavanagh. 1938. Thiamin and growth of *Pythium butleri. Bull. Torrey Bot. Club* **65**: 453–461.

Rogers, T. O., and H. C. Lichstein. 1969a. Characterization of the biotin transport system in *Saccharomyces cerevisiae. J. Bacteriol.* **100**: 565–572.

Rogers, T. O., and H. C. Lichstein. 1969b. Regulation of biotin transport in *Saccharomyces cerevisiae. J. Bacteriol.* **100**: 565–572.

Rozenfeld, S. M., and E. N. Disler. 1971. Characteristics of the pool of free intracellular amino acids in the thiamine heterotrophic yeast *Candida lipolytica. Mikrobiologiya* **40**: 218–222.

Samsonova, I. A., D. Birnbaum, and F. Bottcher. 1974. Growth of nicotinic acid deficient mutants of *Rhodotorula glutinis* on *n*-alkanes in the absence of exogenous nicotinic acid. *Mikrobiologiya* **43**: 118–120.

Sato, T., A. Maekawa, T. Suzuki, and Y. Sashashi. 1966. Utilization of hydrocarbons for the formation of riboflavin and its coenzyme in *Eremothecium ashbyii. Vitamins (Kyoto)* **34**: 542–545.

Schlee, D. 1977. Changes in riboflavin overproduction in the yeast *Pichia guilliermondii* under the effect of branched amino acids. *Ukr. Biokhim. Zh.* **49**: 91–96.

Schlee, D., K. Zur Heiden, W. Fritchie, and H. Reinbothe. 1970. Purine metabolism and riboflavin formation in microorganisms: III. Growth and riboflavin overproduction of *Candida guilliermondii* (Cast.) Lang and G. *Biochem. Physiol. Pflanzen.* **161**: 459–468.

Scott, W. A. 1977. Unsaturated fatty acid mutants of *Neurospora crassa. J. Bacteriol.* **130**: 1144–1148.

Scott, W. A., C. Gabrielides, and J. Zrike. 1978. Identification of membrane lipid components essential to adenylate cyclase function. *Fed. Proc.* **37**: 1724.

Shavlovskii, G. M., V. E. Kashchenko, L. V. Koltun, E. M. Logvinenko, and A. E. Zakal'skii. 1977.

Regulation of the synthesis of GTP-cyclohydrolase participating in yeast flavinogenesis by iron. *Mikrobiologiya* **46**: 578–580.

Shavlovskii, G. M., E. M. Logvinenko, R. Benndorf, L. V. Koltun, V. E. Kashchenko, A. E. Zakal'skii, D. Schlee, and H. Reinbothe. 1980. First reaction of riboflavin biosynthesis: Catalysis by a guanosine triphosphate cyclohydrolase from yeast. *Arch. Microbiol.* **124**: 255–259.

Shcherbina, S. M. 1973. Effect of different exogenous fatty acids on biotin deprived death of *Saccharomyces cerevisiae. Agric. Biol. Chem.* **42**: 233–240.

Shimada, S., H. Kuraishi, and K. Aida. 1978. Effect of exogenous fatty acids on biotin deprived death of *Saccharomyces cerevisiae. Agric. Biol. Chem.* **42**: 233–240.

Shukla, D. S., and S. K. Sarkar. 1972. Effect of vitamins on the growth and sporulation of *Botryodiplodia theobromae* in culture. *Indian Phytopathol.* **25**: 40–43.

Sibirnyi, A. A., G. M. Shavlovskii, G. P. Ksheminskaya, and A. G. Orlovshaya. 1977a. Riboflavin transport in the cells of riboflavin-dependent yeast mutants. *Mikrobiologiya* **46**: 376–378.

Sibirnyi, A. A., G. P. Shavlovskii, G. P. Ksheminskaya, and A. G. Orlovskaya. 1977b. Active transport of riboflavine in the yeast *Pichia guilliermondii: Detection and some properties of cryptic riboflavine permease. Biochemistry* **42**: 1451—1460.

Sicho, V., and B. Kralova. 1968. Effect of antivitamins on the microbiological biosynthesis of vitamins: II. Relation between amounts of thiamine, thiamine monophosphate and thiamine diphosphate in *Saccharomyces cerevisiae* under normal conditions and with oxythiamine in the culture medium. *Biochim. Biophys. Acta* **165**: 459–462.

Simpson, B., and A. J. Hayes. 1978. Growth of *Marssonina brunea. Trans. Brit. Mycol. Soc.* **70**: 249–256.

Sladky, Z., and V. Tichy. 1974. Stimulation of the formation of fruiting bodies of the fungus *Lentinus tigrinus* (Bull.) Fr. by growth regulators. *Biol. Plant (Prague)* **16**: 436–443.

Smashevskii, N. D. 1967. Stimulation of the growth of different yeasts by β-alanine and pantothenic acid in the presence of asparagine *Prikl. Biokhim. Mikrobiol.* **3**: 55–58.

Smith, S. W., and R. L. Lester. 1974. Inositol phosphorylceramide, a novel substance and the chief member of a major group of yeast sphingolipids containing a single inositol phosphate. *J. Biol. Chem.* **249**: 3395–3405.

Soda, K., M. Moriguchi, and K. Tanizawa. 1975. Regulation of the activity of microbial kynureninase by transmission of the enzyme bound coenzyme. *Acta Vitaminol. Enzymol.* **29**: 335–338.

Stephen, R. C., and K. K. Fung. 1971. Vitamins requirements of the fungal endophytes of *Arundina chenensis. Can. J. Bot.* **49**: 411–415.

Stone, A. B., and D. Wilkie. 1974. Cellular mitochondrial effects of folate antagonism by pyrimethamin in *Saccharomyces cerevisiae. J. Gen. Microbiol.* **83**: 283–293.

Sullivan, J. L., and A. G. Debusk. 1973. Inositol-less death in *Neurospora* and cellular ageing. *Nature New Biol.* **243**: 72–74.

Tachibana, S. and J. Siode. 1971. Effect of thiamine on the L-malate fermentation by *Schizophyllum commune. J. Vitaminol.* **17**: 215–218.

Tan, K. K. 1978. Light-induced fungal development. In J. E. Smith and D. R. Berry (Eds.), *The Filamentous Fungi*, Vol. 3, pp. 334–357. New York: Wiley.

Tani, Y., Y. Tama, and T. Kamihara. 1979. Morphological change in *Candida tropicalis* pK 233 caused by ethanol and its prevention by myo-inositol. *Biochem. Biophys. Res. Commun.* **91**: 351–355.

Teh, J. S. 1974. Toxicity of short-chain fatty acids and alcohols towards *Cladosporium resinae. Appl. Microbiol.* **28**: 840–844.

Todosuchuk, S. R., D. M. Isakova, and I. E. Kvasnikov. 1976. Limitation of the growth of *Candida tropicalis* by biotin as a function of the source of carbon nutrition and temperature of culturing. *Mikrobiologiya* **45**: 372–375.

Tokuyama, T., H. Kuraishi, K. Aida, and T. Uemura. 1973. Hydrogen sulfide evolution due to pantothenic acid deficiency in the yeast requiring this vitamin, with special reference to the effect of adenosine triphosphate on yeast cysteine disulfhydrase. *J. Gen. Appl. Microbiol.* **19**: 439–466.

Tomita, I., S. I. Saito, and T. Ozawa. 1974. Coenzyme analog inhibition in the reconstitution of yeast pyruvate decarboxylase. E. C. 4.1.1.1 and transketolase. *Biochem. Biophys. Res. Commun.* **57**: 78–84.

Tomita, I., Y. Satou, T. Ozawa, and S. I. Saito. 1973. Studies on the decarboxylase of pyruvate: II. Kinetic studies on the binding of thiamine pyrophosphate to pyruvate decarboxylase. *Chem. Pharm. Bull.* **21**: 252–255.

Tschabold, E. 1967. Physiology of sexual reproduction in *Hypomyces solani* f. *cucurbitae*: IV. Influence of flavin inhibitors on perithecium formation. *Phytopathology* **57**: 1140–1141.

Uzuka, Y., T. Naganuma, and K. Tanaka. 1974. Influence of culture pH on the initiation of biotin synthesis in a strain of *Lipomyces starkeyi*. *J. Gen. Appl. Microbiol.* **20**: 197–206.

Van Driel, R. 1981. Binding of the chemoattractant folic acid by *Dictyostelium discoideum* cells. *Eur. J. Biochem.* **115**: 391–396.

Vatsala, T. M., K. Shanmugasundaram, and E. R. B. Shanmugasunderam. 1976. Role of nicotinic acid in pyridoxine biosynthesis in *Aspergillus nidulans*. *Biochem. Biophys. Res. Commun.* **72**: 1570–1575.

Vir, S., and J. S. Grewal. 1973. Vitamin and trace element requirements of *Fusarium caeruleum* the causal agent of guar wilt. *Indian Phytopathol.* **26**: 274—278.

Wainwright, T. 1970. Hydrogen sulfide production by yeast under conditions of methionine, pantothenate or vitamin B_6 deficiency. *J. Gen. Microbiol.* **61**: 107–119.

Waller, J. R. 1970. Amino acid-induced inhibition and stimulation of *Saccharomyces carlsbergensis*: I. Variation in the response to pantothenic acid induced by casein hydrolysate. *App. Microbiol.* **20**: 857–860.

Weigl, E., M. Hejtmanek, N. Hejtmankova, J. Kunert, and K. Lenhart. 1975. Biochemical mutants of *Trichophyton mentagrophytes*. *Acta. Univ. Palack Olomuc. Fac. Med.* **74**: 69–82.

West, D. 1975. Effects of pH and biotin on a circadian rhythm of conidation in *Neurospora crassa*. *J. Bacteriol.* **123**: 387–389.

Whaley, J. W., and H. L. Barnett. 1963. Parasitism and nutrition of *Gonatobotryum simplex*. *Mycologia* **55**: 199–210.

Wilkins, A. S., and G. Reynolds. 1979. The development of sporulation competence in *Physarum polycephalum*. *Dev. Biol.* **72**: 175–181.

Witt, I., and B. Neufang. 1970. Studies on the influence of thiamine on the synthesis of thiamine pyrophosphate-dependent enzymes in *Saccharomyces cerevisiase*. *Biochem. Biophys. Res. Commun.* **25**: 340–345.

Wurster, B., and K. Schubiger. 1977. Oscillations and cell development in *Dictyostelium discoideum* stimulated by folic acid pulses. *J. Cell. Sci.* **27**: 105–114.

Wurster, B., and K. Schubiger, and P. Brachet. 1979. Cyclic GMP and cyclic AMP changes in response to folic acid pulses during cell development of *Dictyostelium discoideum*. *Cell Diff.* **8**: 235–242.

Yamagata, S., and K. Takeshima. 1976. O-acetylserine and O-acetylhomoserine sulfhydrylase of yeast: Further purification and characterization as a pyridoxal enzyme. *J. Biochem.* **80**: 777–785.

Yamaguchi, E. 1974. Effect of biotin insufficiency on the composition and ultrastructure of cell walls of *Candida albicans* in relation to its mycelial morphogenesis. *J. Gen. Appl. Microbiol.* **20**: 217–228.

Yamamoto, S., Y. Minoda, and K. Yamada. 1973. Formation and some properties of the acid phosphatase in *Asperigillus terreus*. *Agric. Biol. Chem.* **37**: 2719–2726.

Yanofsky, C. 1955. Tryptophan and niacin synthesis in various organisms. In W. D. McElroy and H. B. Glass (Eds.), *Amino Acid Metabolism*, pp. 930–939. Baltimore: Johns Hopkins.

Zhukova, S. V. 1972. Folic acid and biosynthesis of nucleic acids by microorganisms. *Biol. Nauk.* **15**: 94–97.

7

SPORE DORMANCY, ACTIVATION, AND GERMINATION

Spores are units of dispersal that are usually products of both the sexual and asexual cycles in fungi. The spores of some species may be capable of germinating immediately upon being shed from the parent; others enter a period of dormancy before germination can occur. Dormancy and germination are interrelated phenomena, and much research has been focused on mechanisms that cause a spore to become dormant as well as those that activate a spore and thus enable it to germinate. In this chapter we briefly discuss certain aspects of spore dormancy, activation, and germination. A detailed review of these topics has recently been published (Weber and Hess, 1976).

DORMANCY

Dormancy is a rest period in the life cycle of a fungus, a reversible interruption in the phenotypic development of the organism (Sussman, 1966). Some spores are **exogenously dormant**; their failure to germinate is due to unfavorable environmental conditions. In this case, germination will proceed as soon as environmental conditions are satisfactory. In contrast, **endogenously dormant** spores do not germinate even under ideal environmental conditions. Spores of this type require either a period of aging or specific treatments to activate the germination process.

Exogenous Dormancy

In general, fungal spores will not germinate under conditions of unfavorable temperature, moisture, oxygen, or pH, although the absolute requirements vary considerably among species. Some fungi have additional requirements. For example, high intensity white or blue light stimulates germination of *Aspergillus niger* conidia (Khan, 1977), and increased CO_2 levels enhance germination of *A. nidulans* (Trinci and Whittaker, 1968). In other species, such as *Trichophyton ajelloi* (Buchnicek, 1976) and *Colletotrichum lindemuthianum* (Arnold and Rahe, 1976), light and CO_2, respectively, are inhibitory. Care must be taken in doing chemical analyses of dormant spores because even brief (30 minute) exposure to water during harvesting causes spores of species such as *Rhizopus stolonifer* to become metabolically active (Nickerson et al., 1981).

Spores of some species are exogenously dormant because they require various nutrients for germination. For example, conidia of *Penicillium griseofulvum* have an absolute requirement for glucose to initiate germination (Fletcher and Morton, 1970). Oospores of *Pythium hydnosporum* will germinate in distilled water, but the addition of nutrients increases the percentage of germination (Al-Hassan and Fergus, 1973).

An intriguing example of selective nutritional control of spore germination is seen with conidia of *Neurospora crassa* (Farach et al., 1979). In this species germination experiments are usually carried out using a minimal medium containing a carbon source (usually glucose or sucrose), major salts, trace elements, and biotin. Figure 7.1A indicates that the most critical combination of nutrients for germ tube emergence is the carbon source plus the major salts. However, trace elements are additionally required for germ tube growth, as measured by dry weight (Figure 7.1B). From experiments in which individual major salts were deleted from the medium (Table 7.1A) it was determined that the absence of either KH_2PO_4, NH_4NO_3, or $MgSO_4$ leads to a substantial retardation of germination when assayed at 3 hours. However, KH_2PO_4 deprivation has no effect after 5 hours nor does NH_4NO_3 deprivation after 24 hours. The requirement for magnesium is diminished but still apparent even after this longest incubation. Experiments in which individual trace elements are deleted (Table 7.1B) indicate that deprivation of $ZnSO_4$ has the greatest

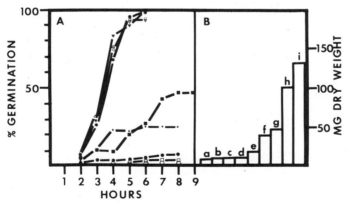

Figure 7.1 Ability of various "deficient" media to support (A) germ tube emergence and (B) germ tube elongation in *Neurospora crassa*. ($\triangle$,f) sucrose + major salts; (*,g) sucrose + major salts + biotin; ($\triangle$,i) sucrose + major salts + trace elements; ($\hollowcircle$,h) minimal medium; ($\bullet$,d) sucrose + trace elements; ($\circ$,b) major salts; ($\blacksquare$,e) sucrose + biotin; ($\star$,c) sucrose; ($\square$,a) sucrose + trace elements + biotin. (Farach et al., 1979)

Table 7.1 The Effect of (A) Major Salts on Germ Tube Emergence and (B) Trace Elements on Germ Tube Elongation of Spores of *Neurospora crassa*

Medium	Percentage Germination			Dry Weight (mg)
	3 h	7 h	24 h	
A. Minimal	81	100	100	194
Sodium citrate	60	100	100	133
Potassium phosphate	5	100	100	20
Ammonium nitrate	22	86	100	16
Magnesium sulfate	36	79	76	18
Calcium chloride	64	100	100	119
B. Minimal			100	220
Zinc sulfate			100	90
Cupric sulfate			100	195
Ferrous ammonium sulfate			100	220
Manganese sulfate			100	192
Sodium molybdate			100	190
Boric acid			100	185
Citric acid			100	195

Source: Farach et al. (1979).

effect on germ tube growth, even though germination itself is unaffected by trace element levels. Previous work with this species (Mirkes, 1974, 1977) indicated that protein and RNA synthesis can be activated by incubating these conidia in a carbon source alone, even though germ tubes do not emerge. Thus, manipulation of the composition of the growth medium permits the study of discrete aspects of germination in this species. Media containing only a carbon source can be used for the study of metabolic activation, magnesium-deficient media for the study of germ tube emergence, and zinc-deficient media for the study of germ tube growth (Farach et al., 1979).

Exogenous dormancy may also be caused by inhibitors present in the environment. Inhibitory compounds are naturally present in plants and may function as resistance factors in preventing spore germination of various plant pathogens. Phenolic compounds appear to be a particularly important class of inhibitors. One group, the hydroxylated cinnamic acid derivatives, are universally present in plants either as monomers or as part of more complex products such a lignin. Stahmann and co-workers (1975) tested the inhibitory effect of 25 cinnamic acid derivatives on the germination of coffee rust uredospores; more than half of the compounds were inhibitory. However, one of these derivatives, ferulic acid, although inhibitory to germination of angiosperm saprophytes actually stimulates germination of soil-inhabiting organisms such as *Penicillium* (Black and Dix, 1976).

Other inhibitory compounds, the **phytoalexins,** are produced by the host plant in response to infection. These are usually complex, multiring phenolics (Figure 7.2), and much interest has been directed to understanding the mechanisms by which they restrict spore germination and growth of various organisms (Keen 1981).

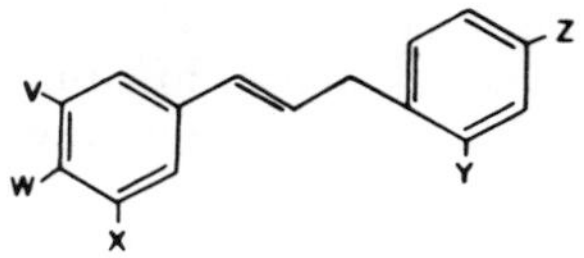

	V	W	X	Y	Z
Resveratrol	OH	H	OH	H	OH
4-prenylresveratrol	OH	Prenyl	OH	H	OH
Oxyresveratrol	OH	H	OH	OH	OH
4-prenyloxyresveratrol	OH	Prenyl	OH	OH	OH
Pterostilbene	OCH₃	H	OCH₃	H	OH

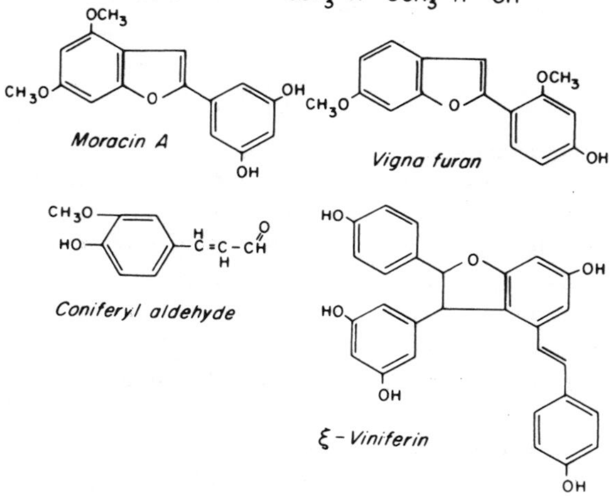

Moracin A

Vigna furan

Coniferyl aldehyde

ξ - Viniferin

Figure 7.2 Recently described phytoalexins. (Keen, 1981)

Spores may also fail to germinate owing to the presence of organisms that are not hosts to fungal pathogens. This is termed **fungistasis** and can often be overcome by sterilizing the medium or by adding nutrients. This suggests that microorganisms such as bacteria may compete with fungal spores for nutrients needed for germination. In some cases, the addition of fungal spores to soil not only inhibits their germination but also causes an increase in the number of soil bacteria. It is hypothesized that bacteria function as a "nutrient sink" and thrive on compounds released from the spore during hydration and which ordinarily would be utilized for germination by the fungus. Spores placed on membrane filters and continuously flushed with water to mimic the removal of nutrients by bacteria germinate very poorly. However, germination is excellent when the flushing ceases or when the spores are instead flushed with a sterile glucose solution (Lockwood, 1977).

Bacteria may also produce compounds that inhibit germination of fungal spores. Volatile hydrocarbons such as allyl alcohol, the plant hormone ethylene, and various unidentified compounds have been shown to be fungistatic. Such compounds may interfere directly with certain key steps in the germination process or indirectly via an effect on nutrient uptake or metabolism (Lockwood, 1977). Many fungicides, particularly the fungal antibiotics, exert their effect on spore germination and do so by a variety of mechanisms (Torgeson, 1969).

Exogenous dormancy appears to be an adaptation to postpone germination until conditions are favorable for survival as well as to ensure effective dissemination of spores and improve the chances for establishment on substrates that can be readily used for completion of the life cycle.

Endogenous Dormancy

Endogenously dormant spores fail to germinate because of internal rather than external factors. There are three general mechanisms proposed to explain this type of dormancy: permeability phenomena, the presence of self-inhibitors, and metabolic impairment (Sussman, 1966).

PERMEABILITY

The observation that various chemicals that alter membrane permeability also stimulate spore germination in some species suggests that certain spores are dormant because nutrients, water, oxygen, or other compounds required for germination cannot enter the cell. For example, treating rust uredospores with nonyl alcohol enhances germination and also increases the leakage of spore constituents into the medium (Maheshwari and Sussman, 1971). Dimethylsulfoxide (DMSO), which also affects membrane permeability, stimulates germination of Lycogala epidendrum (Butterfield, 1968) and swelling of spores of Dictyostelium discoideum (Cotter et al., 1976).

Spores of D. discoideum can be stimulated to germinate by incubating them at 40° C for 5 minutes. Work with this organism suggests that heat activation may also cause changes in membrane permeability. Spores given a heat treatment begin to swell after 60–75 minutes, just as with DMSO, and show increased permeability within 30 minutes as determined by efflux experiments using radioactively labeled spores. An analysis of freeze fracture faces shows a drastic effect of heat on the distribution of particles in the plasma membrane, indicating a subtle change in its architecture and/or functional properties. It is hypothesized that these changes in membrane permeability trigger the sequence of events associated with germination (Hohl et al., 1978).

SELF-INHIBITORS

Evidence has long been accumulating that spores of certain fungi possess compounds that inhibit their own germination: (1) high concentrations of spores germinate less well than low concentrations, (2) washing high concentrations of spores with water stimulates germination, (3) such water extracts are inhibitory to germination, and (4) inhibition of this type is quickly and easily reversed. Although self-inhibitors have been reported for at least 53 fungal species, their precise chemical structures and modes of action are known in only a few instances (Allen, 1976).

The most thoroughly studied self-inhibitors are those from rust uredospores. Table 7.2 indicates that in the many species tested the inhibitors are the cinnamic acid derivatives methyl-cis-3,4-dimethoxycinnamate and methyl-cis-ferulate (Figure 7.3).

Table 7.2   Effects of Methyl-*cis*-3,4-Dimethoxycinnamate on the Germination of Uredospores of Different Rust Fungi.

Rust Species	Host	ED_{50} (ng/ml)	Natural Inhibitor
Uromyces phaseoli	Bean	2.72	MDMC[a]
Puccinia antirrhini	Snapdragon	0.23	MDMC[a]
P. arachidis	Peanut	0.08	MDMC[a]
P. graminis tritici	Wheat	3.7	MF[b]
P. coronata	Oat	0.01	MDMC[a]
P. helianthi	Sunflower	1.65	MDMC[a]
P. striliformis	Wheat	4.0	MDMC[a]
Hemileia vastatrix	Coffee	Insensitive	Unknown
Melampsora lini	Flax	Insensitive	Unknown

Source: Trione (1981).
[a]Methyl-*cis*-3,4-dimethoxycinnamate.
[b]Methyl-*cis*-ferulate.

These compounds are effective in inhibiting germination at 10^{-9} and 10^{-10} M but have no effect once the germ tube has protruded from the spore. Rust uredospores develop a pore in the wall through which the germ tube eventually emerges. Initially this pore is plugged with material that must be hydrolyzed prior to germination. The self-inhibitors present prevent hydrolysis of the plug and thus inhibit the emergence of the germ tube (Macko et al., 1976; Hess et al., 1976).

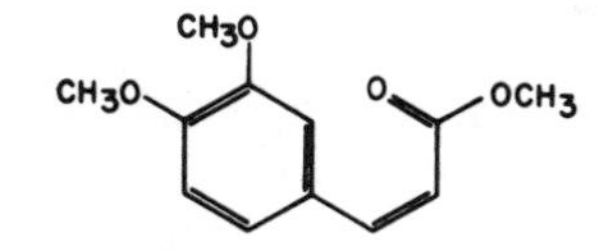

Methyl-cis-3,4-dimethoxycinnamate

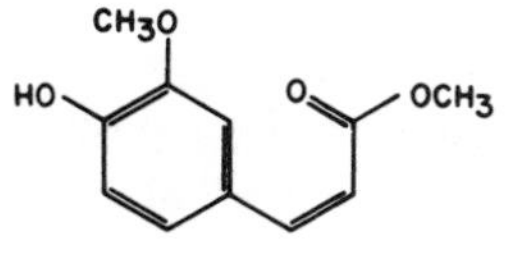

Methyl-cis-ferulate

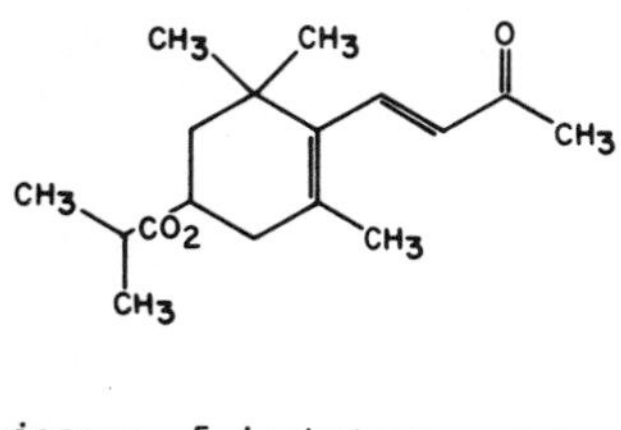

Quiesone, 5-isobutyroxy-β-ionone

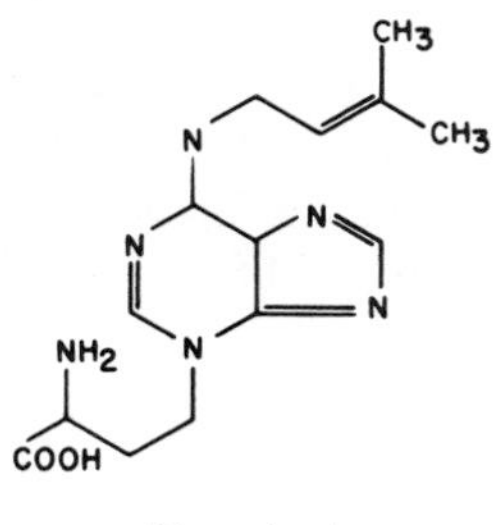

Discadenine

Figure 7.3   The structures of some inhibitors of fungal spore germination. (Trione, 1981)

Uredospores of *Uredo eichhorniae* and *Uromyces pontederiae* exhibit less than 20% germination when incubated in distilled water. Up to 90% germination is obtained when the spores are exposed to water containing ketones such as α-ionone and 2-hexanone or aldehydes such as retinal (vitamin A aldehyde) and its derivative retinol (vitamin A). Because both these fungal species are pathogens of aquatic plants, it would be evolutionarily advantageous for their spores not to germinate solely upon contact with free water but instead remain dormant until a suitable nutrient supply is present. The effects of these germination stimulants are thought to be involved in overcoming endogenous dormancy in the spores (Charudattan et al., 1981).

A self-inhibitor of conidial germination in *Peronospora tabacina* has been identified and named quiesone (Figure 7.3). This compound has the same basic ionone skeleton as abscisic acid, a plant hormone involved in the inhibition of seed and bud germination as well as in the abscission of leaves. Interestingly, quiesone stimulates germination of rust uredospores (Macko et al., 1976; Trione, 1981).

Spores of *D. discoideum* contain N,N-dimethylguanosine (DMG), which is especially effective in inhibiting spore germination prior to swelling. This compound immediately arrests protein synthesis when added to activated spores and slows it when added to spores that have already begun to germinate (Bacon and Sussman, 1973). An even more potent self-inhibitor called discadenine has also been isolated (Figure 7.3), (Abe et al., 1976; Trione, 1981).

Sulfur appears to be a nonspecific self-inhibitor of spore germination in many fungi. Pezet and Pont (1977) noted that elemental sulfur (S_8) is present in numerous self-inhibited and resting structures such as dormant spores and sclerotia. It is hypothesized that elemental sulfur may inhibit germination by competing with the mitochondrial electron transport chain for electrons, thus keeping the respiration rate low.

METABOLIC IMPAIRMENT

Spores that are endogenously dormant may fail to germinate because of the presence of a metabolic block that is not associated with either a self-inhibitor or a permeability problem. Germination may not occur because key enzymes associated with respiration or other pathways of intermediary metabolism are not functioning normally or because RNA and protein synthesis are blocked. In *Phycomyces blakesleeanus*, the activity of trehalase, an enzyme responsible for breaking down stored trehalose, increases tenfold to fifteenfold when the spores are activated by heating (Van Assche et al., 1972). However, enzyme activation is generally associated with spore activation, and it is difficult to determine whether the lack of activity in an ungerminated spore is a cause or a consequence of dormancy. It is also difficult to rule out the possibility of self-inhibitors in such cases (Sussman, 1966).

ACTIVATION

As we have seen, endogenously dormant spores must be activated in some way before germination can occur. Methods to activate spores are diverse, and frequently a combination of treatments is required (Sussman, 1976).

Temperature

We have mentioned the stimulatory effect of heat on the germination of *Dictyostelium discoideum* spores. Heat treatments are commonly used to stimulate germination of many species. Usually temperatures of 40–75° C for a duration varying from a few minutes to several hours are necessary for activation to occur (Table 7.3). Germination may also be facilitated by prior treatment of spores with low temperatures (− 5° to 10° C) for varying periods of time (Sussman, 1966).

Spores of *Phycomyces blaskesleeanus* are endogenously dormant and require heat activation (50° C for 3 minutes). Thermodynamic studies indicate that the mechanism of heat activation in this species is primarily via protein conformational change rather than alterations in membrane structure (Thevelein et al., 1979).

In addition to temperature, there is a nutritional component to the activation process in *P. blaskesleeanus*. Heating the spores for 3 minutes at 44° C results in nearly complete germination if the spores are rapidly transferred to a glucose-asparagine-phosphate culture medium; however, when such spores are transferred instead to water, rapid deactivation occurs (Figure 7.4). This has been described as reversible activation. But if spores are heat-activated for 3 minutes at 50° C, there is little difference in germination between incubation in culture medium and in water—in this case, the spore cannot be deactivated, and irreversible activation has occurred. Deactivation results in decreased RNA and protein synthesis as well as decreased rate of respiration. Glucose was shown to be the key nutrient necessary for germination after a heat treatment of less than 50° C. Thus, at these lower temperatures, a carbon source is necessary to start germination (Van Laere et al., 1980a). In addition, acetate can replace heat as an activator in this species. Activation by acetate can be inhibited by azide, even though azide does not inhibit oxygen uptake by dormant spores. This suggests that acetate metabolism is necessary for spore activation (Van Laere et al., 1980b).

Chemicals

A wide variety of chemicals have been used to activate fungal spores; a representative sampling is listed in Table 7.4. Most of these compounds disrupt lipoprotein bilayers,

Table 7.3 **Optimal Temperatures and Duration of Treatment Required for the Heat-Activation of Selected Fungal Spores**

Organism	Temperature	Duration
Dictyostelium discoideum	40°C	5 min.
Puccinia graminis	40°C	5 min.
Coprinus radiatus	44–46°C	4 h
Mucor miehi	45°C	5 h
Phycomyces blakesleeanus	50°C	3 min.
Neurospora tetrasperma	50–65°C	10–20 min.
Byssochlamys fulva	75°C	5–10 min.

Source: Sussman (1976).

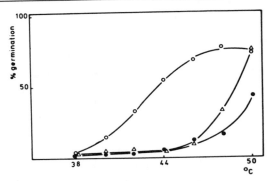

Figure 7.4 Percentage germination of *Phycomyces* spores in culture medium as a function of activation temperature. ○ Immediately after activation. △ After 1 hour in water. ● After 24 hours in water. (Van Laere et al., 1980a)

therefore their mode of action is presumably to increase membrane permeability. Certain species secrete compounds that activate the spores of other species. For example, cultures of *Ceratocystis fagacearum* release volatile alcohols and aldehydes that activate spores of several *Lactarius* species (Oort, 1974). Safflower plants produce volatile polyacetylene compounds that trigger the germination of safflower rust teliospores (Binder et al., 1977).

Aging

Many spores that are endogenously dormant may not respond to activation treatments until a certain period of time has elapsed. Such spores may simply not be old enough to be capable of germination. This phenomenon is termed after-ripening

Table 7.4 Selected List of Fungal Spores Activated by Detergents, Organic Solvents, or Lipids

Organism	Activator
More than 50 slime molds	Detergents
Colletotrichum trifolii	Tween-20
Phycomyces blakesleeanus	Acetic, propionic, and butyric acids
Penicillium frequentans (conidia)	Ethanol, acetone
Neurospora spp. (ascospores)	In low concentrations: furans, thiophenes, pyrrole and aromatic alcohols. In high concentration: ethanol, acetone, and other organic solvents
Coprinus radiatus (basidiosphores)	Furfural and other heterocyclics
Anthraciodea spp.	Quarternary detergents

Source: Sussman (1976).

and has been defined by Sussman (1966) as that part of the dormancy period during which changes occur that lead to germination. After-ripening can be thought of as an evolutionary adaptation to ensure that the spore is fully mature before germination begins. Ayers and Lumsden (1975) demonstrated that the germination of *Pythium* oospores increases from 0% to more than 80% over a six-week period in nonsterile soil extracts; these observations imply that an after-ripening phenomenon is involved.

Activation treatments may accelerate or induce the after-ripening period but apparently cannot substitute for it. Haware and Pavgi (1976) noted that acid or alkali treatments are effective in stimulating the germination of *Protomycopsis* chlamydospores only when these spores were five to six months old. Treating fresh spores in this way had no effect.

Multiple Activation Treatments

Frequently a combination of treatments is required before activation is accomplished. Alternate wetting and drying as well as alternate high and low temperatures are often effective. Mills and Eilers (1973) found that heating basidiospores of *Coprinus radiatus* for 4 hours at 45°C results in 23% germination. Treatment of these spores with chemicals such as furfural, thiophene, or pyrrole yields about 3% germination. However, spores simultaneously treated with heat and chemicals showed germination percentages up to 88%.

GERMINATION

Once factors causing dormancy have been overcome, germination may proceed. Germination can conveniently be divided into two stages: a period of increase in spore size, sometimes called the swelling phase, followed by the protrusion of the germ tube.

During the first phase, spores of most fungi increase approximately twofold to threefold in diameter (Fletcher, 1969). This increase is apparently due both to imbibition and to synthesis of new cellular material. In Figure 7.5a–d are scanning electron micrographs of the time-course of germination of *Mucor rouxii*. These spores undergo roughly a fourfold increase in diameter in 6 hours. Notice that a higher magnification (Figure 7.5e) reveals pieces of wall material on the 6-hour spore surface (Storck and Price, 1977). Bartnicki-Garcia and co-workers (1968) determined that these pieces constitute remnants of the original spore wall that overlays an entirely new wall, which is synthesized as spherical growth proceeds (Figure 7.6). The germ tube that eventually emerges is an extension of this new vegetative wall (Bartnicki-Garcia and Lippman, 1977). The synthesis of new spore wall layers during germination has also been reported for other species and appears to be a common phenomenon in fungi (Akai et al., 1976).

Numerous cytological changes have been observed in fungal spores as swelling and germ tube formation occur (Smith et al., 1976). These include the disappear-

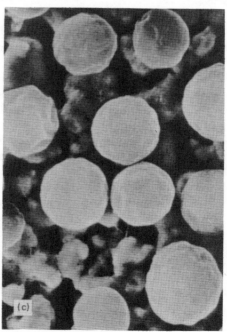

Figure 7.5 Scanning electron micrographs of *Mucor* spores (a) after 0, (b) 2, (c) 4, and (d) 6 hours incubation. All × 2900. (e) Spore after 6 hours incubation, × 14,000. (Storck and Price, 1977)

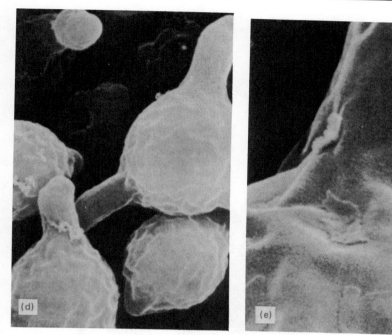

Figure 7.5   (*Continued*)

ance of lipid bodies as well as increases in the extent of endoplasmic reticula, in the size and/or number of mitochondria, in the size and number of nuclei, and in the number of ribosomes. Of particular interest is the accumulation of small vesicles occasionally observed on one side of the spore and at the tips of germ tubes (Figure 7.7). These vesicles appear to be similar to those involved in wall growth and are thought to contain the enzymes necessary for loosening the wall and also the subunits needed for new wall synthesis. One hypothesis is that the spore becomes polarized in some manner, resulting in the accumulation of vesicles in the region of the wall where the germ tube will eventually emerge.

Bartnicki-Garcia and Lippman (1977) have demonstrated such a polarity in germinating spores of *M. rouxii*. In Figure 7.8 are autoradiographs of spores incubated

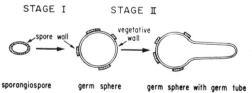

Figure 7.6   Stages of sporangiospore germination in *Mucor rouxii*. (Bartnicki-Garcia and Lippman, 1977)

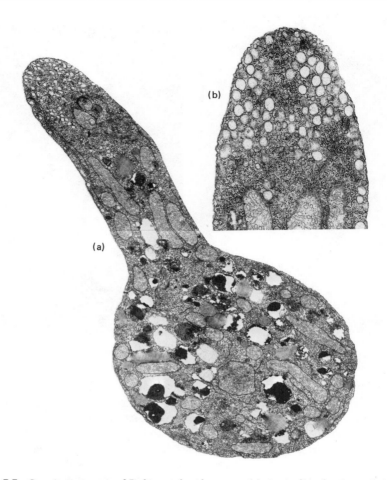

Figure 7.7 Germinating spores of *Pythium aphanidermatum*. (a), A germling showing typical proto-plasmic organization including the cluster of apical vesicles at the tip of the germ tube. × 11,500. (b) Details of the apex of the germ tube. The apical vesicles are in a region of cytoplasm with a lower concentration of ribosomes than the rest of the protoplast. × 31,000. V, apical or Golgi vesicles; FV, "Fingerprint vacuole"; L, lipid inclusion; M, mitochondrion; GA, Golgi apparatus; N, nucleus; ER, endoplasmic reticulum (Grove and Bracker, 1978)

for various periods of time and then given a pulse of tritiated N-acetylglucosamine for incorporation into the cell wall. Notice that although some spores exhibit uniform incorporation of label indicating spherical growth (row A), other spores show pro-gressively greater concentration of silver grains in one hemisphere (rows B and C). All cells with incipient (row D) or long (row E) germ tubes show a heavy accumula-tion of label at the tube apex. Clearly, polarization of wall synthesis has preceded the protrusion of the germ tube. Quantitative analysis indicates that the total rate of wall synthesis does not change during this period; there is only a spatial redistribution of

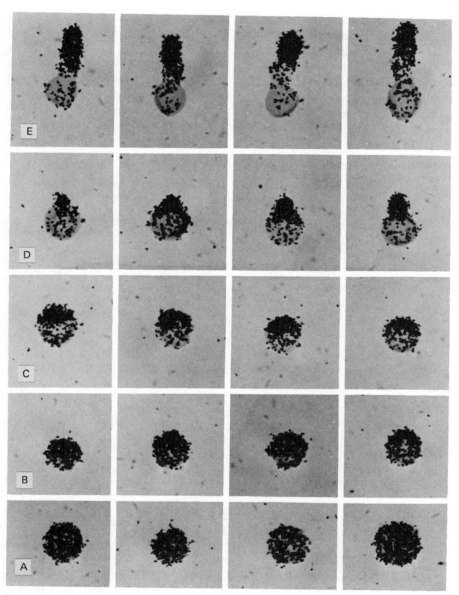

Figure 7.8 Patterns of wall synthesis in a population of spores of *Mucor rouxii* germinating for 5.5 hours. The autoradiographs were deliberately overexposed for 11 days. The cells have been arranged in rows to depict progressive changes in the pattern of wall synthesis from (A) nonpolarized germ tubes to (B and C) increasingly polarized germ spheres to (D) germ spheres with incipient or (E) well-developed germ tubes. The cells in row D have incipient germ tubes (2–6 μm long) seen by direct microscopic observations to follow the outline of the heavy silver grain deposit. ×1100. (Bartnicki-Garcia and Lippman, 1977)

the wall synthesizing machinery in the spore. As with hyphal growth, the timing mechanism that triggers polarization and the mechanisms by which wall synthesis becomes polarized are unknown.

Even from this brief discussion it is obvious that a host of biochemical processes become activated as germination proceeds. Some of the most basic pathways involved are those concerned with the synthesis of DNA, RNA, and protein. In spores of *Rhizopus stolonifer*, RNA and protein contents begin to increase during the first hour of incubation—several hours before any germination actually occurs (Table 7.5). On the other hand, spores of *Botryodiplodia theobromae* exhibit little change

Table 7.5   Content of Polyamines, DNA, RNA, and Protein During Germination of Spores of (top) *Rhizopus stolonifer* and (bottom) *Botyrodiplodia theobromae*

Germination Time (h)	Germination (%)	Dry Weight (pg/spore)	Content (pg/spore)[a]					Micromoles of Spermidine/ Micromoles of RNA Nucleotide
			Spermidine	Putrescine	DNA	RNA	Protein	
0	0	123	0.11	0.00083	0.62	11.5	32.5	0.021
1	0	128	0.16	0.0042	0.62	12.7	35.8	0.028
2	0	134	0.25	0.046	0.64	16.2	45.2	0.033
3	0	158	0.34	0.090	0.71	24.6	60.3	0.031
4	3	212	0.47	0.11	1.02	43.0	79.9	0.025
5	67	298	0.63	0.12	1.39	63.2	108	0.002
6	95	425	0.68	0.085	1.90	93.8	148	0.016

Germination time (h)	Germination (%)	Dry Weight (pg/spore)	Content (pg/spore)[b]				Micromoles of Spermidine/ Micromoles of RNA Nucleotide
			Spermidine	DNA	RNA	Protein	
0	0	1614	0.98	2.8	34.4	351	0.064
1	0	1614	1.11	2.7	35.1	382	0.071
2	0	1589	1.01	2.8	37.3	382	0.061
3	14	1614	1.00	2.9	46.0	412	0.049
4	30	1606	1.38	3.3	61.5	394	0.050
5	90	1614	1.76	4.0	73.7	438	0.053
6	95	1720	1.95	5.0	95.5	501	0.046

Source: Nickerson et al. (1977).

[a]Standard deviations were ±5% for spermidine and putrescine, ±8% for protein, and ±4% for RNA and DNA.

[b]Trace amounts of putrescine were detected at all time periods. Standard deviations were ±5% for spermidine and protein, ±4% for RNA, and ±3% for DNA.

in any of these macromolecules during the first 3 hours (Table 7.5) (Nickerson et al., 1977). Use of radioactive precursors indicates that protein synthesis in B. theobromae actually precedes RNA synthesis during germination (Brambl and Van Etten, 1970). This observation suggests that in this case ungerminated spores have all the necessary RNA (i.e., rRNA, tRNA, mRNA) already present such that protein synthesis can take place without RNA synthesis (Knight and Van Etten, 1976). This preformed RNA has been reported in several fungal species. In B. emersonii germination can take place in the absence of both RNA and protein synthesis and implies the presence of both preformed RNA and preformed protein (Soll et al., 1969; Soll and Sonneborn, 1969, 1972). In Aspergillus nidulans there is a rapid increase in chitin synthetase activity, the amount of chitin synthetase zymogen, and the chitin content of the cell wall just prior to emergence of the germ tube. Experiments using the transcription inhibitor 5-fluorouracil and the translation inhibitor cycloheximide indicate that the chitin synthetase zymogen is synthesized de novo during germination initially using mRNA preformed during conidiogenesis (Ryder and Peberdy, 1979). However, in other species such as Achlya bisexualis, germination requires the concurrent synthesis of both RNA and protein (MacLeod and Horgen, 1979).

The respiration rate of fungal spores also increases dramatically as germination begins. Dormant spores of B. theobromae have mitochondria that lack two of the enzymes involved in respiratory electron transport: cytochrome c oxidase and an oligomycin-sensitive ATPase. During germination the activities of both enzymes increase rapidly. The ATPase is synthesized de novo from preformed mRNA. However, cytochrome c oxidase is assembled from subunits that are stored in the cytoplasm during dormancy and that move into the mitochondria and form a complex with three additional subunits (Brambl, 1980; Wenzler and Brambl, 1981). On the other hand, dormant spores of Neurospora crassa contain all the components necessary for a functional electron transport chain. These preserved components are responsible for the increased respiration rate that begins immediately after the spores are suspended in an incubation medium (Stade and Brambl, 1981).

In spores with stored reserve materials, such as lipid, trehalose, or mannitol, these reserves are broken down and used for energy production and the synthesis of new cellular material. For example, in Achlya sp. total lipid accounts for 10% dry weight of the ungerminated spore. After germination, lipids fall to 6% with half of the initial loss occurring within the first 2 hours (Law and Burton, 1976). In spores of Rhizopus stolonifer, de novo lipid synthesis is necessary for germination (Nickerson and Leastman, 1978). Incubating these spores in cerulenin, a known inhibitor of fatty acid synthesis, inhibits spore swelling, germ tube formation, and apparently mitochondrial biogenesis without affecting protein and RNA synthesis or oxygen uptake.

SUMMARY

Both sexual and asexual reproductive cycles usually involve the germination of spores. Once shed from the parent, spores may be unable to germinate and are said to be dormant. Exogenously dormant spores do not germinate because of unfavorable environmental conditions, including nutrient deficiency and the presence of external inhibitors. Endogenously dormant spores fail to germinate because of permeability phenomena, the presence of endogenous inhibitors, or metabolic impairment. Such spores must be activated by treatments such as chemicals, extremes of temperature, or aging before germination can occur.

Germination involves swelling, a localized polarization and synthesis of new wall material, and the emergence of the germ tube. These are accompanied by numerous changes at both the ultrastructural and biochemical levels.

REFERENCES

Abe, H., M. Uciyama, Y. Tanaka, and H. Saito. 1976. Structure of discadenine, a spore germination inhibitor from the cellular slime mold, Dictyostelium discoideum. Tetrahedron Lett. 42: 3807–3810.

Akai, S., M. Fukutomi, H. Kunoh, and M. Shiraishi. 1976. Fine structure of the spore wall and germ tube change during germination. In D. J. Weber and W. M. Hess (Eds.), The Fungal Spore: Form and Function, pp. 355–410. New York: Wiley.

Al-Hassan, K. K., and C. L. Fergus. 1973. The effects of nutrients and environment on germination and longevity of oospores of Pythium hydnosporum. Mycopathol. Mycol. Appl. 51: 283–297.

Allen, P. J. 1976. Spore germination and its regulation. In R. Heitefuss and P. H. Williams (Eds.), Encyclopedia of Plant Physiology, pp. 51–85. New York: Springer-Verlag.

Arnold, R. M., and J. E. Rahe. 1976. Effects of 15% CO_2 on germination, germ tube elongation and sporulation in cultures of Colletotrichum lindemuthianum. Can. J. Bot. 54: 1044–1048.

Ayers, W. A., and R. D. Lumsden. 1975. Factors affecting production and germination of oospores of three Pythium species. Phytopathology 65: 1094–1100.

Bacon, C. W., and A. S. Sussman. 1973. Effects of the self-inhibitor of Dictyostelium discoideum on spore metabolism. J. Gen. Microbiol. 76: 331–344.

Bartnicki-Garcia, S., and E. Lippman. 1977. Polarization of cell wall synthesis during spore germination of Mucor rouxii. Exp. Mycol. 1: 230–240.

Bartnicki-Garcia, S., S. N. Nelson, and E. Cota-Robles. 1968. Electron microscopy of spore germination and cell wall formation in Mucor rouxii. Arch. Microbiol. 63: 242–255.

Binder, R. G., J. M. Klisiewicz, and A. C. Waiss. 1977. Stimulation of germination of Puccinia carthami teliospores by polyacetylenes from safflower. Phytopathology 67: 472–474.

Black, R. L. B., and N. J. Dix. 1976. Spore germination and germ hyphal growth of microfungi from litter and soil in the presence of ferulic acid. Trans. Brit. Mycol. Soc. 66: 305–311.

Brambl, R. 1980. Mitochondrial biogenesis during fungal spore germination: Biosynthesis and assembly of cytochrome c oxidase in Botryodiplodia theobromae. J. Biol. Chem. 255: 7673–7680.

Brambl, R. M., and J. L. Van Etten. 1970. Protein synthesis during fungal spore germination: V. Evidence that the ungerminated conidiospores of Botryodiplodia theobromae contain messenger ribonucleic acid. Arch. Biochem. Biophys. 137: 424–425.

Buchnicek, J. 1976. Light resistance in geophilic dermatophytes. Sabouraudia 14: 75–80.

Butterfield, W. 1968. Effect of dimethylsulphoxide on the germination of spores of Lycogala epidendrum. Nature 218: 494–504.

Charudattan, R., D. E. McKinney, and K. T. Hepting. 1981. Production, storage, germination, and infectivity of uredospores of Uredo eichhorniae and Uromyces pontederiae. Phytopathology 71: 1203–1207.

Cotter, D. A., J. W. Morin, and R. W. O'Connell. 1976. Spore germination in Dictyostelium discoideum: II. Effects of dimethylsulfoxide on post-activation lag as evidence for the multistate model of activation. Arch. Microbiol. 108: 93–98.

Farach, M. C., H. Farach, and P. E. Mirkes. 1979. Control of development in Neurospora crassa:

Nutrient requirements for conidial germ tube emergence, elongation, and metabolic activation. *Ex. Mycol.* **3**: 240–248.

Fletcher, J. 1969. Morphology and nuclear behavior of germinating conidia of *Penicillium griseofulvum. Trans. Brit. Mycol. Soc.* **53**: 425–432.

Fletcher, J., and A. G. Morton. 1970. Physiology of germination of *Penicillium griseofulvum* conidia. *Trans. Brit. Mycol. Soc.* **54**: 65–81.

Grove, S. N., and C. E. Bracker. 1978. Protoplasmic changes during zoospore encystment and cyst germination in *Pythium aphanidermatum. Exp. Mycol.* **2**: 51–98.

Haware, M. P., and M. S. Pavgi. 1976. Chlamydospore germination in *Protomycopsis* species. *Mycopathologia* **59**: 105–111.

Hess, S. L., P. J. Allen, and H. Lester. 1976. Uredospore wall digestion during germination independent of molecular synthesis. *Physiol. Plant Pathol.* **9**: 265–272.

Hohl, R., M. Buehlmann, and E. Wehrli. 1978. Plasma membrane alterations as a result of heat activation in *Dictyostelium* spores. *Arch. Microbiol.* **116**: 239–244.

Keen, N. T. 1981. Evaluation of the role of phytoalexins. In R. C. Staples and G. H. Toenniessen (Eds.), *Plant Disease Control: Resistance and Susceptibility,* pp. 155–178. New York: Wiley.

Khan, M. 1977. Light stimulation of fungus spore germination. *Z. Pflanzenphysiol.* **81**: 374–376.

Knight, R. H., and J. L. Van Etten. 1976. Characteristics of ribonucleic acids isolated from *Botryodiplodia theobromae* pycnidiospores. *Arch. Microbiol.* **109**: 45–50.

Law, S. W. T., and D. N. Burton. 1976. Lipid metabolism in *Achlya:* Studies of lipid turnover during development. *Can. J. Microbiol.* **22**: 1710–1715.

Lockwood, J. L. 1977. Fungistasis in soils. *Biol. Rev.* **52**: 1–43.

Macko, V., R. C. Staples, Z. Yaniv, and R. Granados. 1976. Self-inhibitors of fungal spore germination. In D. J. Weber and W. H. Hess (Eds.), *The Fungal Spore: Form and Function,* pp. 73–100. New York: Wiley.

MacLeod, H., and P. A. Horgen. 1979. Germination of the asexual spores of the aquatic fungus *Achlya bisexualis. Exp. Mycol.* **3**: 70–82.

Maheshwari, R., and A. S. Sussman. 1971. The nature of cold-induced dormancy in uredospores of *Puccinia graminis* tritici. *Plant Physiol.* **47**: 289–295.

Mills, G. L., and F. L. Eilers. 1973. Factors influencing the germination of basidiospores of *Coprinus radiatus. J. Gen. Microbiol.* **77**: 393–401.

Mirkes, P. E. 1974. Polysomes, ribonucleic acid, and protein synthesis during germination of *Neurospora crassa* conidia. *J. Bacteriol.* **117**: 196–202.

Mirkes, P. E. 1977. Role of the carbon source in the activation of ribonucleic acid synthesis during the germination of *Neurospora crassa* conidia. *Exp. Mycol.* **1**: 271–279.

Nickerson, K. W., S. N. Freer, and J. L. Van Etten. 1981. *Rhizopus stolonifer* sporangiospores: a wet-harvested spore is not a native spore. *Exp. Mycol.* **5**: 189–192.

Nickerson, K. W., and E. Leastman. 1978. Cerulenin inhibition of spore germination in *Rhizopus stolonifer. Exp. Mycol.* **2**: 26–31.

Nickerson, K. W., B. K. McCune, and J. L. Van Etten. 1977. Polyamine and macromolecule levels during spore germination in *Rhizopus stolonifer* and *Botryodiplodia theobromae. Exp. Mycol.* **1**: 317–322.

Oort, A. J. P. 1974. Activation of spore germination in *Lactarius* species by volatile compounds of *Ceratocystis fagacearum. Proc. K. Ned. Akad. Wet. Ser. C., Biol. Med. Sci.* **77**: 301–307.

Pezet, R., and V. Pont. 1977. Elemental sulfur: Accumulation in different species of fungi. *Science* **196**: 428–429.

Ryder, N. S., and J. F. Peberdy. 1979. Chitin synthetase activity and chitin formation in conidia of *Aspergillus nidulans* during germination and the effect of cycloheximide and 5′-fluorouracil. *Exp. Mycol.* **3**: 259–269.

Smith, J. E., K. Gull, J. G. Anderson, and S. G. Deans. 1976. Organelle changes during fungal spore germination. In D. J. Weber and W. M. Hess (Eds.),*The Fungal Spore: Form and Function,* pp. 301–354. New York: Wiley.

Soll, D. R., R. Bromberg, and D. R. Sonneborn. 1969. Zoospore germination in the water mold, *Blastocladiella emersonii:* I. Measurement of germination and sequence of subcellular morphological changes. *Dev. Biol.* **20**: 183–217.

Soll, D. R., and D. R. Sonneborn. 1969. Zoospore germination in *Blastocladiella emersonii*: II. Influence of cellular and environmental variables on germination. *Dev. Biol.* **20**: 218–235.

Soll, D. R., and D. R. Sonneborn. 1972. Zoospore germination in *Blastocladiella emersonii*: IV. Ion control over differentiation. *J. Cell Sci.* **10**: 315–333.

Stade, S., and R. Brambl. 1981. Mitochondrial biogenesis during fungal spore germination: respiration and cytochrome c oxidase in *Neurospora crassa*. *J. Bacteriol.* **147**: 757–767.

Stahmann, M. A., P. Abramson, and L. Wu. 1975. A chromatographic method for estimating growth by glucosamine analysis of diseased tissues. *Biochem. Physiol. Pflanz. BPP.* **168**: 267–276.

Storck, R., and J. S. Price. 1977. Sensitivity of germinating spores of *Mucor rouxii* to chitosanase. *Exp. Mycol.* **1**: 323–338.

Sussman, A. S. 1966. Dormancy and spore germination. In G. C. Ainsworth and A. S. Sussman (Eds.), *The Fungi*, Vol. 2, pp. 733–764. New York: Academic.

Sussman, A. S. 1976. Activators of fungal spore germination. In D. J. Weber and W. H. Hess (Eds.), *The Fungal Spore: Form and Function*, pp. 101–139. New York: Wiley.

Thevelein, J. M., J. A. Van Assche, A. R. Carlier, and K. Heremans. 1979. Heat activation of *Phycomyces blakesleeanus* spores: Thermodynamics and effects of alcohols, furfurals and high pressure. *J. Bacteriol.* **139**: 478–485.

Torgeson, D. (Ed.). 1969. *Fungicides, an Advanced Treatise: Vol. 2. Chemistry and Physiology*. New York: Academic Press.

Trinci, A. P. J., and C. Whittaker. 1968. Self-inhibitor of spore germination in *Aspergillus nidulans*. *Trans. Brit. Mycol. Soc.* **51**: 594–596.

Trione, E. J. 1981. Natural regulators of fungal development. In R. C. Staples and G. H. Toenniessen (Eds.), *Plant Disease Control: Resistance and Susceptibility*, pp. 85–102. New York: Wiley.

Van Assche, J. A., A. R. Carlier, and H. I. Dekeersmaeker. 1972. Trehalase activity in dormant and activated spores of *Phycomyces blakesleeanus*. *Planta* **103**: 327–333.

Van Laere, A. J., J. A. Van Assche, and A. R. Carlier. 1980a. Reversible and irreversible activation of *Phycomyces blakesleeanus* spores. *Exp. Mycol.* **4**: 96–104.

Van Laere, A. J., J. A. Van Assche, and A. R. Carlier. 1980b. Metabolism and chemical activities of *Phycomyces blakesleeanus* spores. *Exp. Mycol.* **4**: 260–268.

Weber, D. J., and W. M. Hess. 1976. (Eds.), *The Fungal Spore: Form and Function*. New York: Wiley.

Wenzler, H., and R. Brambl. 1981. Mitochondrial biogenesis during fungal spore germination: Catalytic activity, composition, and subunit biosynthesis of oligomysin-sensitive ATPase in *Botryodiplodia*. *J. Biol. Chem.* **256**: 7166–7172.

8

GROWTH OF FUNGI

Growth is generally considered to be an increase in the mass of an organism that occurs after a given period of incubation. Growth is the net result of a multitude of molecular, cellular, and morphological changes which are still only incompletely known. Thus, the term "growth" may take on a variety of meanings, as we shall see. In this chapter we first examine methods of measuring growth and the various ways in which growth of fungi can be analyzed. This will be the basis for an overview of the mechanisms of growth of both filamentous and nonfilamentous fungi and some of the factors affecting this complex process.

METHODS OF MEASURING GROWTH

Methods that measure an increase in fungal mass per unit time provide the most direct means of assessing growth. Operationally, however, the morphology of the fungus, culture conditions, and experimental considerations may necessitate the use of other methods. Thus an experimenter may have to base a definition of growth on his or her method of measurement.

Dry Weight

Weighing the dried fungus is the most widely used and often the most convenient method of measuring growth. Tissue is placed in a tared pan, heated for 24 hours at 80°C, allowed to cool, and then weighed. Limitations of this method are that bulk quantities of tissue are required, and it is mostly cell wall material that is measured. In addition, when solid media are used it is often a tedious, if not impossible, process to separate the tissue from the medium. And, obviously, continuous changes in the growth of a specimen cannot be observed.

Linear Extension

This method involves assessment of the linear increase of either individual hyphae or colony diameter per unit time. Linear extension is a nondestructive measure of growth, and thus continued observations can be made. It is certainly the most rapid method when solid media are used. However, the usefulness of this method when measuring colony diameter is often limited by the fact that vertical growth of hyphae is not included, and thus the growth kinetics observed may be different from those obtained using dry weight.

Cell Number

Measurement of cell number is a useful method for determining growth of unicellular fungi such as yeasts. Counts can be made with the aid of a counting chamber such as a hemacytometer or a Sedgwick–Rafter cell. Alternatively, a spectrophotometer can often be used to determine cell number indirectly by measuring culture turbidity. In this case, turbidity can be converted to cell number by using a standard curve of hemacytometer counts plotted against optical density. Care must be taken with this method, however. Borzani and co-workers (1975) found that although the growth rates of *Saccharomyces cerevisiae* measured either by spectrophotometry or by dry weight were identical, the absolute estimates of cell number by the two methods were different.

Concentration of Cell Components

The changes in concentration of certain cellular constituents such as chitin and glucosamine have been used as a measure of fungal growth. This method has been very useful in assessing the growth of plant pathogens. For example, Stahmann et al. (1975) estimated the growth of *Puccinia graminis* var. *tritici* in infected plant tissue. Acid hydrolysis releases glucosamine which can be separated chromatographically, and the amount recovered correlates well with fungal dry weight. In addition, Donald and Mirocha (1977) used the glucosamine content to measure the extent of invasion by pathogens of stored soybeans and corn.

Sterols such as ergosterol and cholesterol are also good indicators of fungal growth. Seitz et al. (1979) note that levels of ergosterol parallel levels of chitin, which in turn parallel levels of dry weight. Furthermore, they note that the ergosterol assay is more sensitive and rapid than the chitin assay as an indicator of growth.

Metabolism

Estimates of metabolic activity are often used to measure fungal growth. Boyle (1977) measured the rate of CO_2 production to determine growth kinetics of *S. lipolytica* in a fermenter. Qualman et al. (1976) successfully monitored the growth of *Trichophyton mentagrophytes* and other filamentous fungi using ^{14}C-glucose incorporation into cell constituents. Such metabolic assays are discussed in detail in Chapter 10.

ANALYSIS OF GROWTH

Studies of the kinetics of fungal growth are often helpful in analyzing how growth of the organism is modified by changes in the environment. Several reviews dealing with growth of filamentous and nonfilamentous fungi have been published by Righelato (1975), Trinci (1978), and van Uden (1971).

Unicellular Fungi

When an inoculum of unicellular organisms such as yeasts is transferred to fresh medium and the growth of the population measured over time, a sigmoidal curve such as that in Figure 8.1 is typically obtained. Initially there is usually a lag phase during which the organism becomes adapted to its new environment. This may involve the induction and/or derepression of the enzymes to metabolize available nutrients. In addition, some organisms may modify the environment by producing CO_2, H^+, or secondary metabolites so that conditions are more favorable for growth.

Once this period of adaptation is over, the population enters the exponential phase. This is the phase of maximal growth for the particular set of environmental conditions, and cell division proceeds at a constant rate called the specific growth rate, denoted μ. The specific growth rate is defined as the amount of tissue produced by a unit amount of organism per unit time. Growth in this phase is autocatalytic: The number of cells (N) at a given time t (together denoted N_t) is an exponential function of the number cells present at the time that this growth phase begins (denoted N_0). Thus,

$$N_t = N_0\, e^{\mu t}$$

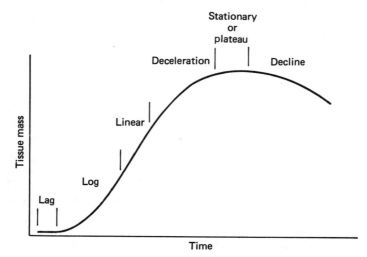

Figure 8.1 A typical growth curve

where e is the base of the natural logarithm (2.71828 . . .). This phase of growth is also called the logarithmic (or log) phase because a plot of log N over this time period gives a straight line having a slope equal to μ. Thus, the specific growth rate is a useful parameter for comparing the growth of different fungi or the growth of one species under different cultural conditions. Obviously, different nutritional conditions may result in different specific growth rates due to differences in the rates of uptake and metabolism.

In batch culture, where no fresh nutrients are supplied, the exponential rate of growth cannot be maintained indefinitely, and a deceleration phase begins. The onset of this phase may be due to some nutrient reaching a limiting concentration, to unfavorable changes in factors such as oxygen tension or pH, or to metabolites excreted by the organism becoming toxic. Eventually, growth of the population slows to the point where there is no net increase in the number of cells, and a plateau or stationary phase is reached. Cessation of growth may be due either to a cessation of cell division or to cells dying at the same rate that new cells are being formed. If these growth-limiting conditions persist, the death rate eventually increases and the population enters a phase of decline.

On the other hand, in a continuous-flow system, organisms can be maintained in exponential phase for an indefinite period of time. In such a system fresh medium is fed into a container at one end, and medium plus organisms are drained off at another. The flow rate (F) is adjusted so that the volume (V) of medium remains constant. Ideally, the cells are continuously exposed to a fixed concentration of nutrients, and waste products and other metabolites do not accumulate.

The F/V ratio is called the dilution rate and is an important regulator of the specific growth rate. If the dilution rate is too slow, fresh nutrients will not be added fast enough nor will wastes be removed in time. The resulting nutrient depletion and waste buildup will cause a slow specific-growth rate. Conversely, if the dilution rate is too fast the continuous supply of fresh nutrients coupled with the lack of toxic wastes result in a rapid growth rate, but all the cells may eventually be washed out. In between is a dilution rate enabling a steady state to be achieved in which the concentrations of organisms and nutrients remain constant over time.

Filamentous Fungi

The growth of many filamentous fungi also exhibits sigmoidal kinetics. For example, the growth curve of *Pycnoporus cinnabarinus* incubated in malt extract broth exhibits lag, log, deceleration, stationary, and decline phases (Figure 8.2). Notice, however, that the shape of the curve is much different when a Czapek's broth basal medium is used. At first glance it may seem surprising that the growth curves of unicellular and filamentous fungi are so similar, particularly the exponential phases. Obviously, the autocatalytic production of new cells observed in yeast does not occur in filamentous fungi. Hyphae grow only at the tips, and thus not all of the hyphal length contributes to growth, as is required for exponential growth. However, the capacity of filamentous fungi to form branches results in new hyphal tips, which enable the colony to grow exponentially. Branches often form at regular intervals, and exponen-

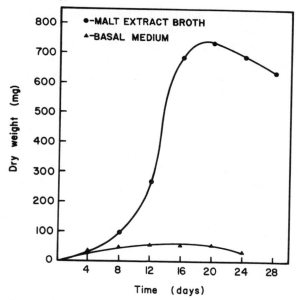

Figure 8.2   Growth curve of *Pycnoporus cinnabarinus* on two different media. (Holler and Brooks, 1980)

tial growth will result provided that the branching rate is proportional to hyphal length. Branching in filamentous fungi is thus physiologically analogous to cell division in unicellular fungi (Trinci, 1978).

Certain mycelial fungi incubated in shake culture grow as pellets of approximately spherical shape. Emerson (1950) demonstrated that exponential growth of *Neurospora cardia* pellets (as measured by dry weight, x) is more of a cubic function than a logarithmic one. The best fit is obtained by the equation

$$x_t^{1/3} = x_0^{1/3} + kt$$

where k is a constant. The different kinetics of pellet growth is due to nutrient depletion in the colony interior. Once the pellet reaches a certain large diameter, the diffusion of nutrients into the center of the tissue mass is too slow to maintain unrestricted growth of the entire mycelium. Thus, growth occurs mainly in the periphery of the pellet, and hyphae in the center may even die.

The kinetics of filamentous fungi growing on the surface of solid media is similar in some ways to that found in pellet growth. Growth of a surface culture in exponential phase can be expressed by the equation

$$x_t = (\pi H d r_0^2)e^{\mu t}$$

where the colony is assumed to grow as a disk of radius r, height H, and density d. Soon, however, conditions toward the center of the colony become unfavorable, and growth in this area slows. Any branches that form in this central zone tend to be narrower, less radially directed, and more meandering than branches in the peripheral zone; this results in effective exploitation of the substrate (Figure 8.3). Growth eventually becomes limited to the peripheral zone, which forms a ring around a nongrowing center. At this point, the increase in the colony radius (r) becomes a linear function as expressed by the equation

$$r_t = a\mu t + r_0$$

where a is the depth of the peripheral zone and μ is the specific growth rate. Experimental measurements confirm that surface-grown cultures have a short exponential phase followed by a long linear phase (Figure 8.4) (Righelato, 1975; Trinci, 1978). A model has been proposed by Trinci (1971) whereby growth kinetics as measured by linear extension can be related to kinetics based on dry weight.

Filamentous fungi can also be grown in continuous-flow systems. However,

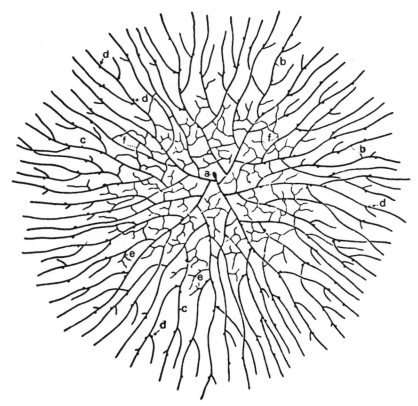

Figure 8.3 Diagram of a dikaryotic mycelium of *Coprinus sterquilinus*. (From Buller, 1958, as presented in Trinci, 1978.)

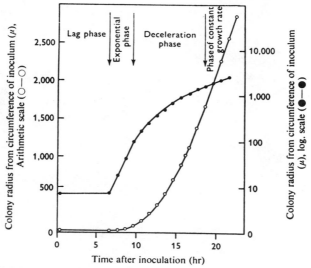

Figure 8.4 Growth of colonies of *Aspergillus nidulans* on a glucose–salts–agar medium at 37° C. (Trinci, 1969)

modifications to the system must be made to prevent the accumulation of mycelia on the walls of the culture vessel and in the outflow lines. Fungi growing as pellets cannot easily be grown in such systems unless new pellets can somehow be formed. As with unicellular fungi, care must be taken in extrapolating data obtained in the "ideal" environment of continuous culture to what occurs in the field.

MECHANISMS OF GROWTH

Considerable progress has been made during the past decade in elucidating many of the biochemical and cellular events involved in growth processes. Several recent reviews of this subject are available, including those by Burnett and Trinci (1979), Cabib (1975), Farkas (1979), Gooday (1978), and Grove (1978). As we shall see, there are certain similarities between the mechanism of growth in filamentous and nonfilamentous fungi even though the resulting growth patterns are different.

Growth of Filamentous Fungi

There is substantial morphological and biochemical evidence that growth in length of filamentous fungi occurs at the hyphal tip. For example, autoradiographic studies have shown that labeled cell wall precursors such as ^{3}H-glucose and ^{3}H-N-acetylglucosamine are incorporated at the apex (Figure 8.5), although label also accumulates farther back where the wall increases in thickness.

The hyphal apex has a distinctive ultrastructure. The 1–2 μm "apical zone" lacks

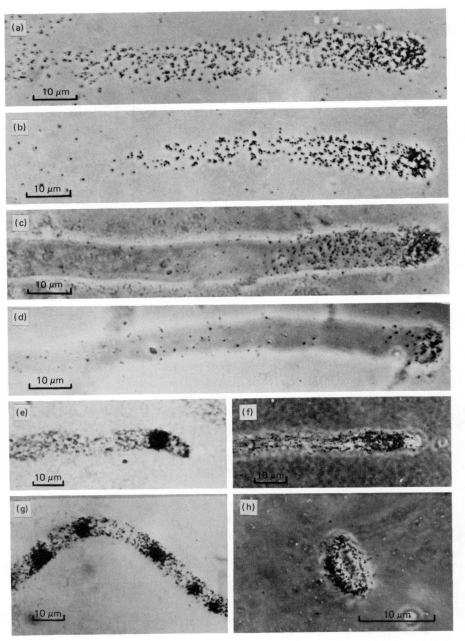

Figure 8.5 Light microscopic autoradiographs of deposition of wall material. (a–f) Growing hyphal tips showing apical incorporation: *Schizophyllum commune* incubated 10 minutes with (a) ³H-glucose, (b) N-(³H)-acetylglucosamine, (f) (³H-glucose followed by a 10-minute chase with (c) (³H)-glucose, (d) N-(³H)-acetylglucosamine. *Phytophthora parasitica* incubated 10 minutes with (e) (³H)-glucose followed by a 10-minute chase with glucose. (Gooday, 1971) (g) *P. parasitica* incubated 1 minute with (³H)-glucose showing subapical formation of cellulose wall ingrowths. (Gooday and Hunsley, 1971) (h) *Neurospora crassa* incubated 10 minutes with N-(³H)-acetylglucosamine showing septal formation. (Hunsley and Gooday, 1974) Scale bar, 10 μm. (Presented in Gooday, 1978.)

all organelles except numerous cytoplasmic vesicles. In some micrographs these vesicles appear to be fusing with the plasma membrane (Figure 8.6). In many Ascomycetes, Basidiomycetes, and Deuteromycetes a dark, iron-hematoxalin staining area called the "Spitzenkörper" has been observed in the apical zone. The function of this structure is not known although there are correlations between its behavior and hyphal growth. For example, it is present only in growing hyphae, and its movements to one side of the hypha precede the turning of the hypha in that direction (Grove, 1978; Gooday, 1978). High-voltage electron microscopy of the hyphal apex of *Fusarium acuminatum* following fixation by freeze-substitution indicates that the Spitzenkörper is composed of a cluster of vesicles surrounding a network of microfilaments. The microfilaments may control the movement of the Spitzenkörper in the apical cell (Howard and Aist, 1980; Howard, 1981).

The cytoplasmic vesicles found at the tip appear to originate at dictyosomes occurring in the subapical zone along with mitochondria, nuclei, and endoplasmic reticula. Although it is often difficult to determine the contents of these vesicles, they have been shown to contain cell wall and plasma membrane precursors, lytic enzymes (e.g., cellulase, β-1,3-glucanase, and proteases), and polysaccharide synthetases such as β-glucan synthetase. In addition, chitin synthetase has been localized in spherical particles called chitosomes (Figure 8.7). These particles have a membranelike shell surrounding a granular interior and are able to synthesize chitin

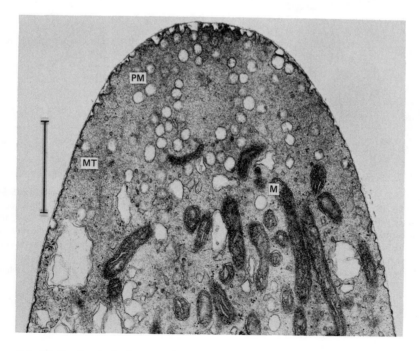

Figure 8.6 Median longitudinal section through the apical zone of a hypha of *Armillaria mellea*. Numerous invaginations of the plasma membrane (*PM*) at the apex may represent sites where vesicles have fused. M, mitochondrion; MT, microtubules. Scale line = 1 μm. (Grove and Bracker, 1970)

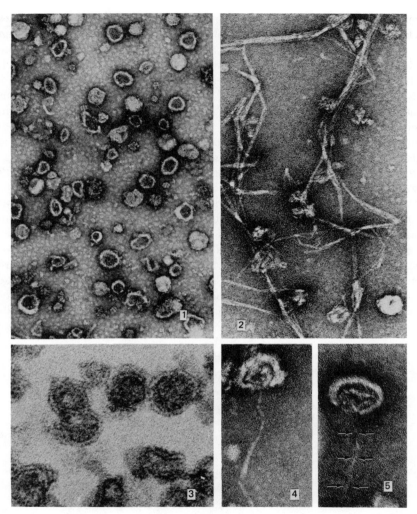

Figure 8.7 Chitosomes. (1) Purified chitosomes of *Mucor rouxii* in a negatively stained preparation. ×95,000. (Bartnicki-Garcia, 1981) □ (2) Chitin microfibrils produced by incubation of chitosomes of *M. rouxii* with substrate (UDP-N-acetylglucosamine) and activators. Negatively stained sample. ×152,000. (After Bartnicki-Garcia et al., 1979) □ (3) Thin-section view of chitosomes of *M. rouxii* fixed with OsO₄/glutaraldehyde. ×380,000. (Bartnicki-Garcia, 1981) □ (4 and 5) Structural connection between chitosomes of *M. rouxii* and chitin microfibrils. A long microfibril is continuous with a fibroid coil inside the partly opened chitosome. ×200,000 and ×280,000, respectively. (After Bracker et al., 1976; Cabib and Farkas, 1971)

microfibrils *in vitro*. Chitosomes have been isolated from all the major taxonomic groups of chitinous fungi and appear to be a means of transporting chitin synthetase to the cell surface (Bartnicki-Garcia et al., 1978).

These observations suggest a model for the mechanism of tip growth (Figure 8.8). Vesicles containing cell wall degrading enzymes are released from dictyosomes, move to the hyphal tip, fuse with the plasma membrane, and release their contents into the wall matrix. The action of these enzymes weakens the microfibrils, and the cell's turgor pressure then expands the wall. Cell wall precursors and the synthetase enzymes for incorporating them into the existing wall are packaged separately at the dictyosome, secreted, and used to rigidify the expanded wall.

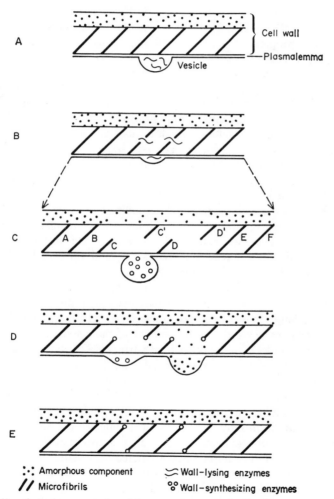

Figure 8.8   Hypothetical representation of the events in a unit of cell wall growth. (Bartnicki-Garcia, 1973)

There are many uncertainties in this model. For example, what causes the vesicles to migrate toward the tip? Cytoplasmic streaming, microtubules, and electrical potential gradients have all been suggested to explain this phenomenon. Also, a key regulatory point in tip growth may be the activation and inactivation of synthetase enzymes already present in the wall; we shall see that such a model has been described for yeasts.

It has been hypothesized that one role of the Spitzenkörper may be to direct the fusion of vesicles with specific sites on the plasma membrane. Microtubules are oriented parallel to the direction of growth and are thought to mediate the long-distance transport of vesicles to the apex and to the Spitzenkörper, if present. Microfilaments in the Spitzenkörper then direct the further movement of these vesicles to the plasma membrane (Howard and Aist, 1980; Howard, 1981).

Tip growth can thus be considered as a dynamic balance between wall lysis, wall synthesis, and turgor pressure. Factors disrupting this balance also alter the resulting pattern of growth. For example, the addition of polyoxin D, a competitive inhibitor of chitin synthetase, to cultures of *Mucor rouxii* results in weakened walls which bulge and burst at the tip (Bartnicki-Garcia and Lippman, 1972). Also, coumarin, an inhibitor of cellulose biosynthesis, supresses growth and branching of *Saprolegnia monoica* (Figure 8.9). On the other hand, growth and branching of this species are also inhibited by glucono-δ-lactone, which inhibits glucanase activity. In this case, the walls are greatly thickened due to decreased lysis (Figure 8.9). Both coumarin and glucono-δ-lactone also modify the cytological organization in the hyphae. Cellular polarity is disturbed, and vesicles accumulate in the subapical zone instead of moving to the tip itself (Fevre and Rougier, 1980).

Hyphal branching can be considered a modified form of tip growth whereby similar types of cytoplasmic vesicles fuse with the plasma membrane along the sides of the hypha. In some species, branches form behind septa. In these cases, vesicles migrating toward the apex may become trapped behind septa, and their accumulation may trigger wall lysis, swelling, and subsequent branch formation (Trinci, 1978). However, in many species there is no correlation between the location of the septum and the branch. Two principal hypotheses have been put forth to explain the pattern of branch formation in these species. Bartnicki-Garcia (1973) suggests that electrical gradients between vesicles and specific portions of the cytoplasm cause a localized accumulation of vesicles. On the other hand, Trinci (1974) argues that vesicles will accumulate whenever the rate of vesicle production exceeds the rate of their incorporation at the apex. That is, once the volume of cytoplasm associated with each hyphal tip (called the "hyphal growth unit") exceeds a critical size, excess vesicles will cause a new branch to form somewhere in the mycelium (Trinci, 1978).

Branches usually form only in the subapical zone. In the older portions of the hypha is a zone of vacuolation containing small amounts of cytoplasm and vacuoles that become larger and more numerous with increasing distance from the apex. This zone may be important in the production of secondary metabolites. The hyphal strand can thus be considered a tube through which the cytoplasm flows toward the growing tip away from the older regions. There are exceptions to this norm, however. Stipe cells of *Coprinus cinereus* exhibit intercalary elongation and incorporate labeled N-acetylglucosamine uniformly along the hypha (Gooday, 1973, 1975). In sporan-

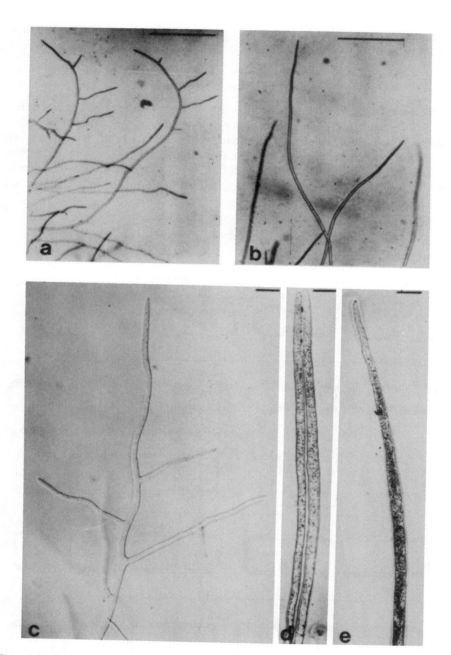

Figure 8.9 Photomicrographs of hyphae of *Saprolegnia monoica*. (a) Branched hyphae grown on control medium for 16 hours. Bar = 0.5 mm. (b) Unbranched hyphae grown for 16 hours in the presence of coumarin (3 m*M*). Bar = 0.5 mm. (c) Branched hyphae grown on control medium for 16 hours. Bar = 50 nm. (d) Unbranched hyphae grown for 16 hours in the presence of gluconolactone (0.33 *M*). Bar = 50 nm. (e) Unbranched hyphae grown for 16 hours in the presence of coumarin (3 m*M*). Bar = 50 nm. (Fevre and Rougier, 1980)

giophores of *Phycomyces blakesleeanus*, the intercalary zone reaches a length of 3 mm (Bergman et al., 1969).

Growth of Unicellular Fungi

Nonfilamentous fungi such as yeasts grow by two basic processes: budding and binary fission. Certain yeasts such as *Saccharomyces cerevisiae* exhibit multipolar budding and form buds over the entire wall surface. On the other hand, bipolar budding yeasts such as *Saccharomycodes ludwigii* always form buds in the same location of the wall (Figure 8.10a). In general, buds form by a balloonlike expansion of one of the parent walls followed by new wall synthesis, septum formation in the constriction

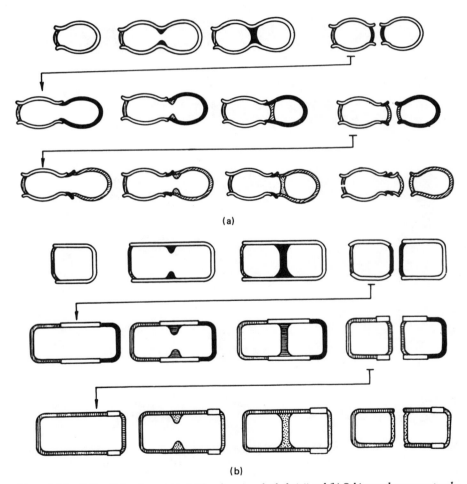

(a)

(b)

Figure 8.10   Cell wall development in (a) *Saccharomycodes ludwigii* and (b) *Schizosaccharomyces pombe*. The sequences illustrate progressive steps in growth and of divisions leading to the formation of cells with three division scars. (Streiblova, 1966, presented in Bartnicki-Garcia and McMurrough, 1971)

between the two cells, and eventual separation of the bud from the parent. Yeasts reproducing by binary fission, such as *Schizosaccharomyces pombe*, elongate at one or both ends, forming a cylinder. Following mitosis, a transverse septum is laid down near the middle; this septum then splits, resulting in two equal-sized daughter cells (Figure 8.10b) (Bartnicki-Garcia and McMurrough, 1971; Johnson et al., 1982). Notice that "binary fission" in fungi is far different from the process with the same name in prokaryotes.

Much progress has been made in elucidating the mechanism of growth in budding yeasts, and we confine our discussion to this group. Two recent reviews have been written on the topic (Cabib 1975; Cabib et al., 1982). There is very little data to explain either the initiation or location of bud formation. It appears that the cell must reach a certain size before it is capable of budding. Numerous cytoplasmic vesicles accumulate very early at the site of the incipient bud and may contain the lytic and synthetase enzymes required for wall growth. Microtubules also aggregate in this area, and these may play a role in orienting the flow of vesicles to the budding site.

Growth of the bud occurs by a combination of apical growth and equatorial enlargement. Autoradiographic studies have demonstrated that glucose is incorporated into glucans more or less uniformly around the bud wall. Mannose, on the other hand, appears to be incorporated first uniformly, then at the tip, and later both at the tip and the sides. Cabib (1975) proposes that a "beam of vesicles" is focused toward the site of bud initiation, resulting in a uniform deposition of mannose in the incipient bud followed by deposition at the tip as soon as apical growth becomes established (Figure 8.11). As the bud enlarges, the beam spreads out and vesicles deposit mannose at the sides as well as at the tip.

Once growth of the bud is complete, the wall thickens at the juncture of the parent and bud (Figure 8.12). A plasma membrane and a primary septum of chitin grow centripetally, and later a secondary septum of glucan–mannan is laid down on either side. When the cells detach, the primary septum remains with the parent as a component of the "bud scar." The remnant of the secondary septum is visible on the bud as a "birth scar."

Chitin appears to be present only in the septum, and much attention has been focused on the mechanism by which chitin synthesis is triggered at a particular location and time in the life cycle. In *Saccharomyces cerevisiae* chitin synthetase is

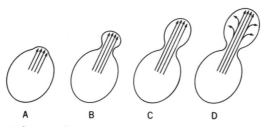

A   B   C   D

Figure 8.11 Schematic diagram of mannan integration into the cell wall of a budding yeast. (Cabib, 1975)

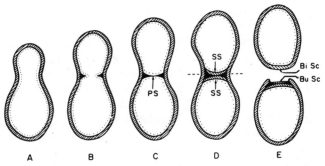

Figure 8.12 Septum formation in budding yeast. The dotted line represents the plasmalemma. The hatched areas are the cell walls. The secondary septa are distinguished by hatching in the opposite direction, although they may well have the same composition as the cell wall. The primary septum is darkened for clarity, although in electron micrographs it is seen as a white area. *PS*, primary septum; *SS*, secondary septum; *Bu Sc*, bud scar; *Bi Sc*, birth scar. (Cabib et al., 1973)

bound to the plasma membrane but is not localized in any one spot. However, most of the enzyme is in an inactive, or zymogen, form that can be activated by a membrane-bound "activating factor" (presumably a protease). The activating factor, in turn, can be inhibited by being bound to another protein in the cytoplasm (Figure 8.13).

Cabib and co-workers hypothesize that vesicles carrying the activating factor fuse with the plasma membrane in the region where the parent and bud join, thereby activating the chitin synthetase in that location (Figure 8.14a). The inhibitor may function as a safety valve to inactivate any protease that spills into other parts of the cell (Cabib, 1975). This localized chitin synthetase activation during septum formation may be replaced by a generalized activation of other polysaccharide synthetases during bud growth to form the spherical cell shape (Farkas, 1979). For example, β-glucan synthetase is present on the inner side of the plasma membrane and becomes active in the bud during growth (Figure 8.14b). Both ATP and GTP stimulate the activity of this synthetase and may be important regulators of the budding process (Cabib et al., 1981).

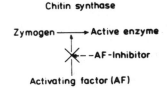

Figure 8.13 Scheme for activation of yeast chitin synthase as proposed by Cabib et al. (1973).

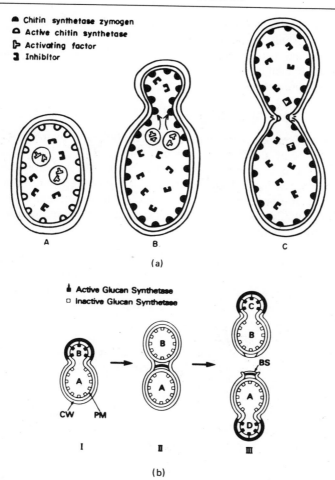

Figure 8.14 Hypothetical schemes for bud formation in yeast. (a) Scheme for the initiation of chitin synthesis. (Adapted from Cabib and Farkas, 1971) (b) Scheme for the role of glucan synthetase. The growing areas of the cell wall are filled in black. A, mother cell; B, bud; C and D, new buds; CW, cell wall; PM, plasma membrane; BS, bud scar. For simplicity, growth of the cell wall is pictured as occurring on the whole surface of the bud. (Shematek et al., 1980)

Dimorphic Growth

Many fungi have the capability of changing their growth habit between mycelium and yeast, depending on the environmental conditions. This ability is called dimorphism and is characteristic of species in the Zygomycetes, Ascomycetes, Basidiomycetes, and Deuteromycetes. Several of the fungal pathogens of humans exhibit dimorphism, with the yeast phase being parasitic and the mycelial phase saprophytic. Thus, from a practical point of view there is considerable interest in understanding the mechanism of dimorphic interconversion. In addition, dimorphism provides a model system for studying morphogenesis in fungi.

Dimorphic fungi can generally be divided into three groups depending on the types of environmental factors triggering interconversion (Romano, 1966). We shall briefly discuss representatives of each group.

TEMPERATURE-DEPENDENT DIMORPHISM

Dimorphism in *Paracoccidioides brasiliensis* is controlled by temperature. This species grows as a yeast at 37°C and as a mycelium at 20°C. One major difference between the two forms is the composition of the cell wall. In yeastlike cells the major polysaccharide is α-1,3-glucan whereas in mycelia it is β-1,3-glucan. Chitin and protein are also present. When the incubation temperature is shifted from 37° to 20°C, the synthesis of α-glucan decreases and the synthesis of β-glucan increases. In addition, the activity of the enzyme β-1,3-glucanase is higher in the mycelial form than in the yeast form and increases during the 37° to 20°C shift. On the other hand, the activity of α-1,3-glucanase is very low in both forms (Kanetsuna and Carbonell, 1970; Kanetsuna et al., 1972; Flores-Carreon et al., 1979). These observations are consistent with a model produced by Kanetsuna and co-workers (1972) (Figure 8.15). The yeast-phase cell contains α-glucan in an outer layer and chitin in an inner layer. β-glucan is presumed to be located at discrete budding sites in the wall, as in true yeasts such as *S. cerevisiae*. At 37°C, β-glucanase activity weakens the β-glucan at a budding site, and the weakened area balloons out as a bud. The relatively low activities of α- and β-glucanase result in a uniform weakening of the new wall, and spherical growth occurs.

At 20°C the increased activity of β-glucanase causes continued expansion at the budding site and the onset of tip growth. Apical growth is maintained by wall lysis due to enhanced β-glucanase activity and by wall synthesis due to increased synthesis of β-glucans. The resulting wall is largely β-glucan with protein and chitin interwoven in an unlayered manner.

TEMPERATURE AND NUTRIENT-DEPENDENT DIMORPHISM

Histoplasma capsulatum also grows as a yeast at 37°C, and mycelial growth results when the temperature is decreased to 25°C. However, an upward temperature shift is not sufficient to cause the conversion from mycelium to yeast. Other factors are required, such as the proper cellular redox potential or the proper concentration of sulfhydryl groups. In fact, mycelial-to-yeast conversion can occur even at 25°C if the medium contains cysteine or cystine or has a low redox potential (Maresca et al., 1977; Rippon, 1968).

The role of sulfhydryl groups has been investigated the most thoroughly. Yeast and mycelial phases each take up ^{35}S-cystine at comparable rates at both 37° and 25°C. In addition, mutants having a yeast phase that is unable to take up cystine can still form mycelia. Thus, cystine uptake is not required for the transition to the yeast phase nor for its maintenance. However, a cystine reductase that catalyzes the reduction of cystine to cysteine has been localized at the plasma membrane. This

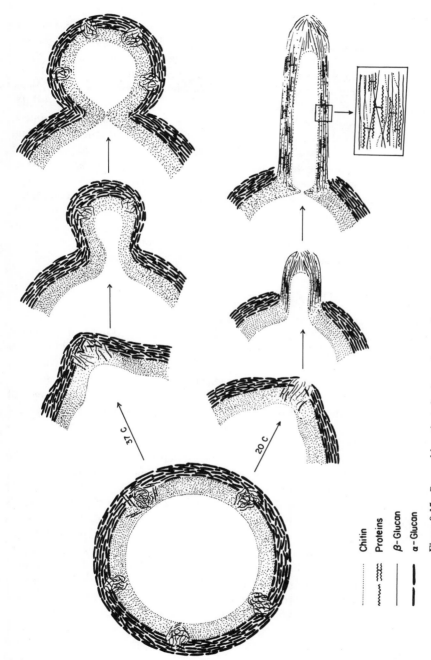

Chitin

Proteins

β- Glucan

α - Glucan

Figure 8.15 Proposed hypothesis for the production of yeastlike and mycelial forms of *Paracoccidioides brasiliensis*. (Kanetsuna et al., 1972)

249

enzyme appears during the transition from 25° to 37°C and is present thereafter in yeast cells. The enzyme seems to be necessary for the maintenance of the yeast phase. Cells treated with an inhibitor of cystine reductase remain mycelial even at 37°C. Thus, this enzyme may cause an increase in the level of cellular -SH groups, which are required for the transition of mycelium to yeast (Maresca et al., 1978).

Studies of cellular respiration rates during the transition from mycelium to yeast have shown that there are three distinct metabolic changes. Following the temperature increase there is a rapid decrease in cell respiration and in the intracellular levels of cysteine and other amino acids. This may be due to a temperature-induced increase in cellular redox potential. When the second stage is reached, respiration has completely stopped. It is at this stage that exogenous cysteine is required. Studies with isolated mitochondria show that cysteine activates mitochondrial respiration. During the third stage the respiration rate rises, intracellular concentrations of cysteine and other amino acids increase, and the morphological transition is completed. It is thought that the requirement for cysteine to activate respiration and to maintain the yeast phase ensures that random temperature increases will not cause the morphological conversion and that the yeast phase will occur only in an appropriate host where cysteine is available (Maresca et. al., 1981).

NUTRIENT-DEPENDENT DIMORPHISM

Dimorphism in species such as *Candida albicans* and *M. rouxii* is regulated by environmental conditions other than temperature. This is called nutrient-dependent dimorphism. For example, *C. albicans* grows as a yeast at either 37° or 25°C when incubated in a glucose medium. However, the mycelial phase forms when this species is transferred to a less readily metabolizable carbon source such as starch or glycogen. This shift from yeast to mycelium can be inhibited by the addition of cysteine, indicating the involvement of sulfhydryl groups (Romano, 1966). Micromolar concentrations of zinc inhibit mycelium formation at 25°C but not at 37°C (Bedell and Soll, 1979).

Various species of *Mucor* also exhibit nutrient-dependent dimorphism, but the key environmental factors appear to be carbon dioxide and hexoses. For example, *M. rouxii* grows as a mycelium when incubated either aerobically or anaerobically in the absence of CO_2. However, it grows as a yeast when the CO_2 level reaches 0.3 atm under anaerobic conditions or 0.9 atm under aerobic conditions (Bartnicki-Garcia and Nickerson, 1962). The hexose concentration is also an important factor. Under anaerobic conditions and 0.3 atm CO_2, tissue incubated on a medium containing a low glucose concentration develops as a mycelium, whereas on a high glucose concentration yeastlike cells form. In fact, CO_2 is not required at all for yeast development if the glucose level is high enough. Fructose, mannose, and galactose (in decreasing order of effectiveness) also stimulate yeast development. Under aerobic conditions only mycelia form, regardless of the hexose levels. It is attractive to hypothesize that high hexose levels might stimulate CO_2 production but this is not the case. One suggestion is that hexose inhibits hyphal development by inhibiting in some way the initiation of apical growth (Bartnicki-Garcia, 1968).

More recently, cyclic AMP has been implicated in *Mucor* dimorphism. In *M.*

racemosus the addition of dibutyryl cAMP induces yeastlike development in hyphae grown aerobically. In addition, endogenous cAMP levels are high in yeast-phase cells and low in mycelia (Larsen and Sypherd, 1974). Because cell walls of yeast and mycelia are chemically similar, it has been hypothesized that cAMP may affect the pattern of wall synthesis. Indeed, there is evidence that in M. *racemosus* cAMP alters the rate of chitin-plus-chitosan synthesis in the cell wall (Domek and Borgia, 1981). Other possible roles for cAMP in dimorphism include regulation of gene expression, activation of protein kinases involved in wall biosynthesis, and interaction with microtubules and microfilaments, thereby regulating the flow of precursors to the sites of wall synthesis (Stewart and Rogers, 1978).

FACTORS AFFECTING GROWTH

Considering the complexity of the processes involved in growth it is not surprising that fungal growth is affected by a variety of environmental factors. Indeed, we have been discussing the effects of nutritional factors on growth throughout much of this book. However, in this section we concentrate on certain gross factors such as water, temperature, pH, irradiation, aeration, and CO_2 (already mentioned in reference to dimorphism), and chemicals such as pesticides and herbicides, which have been found to affect the growth of fungi.

Water

Water is required for the growth of fungi just as for other organisms. Fungi usually require a film of water around their cells through which nutrients and enzymes diffuse. Species causing dry rot can ramify through very dry regions because water is transported through the hyphae from moist regions and also because water is generated as a by-product in metabolic reactions. However, too much water can be detrimental as well. For example, most filamentous fungi do not sporulate in submerged culture, and growth can be inhibited if submergence causes anaerobiosis.

Growth of fungi can usually be prevented by drying the substrate, a method that the food industry often uses to prevent spoilage. The water content of fungi can also be reduced by making the osmotic potential outside the cell more negative than inside. For example, sugar or salt can be added to containers of meat, fruit, or jam to prevent fungal growth. Sugar concentrations of 50–70 percent or salt concentrations of 20–25 percent are usually effective. However, osmophillic yeasts such as *Saccharomyces rouxii* and *S. mellis*, as well as filamentous species such as *Aspergillus glaucus*, grow well in severe osmotic environments (Eggins and Allsopp, 1975). The effect of osmotic stress on fungal growth and development was discussed in relation to sodium and potassium utilization in Chapter 5.

Temperature

Fungi differ markedly in their temperature optima. In Figure 8.16 is shown the growth of six marine fungi on four different agar media at various temperatures. In

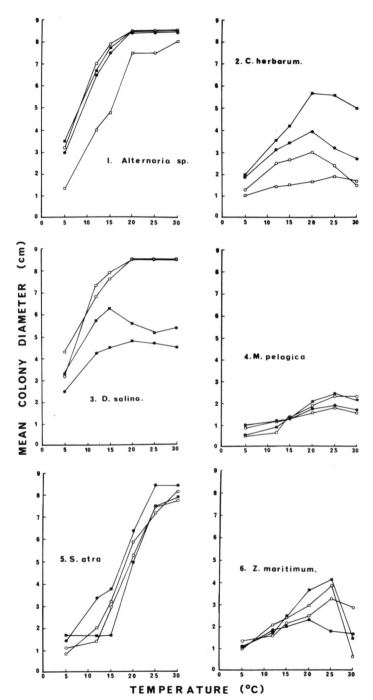

Figure 8.16 Effect of temperature and culture medium on the radial growth of (1) *Alternaria* sp., (2) *Cladosporium herbarum*, (3) *Dendryphiella salina*, (4) *Monodictys pelagica*, (5) *Stachybotrys atra*, (6) *Zalerion maritimum*. ● 2% malt extract-distilled water agar; ■ glucose-yeast extract-seawater; ○ Johnson-Sparrow seawater medium; □ seawater agar. Incubation period 14 days. (Curran, 1980)

most cases, growth is poorest at 5°C and increases with increasing temperature either to a peak or a plateau. Notice that *Alternaria* sp., *Stachybotrys atra*, and *Dendryphiella salina* grow well at 30°C, but this temperature is supraoptimal for the other species (Curran, 1980).

Certain fungi require temperatures above 20°C and often thrive at 50°C and above. These thermophilic fungi are found in garden composts, manure piles, and stored agricultural and forest products. They are also involved in spontaneous combustion of materials and in the diseases of animals (Prodromou and Chapman, 1974). In Figure 8.17 are the growth curves for two such thermophilic species at various temperatures. Both *Mucor miehei* and *Sporotrichum thermophile* have temperature optima in the 40–45°C range (Chapman, 1974).

Although some fungi can survive in extremely cold environments, growth at low temperatures is minimal for most fungi and is a key consideration in food preservation. The temperature at which fungi are stored has an effect on subsequent growth. Raudaskoski et al. (1976) noted that the storage of *Gyromitro esculenta* at subfreezing

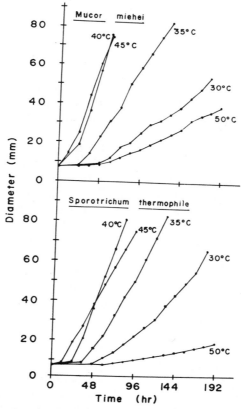

Figure 8.17 Increase in diameter of *Mucor miehei* and *Sporotrichum thermophile* on YpSs agar at different temperatures. Each point represents the average of two colony diameters in each of four replicates. (Chapman, 1974)

temperatures resulted in reduced vigor when it was subsequently incubated at 15–23°C, the temperature range for maximal growth. Cryogenic storage has been used to attempt the long-term maintenance of fungi. Hwang et al. (1976) found that several strains remained viable after nine years in cryogenic storage.

Effects of pH

The optimum pH for the growth of fungi varies with the strain or species and the nutritional environment. In Figure 8.18 are shown the effects of pH on growth of the six marine species mentioned above. Notice that the shapes of the curves vary considerably. *Dendryphiella salina* grows best at pH 4–6, *Cladosporium herbarium* at pH 5–6, *Monodictys pelagica* at pH 6, *Alternaria* sp. at pH 6–7, and the others at basic pH values. In previous chapters we have emphasized the role of pH in the uptake of various nutrients, and thus these growth responses are understandable in terms of differences in uptake rates. The observation that the maximum specific growth rate of *Trichoderma viride* hyphae is linearly correlated with the H^+ concentration (Brown and Halsted, 1975) is consistent with the effect of pH on nutrient uptake.

Visible and Nonvisible Irradiation

The effects of various forms of radiation on development and reproduction in fungi have been given considerable emphasis over the years (e.g., Tan, 1978); there is less information on the effects of radiation on the growth of either filamentous or nonfilamentous species. Light has been shown to stimulate the growth of certain species, to inhibit others, and to have no effect on still others. Observe in Figure 8.19 the various effects of light on the growth of four strains of *Candida albicans*. Light of all wavelengths tested inhibits growth of strain SA#1. In strain 3C, light either is inhibitory or has no effect. In the other two strains, black (UV) light is inhibitory but all other wavelengths stimulate growth (Saltarelli and Coppola, 1979).

It is difficult to pinpoint the mechanisms by which light exerts its effects. In the *C. albicans* strains mentioned above, light generally inhibits protein synthesis and stimulates carbohydrate synthesis. In *Blastocladiella emersonii*, the light-induced stimulation of growth is correlated with an enhancement of polysaccharide synthesis and a decrease in the activity of glucose-6-phosphate dehydrogenase activity (Goldstein and Cantino, 1962). In *Penicillium* spp. the inhibition of growth by light is correlated with a decrese in cell wall synthesis (Chebotarev and Zemlyanukhin, 1974). However, it is difficult to determine if these effects of light are direct or indirect.

The effects of nonvisible irradiation on growth have been studied in several species. For example, γ-irradiation (1.6–4.1 rad/hour) markedly stimulates growth and production of organic acids in *Aspergillus niger*. Low levels of γ-irradiation also increase growth as well as aflatoxin production in *A. flavus*, but high levels are inhibitory (Kuzin et al., 1976).

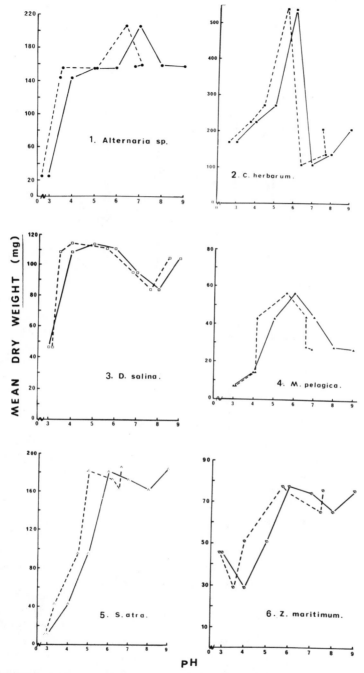

Figure 8.18 Effect of pH on growth of (1) *Alternaria* species, (2) *Cladosporium herbarium*, (3) *Dendryphiella salina*, (4) *Monodictys pelagica*, (5) *Stachybotrys atra*, (6) *Zalerion maritimum*. Incubation period 14 days. (Curran, 1980)

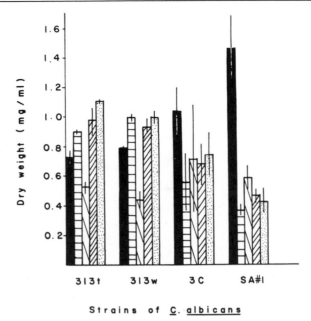

Figure 8.19 Effect of light and dark environments on cell growth of strains of *Candida albicans* as indicated by dry weight. ■ dark, ▤ blue, ◱ blacklite, ◪ Chroma 50 = sunlight equivalent, ▦ pink light. (Saltarelli and Coppola, 1979)

Aeration

Growth of fungi usually increases with increasing aeration. For example, the dry weight of *Penicillium expansum* incubated in shake culture is considerably greater than in static culture at all stages of the growth curve (Figure 8.20). Note also in this graph that the activity of polygalacturonase is also greater in shake culture, suggesting that aeration favors the production and/or secretion of this enzyme (Woodhead and Walker, 1975).

The level of oxygen needed for optimal growth may be quite low. Particularly striking is the observation that growth of *Candida boidinii* is greater at 0.008 atm O₂ than at 0.08 atm (Ivanov and Podgorskii, 1977).

The effect of aeration on growth is related to quantitative and qualitative changes in the physiology of the affected fungus. As we shall see in Chapter 10, increased oxygen levels result in greater rates of cellular respiration and thus greater energy production for cellular activities. In *C. mycoderma* incubated under conditions of decreasing oxygen levels, growth slows and the cyanide-resistant pathway of electron transport appears in the cells. At the same time the content of cytochrome a decreases while that of cytochromes b and c increase (Lozinov et al., 1977). The responsiveness of intracellular cytochromes to reduction upon the addition of glucose to cells of *Saccharomyces cerevisiae* also depends on the level of aeration. Increased

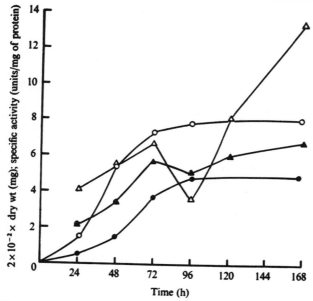

Figure 8.20 Time-course of growth and extracellular enzyme production by *Penicillium expansum* under different conditions of aeration: Dry weight in static (●) and shake (○) cultures; polygalacturonase activity in static (▲) and shake (△) cultures. (Woodhead and Walker, 1975)

cytochrome reduction is seen in aerated cell suspensions but not in nonaerated ones (Israelstam, 1973).

Levels of Carbon Dioxide

Fungi are often exposed to elevated levels of carbon dioxide either as by-products of their own metabolism or from the by-products of other organisms in their environment. As with the other factors we have discussed, the specific effect of CO_2 varies with the species. Accumulation of CO_2 in the culture medium inhibits growth of *Candida utilis*. This inhibition is reversible and can be eliminated either by increasing the aeration or by varying the pH of the medium (Shkidchenko, 1976). High CO_2 concentrations also inhibit the growth of *Phymatotrichum omnivorum* (Gunasekaran, 1973). The inhibitory effects of gases emitted from various *Trichoderma* species and other fungi are attributed in part to CO_2 (Tamimi and Hutchinson, 1975).

On the other hand, stimulatory effects of carbon dioxide on fungal growth have been reported. For example, growth of *Blastocladia ramosa* and *B. pringsheimi* is enhanced by high concentrations of CO_2 (Held et al., 1969); and *Verticillium albo-atrum* requires CO_2 for growth when glucose or glycerol is the sole carbon source (Hartman et al., 1972).

The role of CO_2 as a carbon source is discussed in Chapter 4, and biochemical pathways involved in CO_2 utilization are described in Chapter 10.

Non-Nutrient Chemicals

The considerable importance of fungi as parasites, agents of biodeterioration, mycotoxin producers, and allergens probably accounts for the keen interest in chemicals not ordinarily considered as nutrients which inhibit their growth. Organic and inorganic compounds from various sources have been studied for their effects on fungal growth.

The fungicides triforine and triarimol inhibit growth of many species via an inhibition of ergosterol biosynthesis (Sherald et al., 1973). The inhibitory effect of the polyene antibiotic nystatin on fungi such as *Colletotrichum lagenarium* is a consequence of binding to membrane sterols, which causes increased permeability, the leakage of important cellular constituents, and cell death. Members of the Oomycetes lack membrane sterols and thus are generally insensitive to polyene antibiotics (Fowlks et al., 1967). Inorganics such as cyanide and azide impair the biogenesis and proper functioning of mitochondria (Lyr, 1977). The inhibitory effects of heavy metals such as copper on growth have already been discussed in Chapter 5.

Various compounds applied to plants to control pests may also affect fungi. Pesticides, herbicides, and insecticides have been shown to inhibit the growth of some fungi, but they may have no effect or may even be stimulatory (Schlueter, 1977). For example, the nematocide ethoprop is inhibitory to the growth of *Sclerotium rolfsii* and *Rhizoctonia solani* when present in potato–dextrose agar at concentrations of 10–100 ppm, but *Trichoderma* spp., *Rhizopus stolonifer*, and various *Aspergillus* species are not affected (Rodriguez-Kabana et al., 1976). In addition, the herbicides treflan, basagran, and avadex as well as the insecticides metasystox 55, nicotine, and hostathion stimulate growth and virulence of the bean pathogen *Fusarium solani* f. sp. *phaseoli* (Mussa and Russell, 1977a,b).

Extracts from plants such as corn, apple, or pepper are often potent inhibitors of fungal growth. In addition, extracts from certain fungal species are inhibitory to other fungi. For example, extracts from *Trichoderma viride* are inhibitory to growth of *Fomes annosus*. In this case, the potency of the extracts as inhibitors is affected by the substrate on which *T. viride* is grown (Sierota, 1977). Extracts of *Armillaria mellea* may be self-inhibitory under certain conditions. There is evidence that the inhibitory compounds include phenols, a class of compounds inhibitory to many plants and fungi (Vance and Garraway, 1973).

There is very little information as to the precise mode of action of any of these compounds.

SUMMARY

Growth is defined as an increase in mass occurring after a given period of incubation. In fungi as with other microorganisms, the various methods used to measure growth include dry weight, linear extension, cell number, change in the concentration of various cell components, and estimates of metabolic activity.

Growth curves provide a convenient means of analyzing the growth of both filamentous and nonfilamentous fungi, and mathematical models for various patterns of growth have been developed. In general, these curves are similar in being composed of lag, log, linear, deceleration, stationary, and decline phases. The similarity between the curves of unicellular and mycelial fungi is due to hyphal branching in the latter producing approximately the same effect as autocatalytic growth in the former. The kinetics of growth of fungi incubated on solid media resemble those of pellet growth in certain species incubated in liquid media; in both cases nutrient depletion in the colony center limits growth.

Growth of filamentous fungi occurs at the hyphal tip and at the tips of branches. Growth is the result of interactions among vesicles produced at dictyosomes and containing (1) lytic enzymes that weaken the existing wall and (2) synthetase enzymes that catalyze the incorporation of (3) wall precursors into the weakened wall stretched by the cell's turgor pressure. The mechanism of growth of unicellular fungi such as yeasts is less well known but appears to involve many of the same features found in filamentous fungi that interact to yield instead spherical growth. Some of the enzymes, including chitin synthetase and β-glucan synthetase, are bound to the plasma membrane and undergo localized activation.

Many species can change their growth habit between hyphae and unicell and thus are said to exhibit dimorphism. The interconversion of these growth forms can be stimulated by temperature alone, as in *Paracoccidioides brasiliensis*, by temperature in combination with various nutrients, as in *Histoplasma capsulatum*, or by nutrients alone, as in *Candida albicans* and *Mucor rouxii*. Although the mode of action of these environmental factors is largely unknown, the net result is a change in the pattern of wall lysis and synthesis.

Growth is affected by a host of environmental factors including water, temperature, pH, irradiation, aeration, CO_2, and chemicals. The type of response by fungi, whether positive or negative, to those factors depends on the species. Awareness of these factors and an understanding of the nature of their effects can be useful to mycologists, physiologists, and plant pathologists.

REFERENCES

Bartnicki-Garcia, S. 1968. Control of dimorphism in *Mucor* by hexoses: Inhibition of hyphal morphogenesis. *J. Bacteriol.* **96**: 1586–1594.

Bartnicki-Garcia, S. 1973. Fundamental aspects of hyphal morphogenesis. *Symp. Soc. Gen. Microbiol.* **23**: 245–267.

Bartnicki-Garcia, S. 1981. Role of chitosomes in the synthesis of fungal cell walls. *Microbiology* **1981**: 238–241.

Bartnicki-Garcia, S., C. E. Bracker, E. Reyes, and J. Ruiz-Herrera. 1978. Isolation of chitosomes from taxonomically diverse fungi and synthesis of chitin microfibrils in vitro. *Exp. Mycol.* **2**: 173–192.

Bartnicki-Garcia, S., and E. Lippman. 1972. The bursting tendency of hyphal tips of fungi: Presumptive evidence for a delicate balance between wall synthesis and wall lysis in apical growth. *J. Gen. Microbiol.* **73**: 487–500.

Bartnicki-Garcia, S., and I. McMurrough. 1971. Biochemistry of morphogenesis in yeasts. In A. H. Rose and J. S. Harrison (Ed.), *The Yeasts*, Vol. 2, pp. 441–491. New York: Academic.

Bartnicki-Garcia, S., and W. J. Nickerson. 1962. Induction of yeastlike development in *Mucor* by carbon dioxide. *J. Bacteriol.* **84**: 829–840.

Bartnicki-Garcia, S., J. Ruiz-Herrera, and C. E. Bracker. 1979. Chitosomes and chitin synthesis. In J. H. Burnett and A. P. J. Trinci (Eds.), *Fungal Walls and Hyphal Growth*, pp. 149–168. London: Cambridge University Press.

Bedell, G. W., and D. R. Soll. 1979. Effects of low concentrations of zinc on the growth and dimorphism of *Candida albicans:* Evidence for zinc-resistant and -sensitive pathways for mycelium formation. *Infect. Immun.* **26**: 348–354.

Bergman, K., P. V. Burke, E. Cerda-Olmedo, C. N. David, M. Delbruck, K. W. Foster, E. W. Goodell, M. Heisenberg, G. Meissner, M. Zalokar, D. S. Dennison, and W. Shropshire. 1969. *Phycomyces*. *Bact. Rev.* **33**: 99–157.

Borzani, W., M. L. R. Vairo, and R. E. Gregori. 1975. Influence of cell concentration measurement method on the value of specific growth rate. *Biotechnol. Bioeng.* **17**: 461–462.

Boyle, D. T. 1977. A rapid method for measuring specific growth rate of microorganisms. *Biotechnol. Bioeng.* **19**: 297–300.

Bracker, C. E., J. Ruiz-Herrera, and S. Bartnicki-Garcia. 1976. Structure and transformation of chitin synthetase particles (chitosomes) during microfibril synthesis in vitro. *Proc. Natl. Acad. Sci. USA* **73**: 4570–4574.

Brown, D. E., and D. J. Halsted. 1975. The effect of acid pH on the growth kinetics of *Trichoderma viride*. *Biotechnol. Bioeng.* **17**: 1199–1210.

Buller, A. H. R. 1958. Further observation on the Coprini together with some investigations on social organization and sex in the Hymenomycetes. *Researches in Fungi*, Vol. 4. New York: Hafner.

Burnett, J. H., and A. P. J. Trinci (Eds.). 1979. *Fungal Walls and Hyphal Growth.* Cambridge University Press.

Cabib, E. 1975. Molecular aspects of yeast morphogenesis. *Ann. Rev. Microbiol.* **29**: 191–214.

Cabib, E., and V. Farkas. 1971. The control of morphogenesis: An enzymatic mechanism for the initiation of septum formation in yeast. *Proc. Natl. Acad. Sci. USA* **68**: 2052–2056.

Cabib, E., V. Farkas, R. E. Ulane, and R. Bowers. 1973. Yeast septum formation as a model system for morphogenesis. In J. R. Villanueva, I. Garcia-Acha, S. Gascon, and F. Uruburu (Eds.), *Yeast, Mould and Plant Protoplasts*, pp. 105–116. New York: Academic.

Cabib, E., R. Roberts, and B. Bowers. 1982. Synthesis of the yeast cell wall and its regulation. *Ann. Rev. Biochem.* **51**: 763–793.

Cabib, E., E. M. Shematek, J. A. Braatz, and H. Kawai. 1981. Biosynthesis of the major structural polysaccharides of the yeast cell wall, β-(1-3)-glucan and its regulation. *Microbiology* **1981**: 235–237.

Chapman, E. S. 1974. Effect of temperature on growth rate of seven thermophilic fungi. *Mycologia* **66**: 542–546.

Chebotarev, L. N., and A. A. Zemlyanukhin. 1974. Prolonged effect of visible light on the structure of fungal cell walls. *Mikrobiologiya* **43**: 853–856.

Curran, P. M. T. 1980. The effect of temperature, pH, light and dark on the growth of fungi from Irish coastal waters. *Mycologia* **72**: 350–358.

Domek, D. B., and P. T. Borgia. 1981. Changes in the rate of chitin-plus-chitosan synthesis accompanying morphogenesis of *Mucor racemosus*. *J. Bacteriol.* **146**: 945–951.

Donald, W. N., and C. J. Mirocha. 1977. Chitin as a measure of fungal growth in stored corn and soybean seed. *Cereal Chem.* **54**: 466–474.

Eggins, H. O. W., and D. Allsopp. 1975. Biodeterioration and biodegradation by fungi. In J. E. Smith and D. R. Berry (Eds.), *The Filamentous Fungi*, Vol. 1, pp. 301–319. New York: Wiley.

Emerson, S. 1950. The growth phase in *Neurospora* corresponding to the logarithmic phase in unicellular organisms. *J. Bacteriol.* **60**: 221–223.

Farkas, V. 1979. Biosynthesis of cell walls of fungi. *Microbiol. Rev.* **43**: 117–144.

Fevre, M. and M. Rougier. 1980. Hyphal morphogenesis of *Saprolegnia:* Cytological and biochemical effects of coumarin and glucono-δ-lactone. *Exp. Mycol.* **4**: 343–361.

Flores-Carreon, A., A. Gomez-Villanueva, and G. San Blas. 1979. β-1,3-glucanase and dimorphism in *Paracoccidioides brasiliensis*. Antonie van Leeuwenhoek **45**: 265–274.

Fowlks, E. R., C. Leben, and J. F. Snell. 1967. Sterols in relation to the influence of nystatin on *Pythium aphanidermatum* and *Colletotrichum lagenarium*. *Phytopathology* **57**: 246–249.

Goldstein, A., and E. C. Cantino. 1962. Light-stimulated polysaccharide and protein synthesis by synchronized single generations of *Blastocladiella emersonii*. *J. Gen. Microbiol.* **28**: 689–699.

Gooday, G. W. 1971. An autoradiographic study of hyphal growth of some fungi. *J. Gen. Microbiol.* **67**: 125–133.

Gooday, G. W. 1973. Activity of chitin synthetase during development of fruit bodies of the toadstool *Coprinus cinereus*. *Biochem. Soc. Trans.* **1**: 1105–1107.

Gooday, G. W. 1975. The control of differentiation in fruit bodies of *Coprinus cinereus*. *Rep. Tottori Mycol. Inst. (Jpn)* **12**: 151–160.

Gooday, G. W. 1978. The enzymology of hyphal growth. In J. E. Smith and D. R. Berry (Eds.), *The Filamentous Fungi*, Vol. 3, pp. 51–77. New York: Wiley.

Gooday, G. W., and D. Hunsley, 1971. Cellulose wall ingrowths in *Phytophthora parasitica*. *Trans. Brit. Mycol. Soc.* **57**: 178–179.

Grove, S. N. 1978. The cytology of hyphal tip growth. In J. E. Smith and D. R. Berry (Eds.), *The Filamentous Fungi*, Vol. 3, pp. 28–50. New York: Wiley.

Grove, S. N., and C. E. Bracker. 1970. Protoplasmic organization of hyphal tips among fungi: Vesicles and Spitzenkörper. *J. Bacteriol.* **104**: 989–1009.

Gunasekaran, M. 1973. Physiological studies on *Phymatotrichum omnivorum*: V. Effect of interactions of carbon dioxide, minerals and carbon source on growth in vitro. *Mycopathol. Mycol. Appl.* **50**: 249–253.

Hartman, R. E., N. T. Keen, and M. Long. 1972. Carbon dioxide fixation by *Verticillium albo-atrum*. *J. Gen. Microbiol.* **73**: 29–34.

Held, A., R. Emerson, M. S. Fuller, and F. H. Gleason. 1969. *Blastocladia* and *Aqualinderella:* Fermentative water molds with high carbon dioxide optima. *Science* **165**: 706–709.

Holler, J. R., and J. C. Brooks. 1980. Nutritional studies of *Pyconoporus cinnabarinus*. *Mycologia* **72**: 329–337.

Howard, R. J. 1981. Ultrastructural analysis of hyphal tip cell growth in fungi: Spitzenkörper, cytoskeleton and endomembranes after freeze-substitution. *J. Cell Sci.* **48**: 89–103.

Howard, R. J. and J. R. Aist. 1980. Cytoplasmic microtubules and fungal morphogenesis: Ultrastructural effects of methyl benzimidazole-2-ylcarbamate determined by freeze-substitution of hyphal tip cells. *J. Cell Biol.* **87**: 55–64.

Hunsley, D., and G. W. Gooday. 1974. The structure and development of septa in *Neurospora crassa*. *Protoplasma* **82**: 125–146.

Hwang, S. W., W. F. Kwolek, and W. C. Haynes. 1976. Investigation of ultralow temperature for fungal cultures: III. Viability and growth rate of mycelial cultures following cryogenic storage. *Mycologia* **68**: 377–387.

Israelstam, G. F. 1973. Response of nicotinamide adenine dinucleotide and cytochromes to aerobiosis and anaerobiosis on addition of glucose to *Saccharomyces cerevisiae*. *Biol. Plant (Prague)* **15**: 335–338.

Ivanov, V. N., and V. S. Podgorskii. 1977. Growth energetics of methanol-oxidizing yeast. *Mikrobiologiya* **46**: 203–209.

Johnson, B. F., G. B. Calleja, B. Y. Yoo, M. Zuker, and I. J. McDonald. 1982. Cell division: Key to cellular morphogenesis in the fission yeast, *Schizosaccharomyces*. *Int. Rev. Cytology* **75**: 167–208.

Kanetsuna, F., and L. M. Carbonell. 1970. Cell wall glucans of the yeast and mycelial forms of *Paracoccidioides brasiliensis*. *J. Bacteriol.* **101**: 675–680.

Kanetsuna, F., L. M. Carbonell, I. Azuma, and Y. Yamamura. 1972. Biochemical studies on the thermal dimorphism of *Paracoccidioides brasiliensis*. *J. Bacteriol.* 110: 208–218.

Kuzin, A. M., A. N. Nikitina, S. S. Yarov, and V. N. Primak. 1976. Stimulatory effect of chronic weak gamma-irradiation on the growth and development of *Aspergillus niger*. *Radiobiologiya* 16: 70–72.

Larsen, A. D., and P. S. Sypherd. 1974. Cyclic adenosine 3',5'-monophosphate and morphogenesis in *Mucor racemosus*. *J. Bacteriol.* 117: 432–438.

Lozinov, A. B., S. S. Eremina, and G. V. Sokolov. 1977. The effect of oxygen concentration on the physiological state of cells on the respiratory chain of *Candida mycoderma* yeast. *Mikrobiologiya* 46: 878–884.

Lyr, H. 1977. Mechanism of action of fungicides. In J. G. Horsfall and E. B. Cowling (Eds.), Plant Disease, Vol. 1, pp. 239–261. New York: Academic.

Maresca, B., E. Jacobson, G. Medoff, and G. Kobayashi. 1978. Cystine reductase in the dimorphic fungus *Histoplasma capsulatum*. *J. Bacteriol.* 135: 987–992.

Maresca, B., A. M. Lambowitz, V. B. Kumar, G. A. Grant, G. S. Kobayashi, and G. Medoff. 1981. Role of cysteine in regulating morphogenesis and mitochondrial activity in the dimorphic fungus *Histoplasma capsulatum*. *Proc. Natl. Acad. Sci. USA* 78: 4596–4600.

Maresca, B., G. Medoff, D. Schlessinger, and G. S. Kobayashi. 1977. Regulation of dimorphism in the pathogenic fungus *H. capsulatum*. *Nature* 266: 447–448.

Mussa, A. E. A., and P. E. Russell. 1977a. Influence of growth medium and radiation on sporulation of *Fusarium solani* f. sp. *phaseoli*. *Trans. Brit. Mycol. Soc.* 68: 462–464.

Mussa, A. E. A., and P. E. Russell. 1977b. The influence of pesticides and herbicides on the growth and virulence of *Fusarium solani* f. sp. *phaseoli*. *J. Agric. Sci.* 88: 705–709.

Prodromou, M. C., and E. S. Chapman. 1974. Effects of nitrogen sources at various temperatures on *Papulospora thermophila*. *Mycologia* 66: 876–880.

Qualman, S. J., H. E. Jones, and W. M. Artis. 1976. An automated radiometric microassay of fungal growth: Quantitation of growth of *Trichophyton mentagrophytes*. *Sabouraudia* 14: 287–297.

Raudaskoski, M., K. Pohjola, and I. Saarvanto. 1976. Effect of temperature and light on the mycelial growth of *Gyromitra esculenta* in pure culture. *Karstenia* 16: 1–5.

Righelato, R. C. 1975. Growth kinetics of mycelial fungi. In J. E. Smith and D. R. Berry (Eds.), *The Filamentous Fungi*, Vol. 1, pp. 79–103. New York: Wiley.

Rippon, J. W. 1968. Monitored environment system to control cell growth, morphology and metabolic rates in fungi by oxidation reduction potential. *Appl. Microbiol.* 16: 114–121.

Rodriguez-Kabana, R., P. A. Backman, and P. S. King. 1976. Antifungal activity of the nematocide ethoprop. *Plant Dis. Rep.* 60: 255–259.

Romano, A. H. 1966. Dimorphism. In G. C. Ainsworth and A. S. Sussman (Eds.), *The Fungi*, Vol. 2., pp. 181–209. New York: Academic.

Saltarelli, C. G., and C. P. Coppola. 1979. Effect of light on growth and metabolite synthesis in *Candida albicans*. *Mycologia* 71: 773–785.

Schlueter, K. 1977. Vapor action of the systemic fungicide triadimefon. *Z. Pflanzenkr. Pflanzenschutz*, 84: 612–614.

Seitz, L. M., D. B. Sauer, R. Burroughs, H. E. Mohr, and J. D. Hubbard. 1979. Ergosterol as a measure of fungal growth. *Phytopathology* 69: 1202–1204.

Shematek, E. M., J. A. Braatz, and E. Cabib. 1980. Biosynthesis of the yeast cell wall: I. Preparation and properties of β-(1-3)-glucan synthetase. *J. Biol. Chem.* 255: 888–894.

Sherald, J. L., N. N. Ragsdale, and H. D. Sisler. 1973. Similarities between the systemic fungicides triforine and triarimol. *Pestic. Sci.* 4: 719–727.

Shkidchenko, A. N. 1976. Effect of carbon dioxide on growth of *Candida utilis* during continuous chemical cultivation. *Mikrobiologiya* 45: 57–61.

Sierota, Z. H. 1977. Inhibitory effect of *Trichoderma viride* Pers. ex. Fr. filtrates on *Fomes annosus* (Fr.) Cke. in relation to some carbon sources. *Eur. J. For. Pathol.* 7: 164–172.

Stahmann, M. A., P. Abramson, and L. C. Wu. 1975. A chromatographic method for estimating fungal growth by glucosamine analysis of diseased tissues. *Biochem. Physiol. Pflanz.* 168: 267–276.

Stewart, P. R., and P. J. Rogers. 1978. Fungal dimorphism: A particular expression of cell wall morphogenesis. In J. E. Smith and D. R. Berry (Eds.), *The Filamentous Fungi*, Vol. 3, pp. 164–196. New York: Wiley.

Streiblova, E. 1966. Thesis; Czechoslovakia Academy of Sciences, Prague.

Tamimi, K. M., and S. A. Hutchinson. 1975. Differences between the biological effects of culture gases from several species of *Trichoderma*. *Trans. Brit. Mycol. Soc.* **64**: 455–464.

Tan, K. K. 1978. Light-induced fungal development. In J. E. Smith and D. R. Berry (Eds.), *The Filamentous Fungi*, Vol. 3, pp. 334–357. New York: Wiley.

Trinci, A. P. J. 1969. A kinetic study of the growth of *Aspergillus nidulans* and other fungi. *J. Gen. Microbiol.* **57**: 11–24.

Trinci, A. P. J. 1971. Influence of the width of the peripheral growth zone on the radial growth rate of fungal colonies on solid media. *J. Gen. Microbiol.* **67**: 325–344.

Trinci, A. P. J. 1974. A study of the kinetics of hyphal extension and branch initiation of fungal mycelia. *J. Gen. Microbiol.* **81**: 225–236.

Trinci, A. P. J. 1978. The duplication cycle and vegetative development in moulds. In J. E. Smith and D. R. Berry (Eds.), *The Filamentous Fungi*, Vol. 3, pp. 132–163. New York: Wiley.

Vance, C. P., and M. O. Garraway. 1973. Growth stimulation of *Armillaria mellea* by ethanol and other alcohols in relation to phenol concentration. *Phytopathology* **63**: 743–748.

van Uden, N. 1971. Kinetics and energetics of yeast growth. In A. H. Rose and J. S. Harrison (Eds.), *The Yeasts*, Vol. 2, pp. 75–118. New York: Academic.

Woodhead, S., and J. R. L. Walker. 1975. The effects of aeration on glucose catabolism in *Penicillium expansum*. *J. Gen. Microbiol.* **89**: 327–336.

9

REPRODUCTION

Reproduction results in the formation of new individuals. **Sexual reproduction** occurs via the fusion of compatible nuclei followed by meiotic division of the resulting zygote. If such fusions and meiosis are not involved, reproduction is **asexual**. Reproduction in fungi results not only in the formation of new individuals but may also give rise to agents of dispersal and structures which resist harsh environmental conditions. Sexual reproduction provides a mechanism for increasing genetic variation in populations. Fungi as a group exhibit an astonishing variety of reproductive mechanisms. Differences in sexual reproduction form the basis for fungal taxonomy; asexual structures are important in distinguishing members of the class Deuteromycetes. Most fungi reproduce by forming spores, which may be produced sexually or asexually or both depending on the species. In sexual reproduction, new individuals arise from structures such as zygospores, oospores, ascospores, and basidiospores. Asexual reproduction may occur by simple fragmentation of hyphae (this is also called vegetative reproduction) or by the formation of specialized structures called arthrospores, zoospores, sporangiospores, or conidia. In yeasts, asexual reproduction occurs by budding or fission.

Because of the diversity in reproductive mechanisms exhibited by fungi it is difficult to draw generalizations concerning the physiology of reproduction. The same environmental factors such as temperature, light, pH, aeration, and nutrients shown to have varied effects on fungal growth (Chapter 8) also have varied effects on both sexual and asexual reproduction. However, in her review Hawker (1966) proposes five generalizations concerning the effects of environmental stimuli on the initiation of the reproductive stage and on the development of reproductive structures: (1) Conditions favoring maximum mycelial growth may or may not also favor asexual reproduction but are usually unfavorable to sexual reproduction. (2) The range of any particular external condition allowing sporulation is usually narrower than that permitting mycelial growth and is sometimes narrower for sexual than for asexual reproduction. (3) Sporulation, and in particular, sexual reproduction, usually requires a higher threshold level of nutrients than does mycelial growth, but sporulation is usually inhibited at concentrations considerably below the maximum for growth. (4) Conditions favoring initiation of reproduction are not necessarily equally favorable for later stages in the development and maturation of reproductive bodies. (5) Since the conditions favoring vegetative growth, asexual sporulation, and sexual reproduction differ, it is theoretically possible to control the type of growth by suitable manipulation of the environment.

We shall use examples from the literature to illustrate asexual and sexual reproduction in fungi.

ASEXUAL REPRODUCTION

In this section we focus attention on the two most common types of asexual reproduction in fungi, the production of sporangiospores and conidia.

Zoosporogenesis in *Blastocladiella emersonii*

Members of the lower fungal groups such as Myxomycetes, Chytridiomycetes, Oomycetes, and Zygomycetes produce spores via the formation of a sporangium followed by cleavage of the enclosed cytoplasm into uninucleate spores. To illustrate this type of reproduction we shall discuss the asexual cycle of one of the most thoroughly studied of all fungi, *Blastocladiella emersonii*.

The asexual life cycle of *B. emersonii*, a member of the Chytridiomycetes, is diagrammed in Figure 9.1 and illustrated photographically in Figure 9.2. The following discussion is based on reviews by Lovett (1975) and Cantino and Mills (1976).

Zoospores of *B. emersonii* are highly differentiated cells with a distinctive arrangement of organelles. A posterior flagellum arises from a basal body, which is surrounded by a large mitochondrion (Figure 9.3). On the external side of the mitochondrion is an association of lipid granules and microbodies, presumably glyoxysomes; a double backing membrane lies on the outer side of this mass and frequently surrounds a portion of the mitochondrion as well. The lipid-microbody-mitochondrion association is called the side body, or symphyomicrobody, complex and may play a role in oxidizing lipids to obtain energy for flagellar motion. On top of the nucleus is a large nuclear cap filled with ribosomes. The remaining cytoplasm contains glycogen granules, small vesicles, and membrane-bound, DNA-containing structures called gamma particles.

The zoospores respire rapidly, utilizing endogenous reserves. Most of the resulting ATP is used to maintain motility rather than to biosynthesize macromolecules; there is negligible RNA and protein synthesis. Zoospores may remain active for several hours but eventually they retract their flagella and encyst. Encystment may be triggered by low temperatures, by exposure to monovalent cations, or by dilution of the population (Truesdell and Cantino, 1971). The entire encystment process occurs within 10 to 15 minutes and involves a striking cellular reorganization, including subdivision of the mitochondrion, disruption of the nuclear cap membrane and the resulting dispersal of ribosomes, appearance of dictyosomes, disruption of the side-

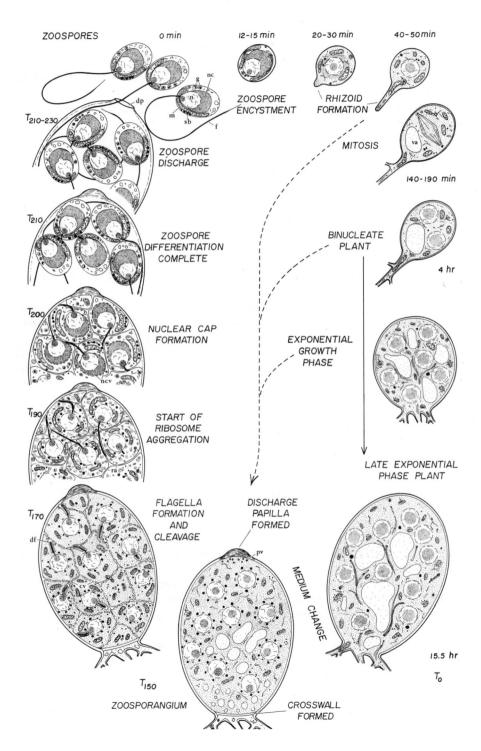

ZOOSPORES   0 min   12-15 min   20-30 min   40-50 min

ZOOSPORE ENCYSTMENT

RHIZOID FORMATION

$T_{210-230}$

ZOOSPORE DISCHARGE

MITOSIS

140-190 min

T_{210}

ZOOSPORE DIFFERENTIATION COMPLETE

BINUCLEATE PLANT

4 hr

T_{200}

NUCLEAR CAP FORMATION

EXPONENTIAL GROWTH PHASE

T_{190}

START OF RIBOSOME AGGREGATION

LATE EXPONENTIAL PHASE PLANT

T_{170}

FLAGELLA FORMATION AND CLEAVAGE

DISCHARGE PAPILLA FORMED

MEDIUM CHANGE

15.5 hr

T_0

T_{150}

ZOOSPORANGIUM

CROSSWALL FORMED

body complex, appearance of rough ER, and the deposition of new cell-wall material aided by the action of gamma particles which then disappear (Mills and Cantino, 1978, 1979).

The encysted zoospore germinates almost immediately, with the resulting germ tube becoming an anchoring rhizoid, and the phase of maximum growth begins. As the cyst increases in size its nucleolus becomes larger; and rough ER, vacuoles, dictyosomes and large microtubules become more numerous. Clusters of vesicles are found close to the cell wall, and it is postulated that they contain the enzymes and wall subunits necessary for wall synthesis. RNA and protein synthesis begin shortly after the germ tube emerges (Figure 9.4). In addition, mitosis is initiated during this period and occurs approximately every 1.75 hours until an average of 256 nuclei are present in the coenocytic thallus.

If the cells are grown on a peptone–yeast-extract–glucose (PYG) medium, roughly 10 percent of the population will begin to form zoosporangia approximately 16 hours after encystment. However, transfer of the cells to a dilute medium containing only mineral salts will induce sporangial formation in almost all cells simultaneously. Under conditions of crowding or high concentrations of bicarbonate or cations, a thick-walled, resistant sporangium is formed which may persist for some time before spores are produced. Under more favorable growth conditions, however, a thin-walled "ordinary colorless" sporangium is formed in which zoospores develop and are released very rapidly. It is this latter developmental pathway that we will follow in this discussion. Even very young germlings can be induced to differentiate into "mini-sporangia," forming only one zoospore, upon transfer of the cells from the PYG medium to dilute phosphate buffer. The ultrastructure of this microcyclic sporogenesis has recently been described by Barstow and Lovett (1978).

In the production of a sporangium, a septum is laid down separating the upper multinucleate cell from the rhizoidal system. The rhizoids soon lose their organelles either by degradation or by being drawn into the upper cell. In addition, a discharge papilla is formed in the wall of this cell by the action of secretory vesicles. Associated with sporangial differentiation is a slow decrease in the dry weight as well as in the content of RNA and protein (Figure 9.4).

By about 18 hours (Figure 9.1, T_{150}), zoospores are beginning to be formed within the sporangial cytoplasm. Each flagellum arises from a centriole functioning as a basal body. Numerous vesicles coalesce, forming individual nuclear and plasma membranes, thus cleaving the cytoplasm into prespore areas. Mitochondria fuse to

---

Figure 9.1 Zoosporangial cycle of *Blastocladiella emersonii*. The early stages are disproportionately enlarged to show details, and the rhizoid systems are not shown in their entirety. The timing from 0 to 4 hours indicates germination and early growth in synthetic medium; the timing from T_0 to T_{230} represents the time in minutes after the inducing medium shift, starting at 15.5 hours. The broken lines indicate that cells shifted to inducing medium at the respective stages will begin zoosporogenesis and zoospore differentiation but in proportionately smaller sporangia and with proportionately fewer zoospores produced, depending on the number of nuclei present (i.e., 1 to 256). *cv,* cleavage vesicles and furrows; *df,* developing flagella; *dp,* discharge pore; *f,* flagella; *g,* golgi; *m,* mitochondrion; *n,* nucleus; *nc,* nuclear cap; *ncv,* nuclear cap membrane vesicle; *pv,* papilla vesicles; *ra,* ribosome aggregations; *sb,* side body granules of lipid and complex microbody; *va,* vesicle. (Lovett, 1975)

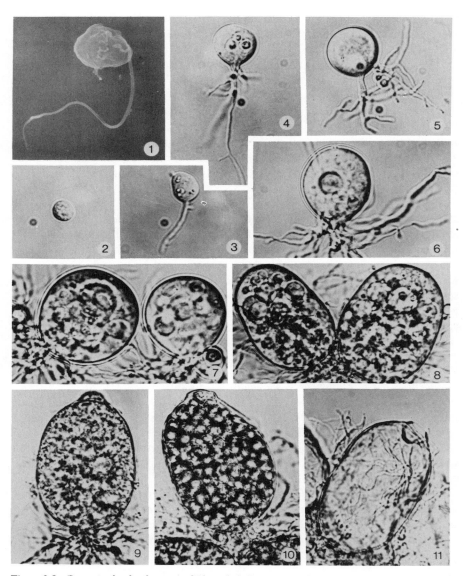

Figure 9.2 Stages in the development of *Blastocladiella emersonii* depicted photographically. (1) Zoospore by scanning electron microscopy. × 2300. (2) Encysted zoospore. × 1070. (3–11) Cells at 3, 6, 9, 12, 15, 18, 21, 23, and 24 hours after encystment, respectively. (3–7) × 1070. (8–11) × 670. (9) Mature thallus with zoosporangium and papilla. (10) Zoosporogenesis. (11) Thallus after release of spores through exit pore. (Cantino and Mills, 1978)

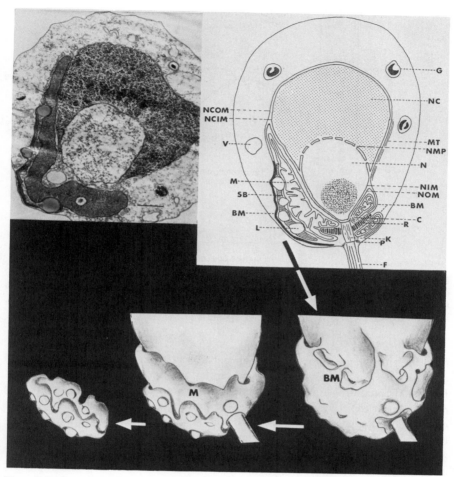

Figure 9.3 Structure of the zoospore of *Blastocladiella emersonii*. Top right, composite diagram of cell components: *NCOM*, nuclear cap outer membrane; *NCIM*, nuclear cap inner membrane; *V*, vacuole; *M*, mitochondrion; *SB*, side-body matrix (symphyomicrobody); *BM*, backing membrane; *L*, lipid globule; *F*, flagellum; *P*, prop; *K*, kinetosome; *R*, rootlet; *C*, centriole; *NOM*, nuclear outer membrane; *NIM*, nuclear inner membrane; *N*, nucleus; *NMP*, nuclear membrane pore; *MT*, microtubule (occurs in triplets); *NC*, nuclear cap; *G*, gamma particle. Top left, longitudinal section through a zoospore fixed in OsO₄ and uranyl acetate. (Truesdell and Cantino, 1970). Bottom, three-dimensional organization (Cantino and Truesdell, 1970) of the side body pictured as if it were being progressively dismantled (from right to left) so as to yield (far left) an isolated side body-lipid complex (i.e., a symphyomicrobody-lipid complex); legends as for top right. (Cantino and Mills, 1976)

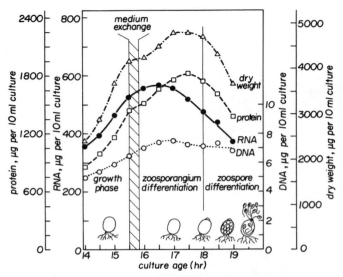

Figure 9.4 Changes in the dry weight, protein, RNA, and DNA content of cells of *Blastocladiella emersonii* before and after induction of zoosporogenesis. The cells were shifted from PYG growth medium to half-strength dilute salts medium at 15.1 to 15.83 hours. (Redrawn from Murphy and Lovett, 1966, as presented in Lovett, 1975.)

form the single large mitochondrion of each zoospore, gamma particles form from other vesicles, and ribosomes begin to cluster around two-thirds of the nucleus. Later, additional vesicles coalesce around the ribosomal mass forming the membrane of the nuclear cap. Once the ribosomes are thus sequestered, translation of mRNA ceases due to the presence of a low molecular weight inhibitor. Lastly, the side-body complex forms by the association of the mitochondrion, lipid droplets, microbodies, and the backing membrane. Zoospore differentiation is now complete and the spores are quickly released from the sporangium upon dissolution of the discharge papilla. The entire cycle is completed within 19 hours.

Development of Conidia

The study of conidial development is complicated by the tremendous variety in the types of conidia produced by fungi. A superbly illustrated review of this topic has been published by Cole and Samson (1979).

There are two principal types of conidial development: blastic and thallic. In blastic development a conidium arises when a portion of the hyphal or conidiophore wall weakens and expands by blowing out in a balloonlike manner (Figure 9.5). In thallic development, on the other hand, the conidium forms by the modification of a preexisting hyphal segment.

The formation of a blastic conidium occurs by a process similar to that involved in the growth of hyphal tips (Figure 9.6). Clusters of large and small vesicles aggre-

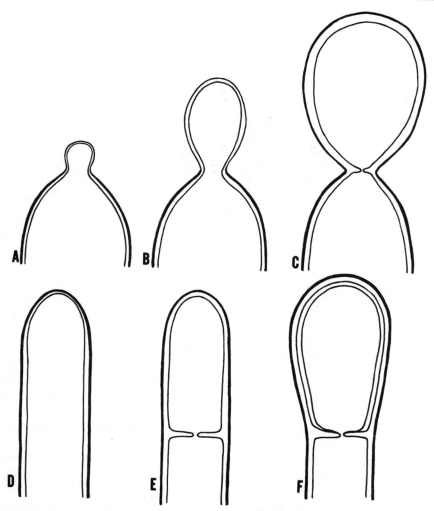

Figure 9.5 Blastic and thallic development. (A–C) In blastic development, a conidium differentiates by blowing out part of the fertile hypha or conidiogenous cell. (D–F) In thallic development, a conidium differentiates by the conversion and disarticulation of a pre-existing segment of a fertile hypha. (Cole and Samson, 1979)

gate at the site of conidial initiation. These vesicles are thought to contain the enzymes and wall subunits necessary for wall growth. Once expansion has begun, the vesicles either disperse, resulting in spherical growth, or remain polarized, yielding a cylindrical cell. Associated with changes in the shape of the developing conidium are changes in the orientation of components called rodlets in the inner wall layer. In *Gonatobotryum apiculatum*, for example, the rodlets in the conidium are arranged in a circumferential pattern during its initial expansion but later change

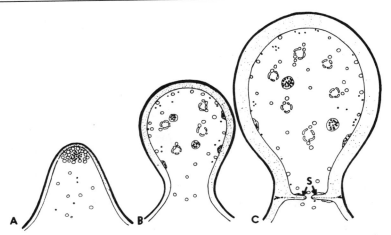

Figure 9.6 Interpretation of protoplasmic differentiation during blastic conidial formation. S, septum. (Cole and Samson, 1979)

so that they are parallel to the long axis during conidial elongation (Figure 9.7). These changes may reflect modifications in the stresses applied to the outer wall layer as growth proceeds.

Structures resembling dictyosomes appear early in the developing conidium initial, and these structures begin to bleb off additional vesicles. The inner wall layer begins to thicken, and a septum forms between the conidial initial and the parent hypha. Migration of one or more nuclei into the conidium coincides with septum formation. If the outer hyphal wall layer breaks during conidial differentiation, development is enteroblastic. On the other hand, if all hyphal wall layers are also found in the conidium at maturity, development is holoblastic. The conidium is finally released from the parent either by a splitting at the septum or by a rupture in the supporting cell. The many different shapes of blastic conidia reflect differences in the rates of wall rigidification and plasticization as well as localized changes in the supply of wall precursors.

In thallic development, whole hyphal segments, either terminal or intercalary, are converted into a single conidium or chains of conidia. Conidiogenesis occurs once apical growth ceases and involves thickening of the inner wall and the development of delimiting septa (Figure 9.8). All hyphal wall layers may be involved in the formation of the conidial wall (holoarthric development) or the outer wall may break, leaving only the inner hyphal wall surrounding the conidium (enteroarthric development). Examples of the wide variety of blastic and thallic conidia are illustrated in Figure 9.9 (Cole and Samson, 1979).

We will use members of the genus *Aspergillus* to illustrate certain aspects of conidial development. The asexual cycle of A. *niger* is shown in Figure 9.10. Swelling of certain hyphal segments gives rise to foot cells, which elongate into conidiophores. The apex of each conidiophore becomes spherical and bears numer-

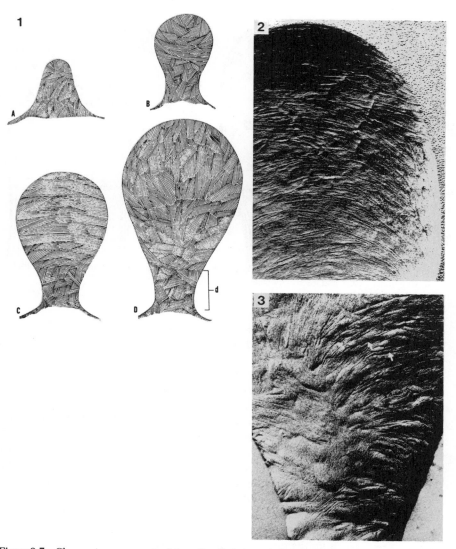

Figure 9.7 Changes in components of the cell wall during conidial development. 1 (A–D): Interpretation of changes in rodlet orientation during blastic conidial development in *Gonatobotryum apiculatum*. 2, 3. Freeze-fractures of a conidial initial (2) and a mature conidium (3) of *Gonatobotryum apiculatum*, revealing different orientations of rodlet fascicles. Arrows in 3 indicate radiating rodlet fascicles. × 52,200 and × 55,000, respectively. (Cole and Samson, 1979)

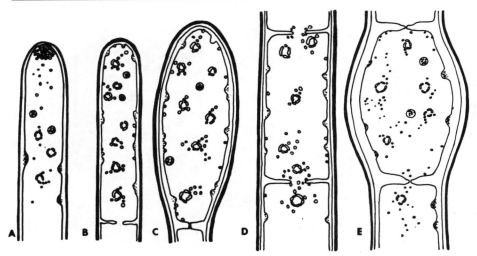

Figure 9.8 Interpretation of protoplasmic and wall differentiation during formation of holothallic, terminal conidia (A–C), and intercalary conidia (D,E). (Cole and Samson, 1979)

ous phialides (Figure 9.11). Conidia break through the tips of the phialides (enteroblastic development) and stick together in chains.

Axelrod and co-workers (Axelrod, 1972; Axelrod et al., 1973) have demonstrated that conidiophores of A. *nidulans* will not be produced until the culture reaches a certain minimum age. Colonies grown in liquid media can be induced to form conidiophores by transferring them to a solid medium. This is most easily accomplished by filtering the colonies onto membrane filters and then placing the filters on absorbent pads soaked in the same medium. Colonies at least 24 hours old will begin forming conidiophores immediately upon transfer (Figure 9.12). Colonies younger than 24 hours at the time of transfer will not form these structures until they have reached an age of 24 hours. That is, the genes for conidiogenesis are capable of being switched on only after the colony is 24 hours old. At this time the organism attains developmental competence. The mechanism by which this occurs is not known. It may involve the sequential turning on of genes or specific gene activation resulting from the interplay of metabolic pathways.

Smith and his colleagues (see Smith, 1978, Smith and Berry, 1974, for reviews) developed a synthetic medium for the production of conidia of A. *niger* in submerged culture and used this to study nutritional factors involved in conidiogenesis. Conidiophores are formed when either the carbon or nitrogen source is limiting. For example, conidiation occurs readily in a basal glucose medium completely lacking in nitrogen. However, conidiation is inhibited when ammonium ions are present at concentrations above 48 mg atom nitrogen/liter. This inhibition must be due to factors other than concentration because inhibition can be overcome by adding various amino acids as well as glyoxylate or intermediates of the Krebs cycle. When

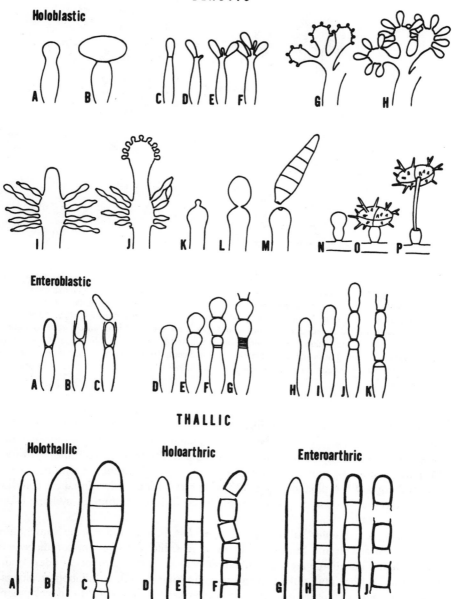

Figure 9.9 Different types of blastic and thallic development. (Cole and Samson, 1979)

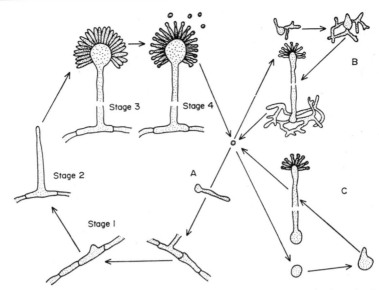

Figure 9.10 Summary of induced morphogenetic sequences leading to conidiophore development in *Aspergillus niger* under submerged agitated coniditons. (A) Sequence of morphological changes in replacement fermenter culture: *Stage 1*, conidiophore initiation; *stage 2*, conidiophore elongation; *stage 3*, vesicle and phialide formation; *stage 4*, conidiospore production. (B and C) Forms of microcycle conidiation. In (B) a branched mycelial system and a mature conidiophore are produced from an enlarged conidium. Treatment consists of incubation at 41° C for 15 hours followed by 30° C for 12 hours. In (C) a mature conidiophore is produced from an enlarged conidium in the complete absence of vegetative development. Treatment consists of incubation at 44° C for 48 hours followed by 30° C for 15 hours. (Smith and Anderson, 1973)

individual amino acids are used in place of ammonium as the nitrogen source, a variety of effects are observed. Normal conidiation occurs with glycine, leucine, isoleucine, serine, citrulline, proline, glutamate, threonine, ornithine, tyrosine, histidine, lysine, aspartate, or asparagine. With arginine, valine, or alanine, development stops at the phialide stage. With phenylalanine, only immature conidiophores develop. No conidiophores are produced with methionine or cysteine (Galbraith and Smith, 1969).

By using a fermenter culture system in which the medium can be sequentially replaced, Anderson and Smith (1971) found that different stages of conidiogenesis have different nutritional requirements. Foot cells are induced to form under conditions of low nitrogen but not when carbon is limiting as well. Conidiophore elongation occurs once the nitrogen source becomes exhausted, but no further development occurs in this medium. Phialide formation occurs when the medium is replaced with one containing a nitrogen source and an intermediate of the Krebs cycle, such as citrate, as the carbon source. However, no conidia are formed unless the colonies are transferred to a medium containing glucose as the carbon source and nitrate as

Figure 9.11 Stages of phialide and conidial development in *Aspergillus* species showing synchrony of conidiogenous cell initiation. 1, *A. fumigatus*; 2, *A. flavus*. Emergence of conidia through ruptured phialide apex (3, 4) and formation of conidial chains (5) are also demonstrated by *A. flavus*. Typical dense clusters of phialides and conidia are shown in 6 for *A. fumigatus. col*, collarette; *cn*, conidial initial. ×2970, ×3960, ×6600, ×1900, ×2160, and ×823, respectively. (Cole and Samson, 1979)

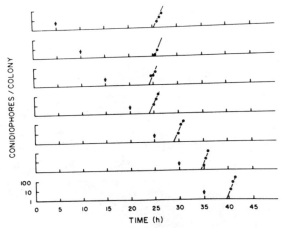

Figure 9.12 Competence of growing colonies of *Aspergillus nidulans* to be induced to develop conidiophores. Colonies of wild-type strain FGSC-4 grown in liquid-nutrient medium at 36° C were induced
by filtration onto membrane filters at the time indicated (↓). Filters were placed on absorbent pads
saturated with the same medium, and colonies were incubated in petri dishes at 36° C. (Axelrod et al.,
1973)

the nitrogen source. In addition, the number of conidia decreases tenfold when the
dilution rate is doubled (Ng et al., 1972).

If conidia of A. *niger* are incubated at increased temperatures, germination is
inhibited to the extent that at 44° C the conidia merely swell instead of producing
germ tubes (Figure 9.10). If after 48 hours at this temperature the conidia are
transferred to a 30° C incubator, they germinate directly into conidiophores. Use of
this microcyclic conidiation may simplify the study of mechanisms underlying conidiophore development in this genus.

SEXUAL REPRODUCTION

The literature abounds with studies of sexual reproduction in the various taxonomic
groups of fungi. To illustrate the physiology of sexual reproduction, we will focus on
four representatives in which hormones have been implicated. These are *Achlya*,
Allomyces, members of the Mucorales, and certain yeasts. Good summaries of this
work have been published by Gooday (1974), Crandall and co-workers (1977), van
den Ende (1976), and Bu'Lock (1976).

Achlya

Raper's pioneering work during the 1940s and 1950s with the oomycete *Achlya* was
the first to demonstrate that hormones are involved in sexual reproduction. Initially
this work was done using A. *ambisexualis* and A. *bisexualis*. These species are
predominantly heterothallic, meaning that a thallus produces either, but not both,

male or female sexual structures, called antheridia and oogonia, respectively. Thus, two compatible thalli are required in order for sexual reproduction to occur. Other species are homothallic; each thallus produces both male and female structures and thus is self-compatible.

In *Achlya* the vegetative thallus is a diploid, branching mycelium. When compatible thalli grow close to each other branched, thin-walled antheridial initials differentiate on the male strain. These specialized hyphae become attracted toward regions of the female thallus which soon differentiate into spherical oogonial initials (Figure 9.13). Once contact is made, cross walls form on each set of initials, delimiting antheridia and oogonia. Meiosis apparently occurs at this time, giving rise to uninucleate haploid fertilization tubes in the antheridia and 1–20 uninucleate haploid oospheres in the oogonia. Plasmogamy occurs between individual fertilization tubes and oospheres, resulting in diploid oospores. These eventually are released from the oogonium, develop germ tubes, and grow into a new mycelium (Raper, 1952; Barksdale, 1966).

In a classic set of experiments, Raper (1940) showed that both the formation of the sex organs and their subsequent interactions are governed by diffusible hormones. He exposed male and female thalli to filtrates of each other in various sequences and combinations and noted whether or not a response occurred. Raper concluded from these experiments that the following series of events took place: (1) The entire sexual process is initiated by the production by the female strain of a substance called hormone A, which induces the formation of antheridial hyphae on the male strain. (2) After induction the male strain secretes a different hormone, hormone B, which induces the formation of oogonial initials on the female strain. (3) After the production of oogonial initials the female strain secretes hormone C, which attracts the antheridial hyphae and delimits the antheridia themselves. (4) The male strain then produces hormone D, which stimulates the delimitation and further development of the oogonia.

Later work, particularly by Barksdale (1963a), has suggested modifications to Raper's hypothesis. For example, there is evidence that hormone A may function both as the attractant of antheridial hyphae and as the stimulus for delimiting antheridia—that is, hormone C may not exist. Because antheridial branches are also attracted to mixtures of amino acids, there is some confusion about this point. There is also some doubt as to the existence of hormone D (Barksdale, 1969; van den Ende, 1976).

Hormones A and B have been isolated, characterized, and given the names antheridiol and oogoniol, respectively (Figure 9.14). Both hormones are steroids. Although steroid hormones are common in animals, *Achlya* is unique in the plant and fungal kingdoms in having steroid-mediated reproduction (Gwynne and Brandhorst, 1980). Two forms of oogoniol with roughly equal activity have been identified. Antheridiol, on the other hand, is quite specific. Several stereoisomers and precursors have been synthesized which have much lower potency than antheridiol itself (Arsenault et al., 1968; Edwards et al., 1969; McMorris et al., 1975). Antheridiol is the most thoroughly studied of these hormones, and we shall briefly discuss its hypothesized mode of action.

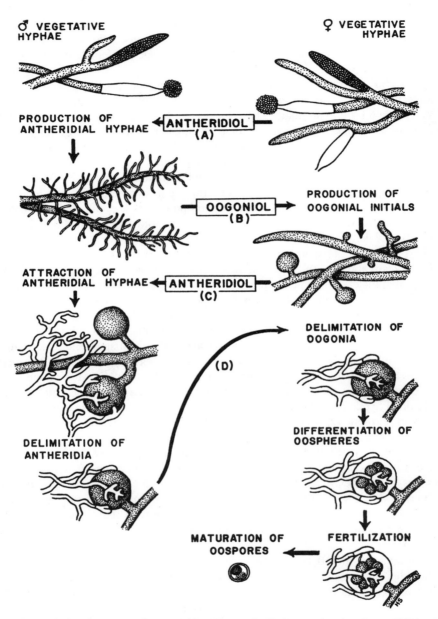

Figure 9.13 Hormones and sex in *Achlya*. (Drawing by H. Stempen based on Raper, 1951.)

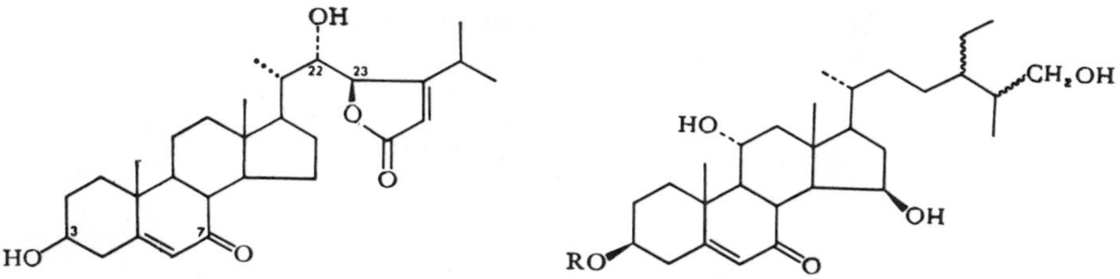

Figure 9.14   The structure of antheridiol (left) and oogoniol (right). (van den Ende, 1976)

Transport of antheridiol into the cell is a key prerequisite for the induction of antheridial branches. Heterothallic strains of *Achlya*, which respond to antheridiol, as well as self-conjugating strains appear to take up antheridiol quite rapidly. On the other hand, strains not readily responding to antheridiol fail to take up the hormone in significant amounts (Table 9.1). As shown, antheridiol is taken up relatively rapidly by the homothallic strains E15 and E22 and by the strongly male strain E87. It is taken up more slowly by strains 9 and 10, which are also capable of producing oogonial initials. Antheridial branching is not induced in strain 10 even at the concentration of 10,000 units per ml. In contrast, branching is induced by as little as 1 unit per ml in strains E87 and E22 (Barksdale, 1963b).

Although antheridiol is taken up at different rates by various strains, work with tritiated antheridiol shows that it binds equally well to mycelia of both male and female strains. This implies that uptake is at least a two-part process in which antheridiol first binds to the cell membrane and then is transported into the cell. Presumably the female strains are unable to accumulate the hormone in sufficient quantities to be effective (Barksdale and Lasure, 1973).

Once inside the cell, antheridiol has been shown to activate the genome, although the mechanism is unknown. Indeed, both transcription and translation are required for the induction of antheridial branches (Timberlake, 1976). Horgen (1977) has indicated that within 30 minutes after antheridiol exposure there is an activation of ribosomal genes and increased rRNA synthesis. This is shortly followed by the acetylation of basic proteins (suggesting gene activation) and increased mRNA and protein synthesis. This latter event coincides with the appearance of antheridial initials. However, analyses of mRNA and protein molecules from control and antheridiol-treated hyphae reveal that antheridiol does not markedly change the type of either mRNA or protein synthesized (Rozek and Timberlake, 1980; Gwynne and Brandhorst, 1980). Perhaps antheridial formation is dependent on a quantitative increase in the rate of RNA synthesis and/or on post-translational modification of proteins already present; differential gene activation does not appear to be involved (Gwynne and Brandhorst, 1980).

One of the proteins whose activity increases in response to antheridiol is cellulase. There is a sharp rise in endogenous cellulase activity at the time of differentiation of antheridial initials (Thomas and Mullins, 1967). The enzyme has been localized in dictyosomes, in vesicles, in the area between the plasma membrane and the cell wall, and outside the cell wall itself (Nolan and Bal, 1974). Electron micrographs

Table 9.1   The Rate of Uptake of Antheridiol by Five
            Strains of *Achlya*

Strain Number	Initial Concentration of antheridiol (units/ml) x 10^4		
	1	3	10
E15 (homothallic	180[a]	450	1600
E22 (homothallic)	180	400	1600
E87 (strongly male)	100	200	660
9	55	130	600
10	30	90	330

Source: Adapted from Barksdale (1963b).

[a]Rate of uptake expressed as units of antheridiol taken up per
mg dry weight per minute. The values given are the mean of
those obtained in three experiments.

(Figure 9.15) depicting freeze-fractured faces of antheridiol-treated *Achlya* hyphae
show aggregates of vesicles adjacent to the cell wall, fusing of vesicles with the
plasma membrane, thinning of the wall in the vicinity of these vesicles, and bulging
of the thinned wall to form an antheridial initial (Mullins and Ellis, 1974). Inhibitory treatments that prevent cellulase induction also prevent the formation of antheridial initials (Kane et al., 1973). These observations provide the basis for the hypothesis that one effect of antheridiol is to increase the activity of cellulase, which is
transported via vesicles to the cell wall where it causes wall loosening at the site of
antheridial initiation (Mullins and Ellis, 1974).

Allomyces

Allomyces is a type of chytrid whose sexual reproduction has been studied in detail.
Members of this genus have a mycelial stage in both the haploid and diploid phases
of the life cycle (Figure 9.16). In the diploid, or sporophyte, phase asexual reproduction occurs by the production of diploid, flagellated zoospores in zoosporangia.
Under harsh environmental conditions such as low water availability or increased
temperature, thick-walled meiosporangia (also called resistant sporangia) are formed
on the diploid thallus (Emerson, 1941). The occurrence of meiosis in these sporangia results in the formation of haploid flagellated zoospores which grow into the
haploid, or gametophyte, thallus. When nutrients such as nitrogen become limiting
(Machlis et al., 1966) or when the temperature is shifted upward (Ojha et al., 1980),
male and female gametangia differentiate one above the other on the haploid thallus.
Allomyces is thus homothallic. Upon dilution of the medium, flagellated male and
female gametes are released. These cells soon fuse, and the resulting zygote grows
into the new sporophyte thallus.

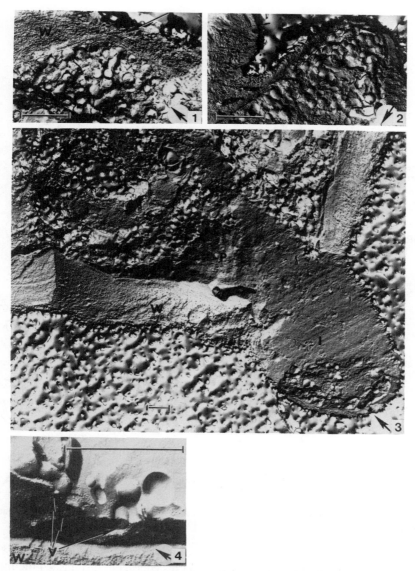

Figure 9.15 Freeze-fracture of *Achlya* hyphae. (1) Longitudinal fracture of a hormone-induced hypha exhibiting an aggregation of vesicles and the associated wall thinning indicated by the arrow. × 19,500. (2) Longitudinal fracture of a hormone-induced hypha exhibiting a bulge at the site of wall thinning. × 17,700. (3) A longitudinal fracture of an antheridiol-induced hyphal initial (I) and the parent hypha. × 8800. (4) An oblique fracture of a hormone-induced hypha exhibiting profiles of exocytosis. × 52,000. (Mullins and Ellis, 1974)

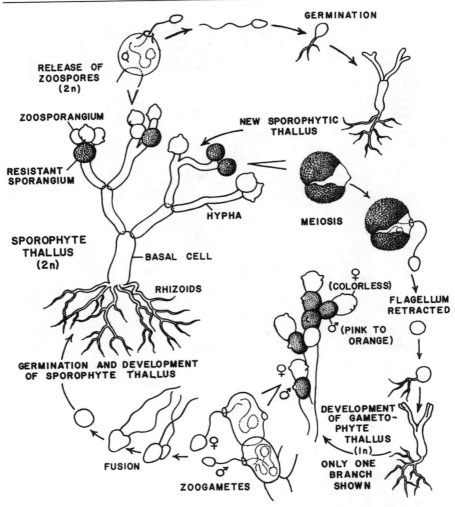

Figure 9.16 Life cycle of *Allomyces arbuscula*. (Drawing by H. Stempen adapted in part from Emerson, 1941.)

In the late 1950s and 1960s Machlis and his colleagues investigated the mechanism by which the male and female gametes come together. They found that within 5 minutes, male gametes are attracted to a piece of dialysis membrane separating the gametes from a solution in which the female gametangia had released their gametes. It appeared that a specific, diffusable, attractant was produced by the female (Machlis, 1958). This bioassay technique was facilitated by the use of mutant strains of *Allomyces* that were largely unisexual. The attractant was named "sirenin" and was found to be a bicyclic sesquiterpendiol (Figure 9.17) (Machlis and Rawitscher-Kunkel, 1967; Machlis et al., 1968; Nutting et al., 1968).

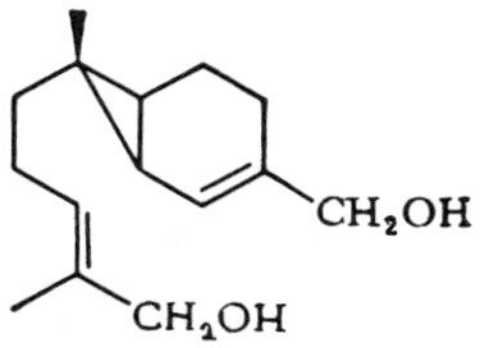

Figure 9.17   The structure of sirenin. (van den Ende, 1976)

Sirenin is effective at very small concentrations (Table 9.2). The optimal concentration for attraction of male gametes appears to be 10^{-8} M, although densely populated suspensions of male gametes are attracted to concentrations as low as 10^{-10} M (Carlile and Machlis, 1965). Sirenin can be secreted by the female gamete for at least 4 to 6 hours, and there is evidence that the hormone is continuously synthesized from enzymes already present in the female gamete at the time of release from the gametangium (Pommerville, 1977).

Little is known about the mechanism by which sirenin causes the attraction of male gametes, although they can rapidly inactivate the hormone. Once the males get very close to a female they are less attracted to it, but mere random motion ensures contact (Ross, 1979). Klapper and Klapper (1977) have reported that male gametes produce a natural inhibitor of sexual attraction, which they name "keerosin," that may function to prevent clustering of the males around the females and thus promote an even distribution of male gametes in the population.

Once contact of the gametes occurs, the male becomes amoeboid and moves over the female until certain regions near the flagellar end are in the proper juxtaposition. As soon as the cells thus "recognize" each other then plasmogamy occurs (Bu'Lock, 1976). This fusion terminates the synthesis and excretion of sirenin (Ross, 1979).

Pommerville (1977) has demonstrated that if males are prevented from fusing, they can produce an attractant for females. Such production may continue for several

Table 9.2   Response Range of Concentrations of Male Gametes of *Allomyces* to Various Concentrations of Sirenin

Male Gametes/ml	Sirenin Concentration (M)[a]							
	0	10^{-10}	10^{-9}	10^{-8}	10^{-7}	10^{-6}	10^{-5}	10^{-4}
10^6	4	35	159	365	∞	∞	∞	71
5×10^5	1	14	72	246	573	∞	∞	20
2×10^5	0	0	22	82	148	413	301	4
10^5	0	0	5	12	58	238	64	2
5×10^4	0	0	1	0	17	169	7	2

Source: Carlile and Machlis (1965)

[a]The figures indicate the number of male gametes settled on 0.13 mm^2 of membrane after 1 h of chemotaxis. ∞ means too great to be counted accurately.

hours and provides a fail-safe mechanism for ensuring that sexual reproduction occurs (Ross, 1979).

Mucorales

An interesting variation on the role of hormones in fungal sexual reproduction is found in the order Mucorales. Members of this order reproduce sexually by the process of gametangial copulation. In heterothallic species the two sexes are morphologically indistinguishable and are called "plus" and "minus." When hyphae of two compatible strains grow close together they each produce specialized hyphae, called zygophores, which become attracted to each other. When the two zygophores come into contact, the tip of each becomes swollen and a cross-wall delimits a terminal cell, called a gametangium, which contains a single haploid nucleus (Figure 9.18). The wall between the gametangia soon dissolves, and the nuclei fuse.

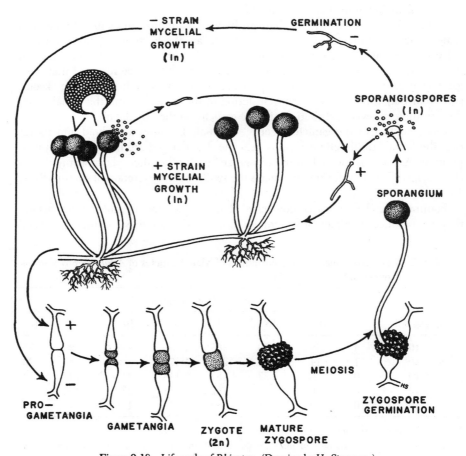

Figure 9.18   Life cycle of *Rhizopus*. (Drawing by H. Stempen.)

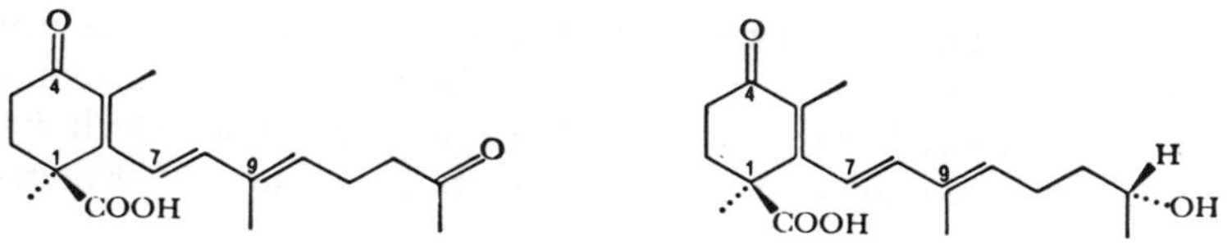

Figure 9.19 Structure of trisporic acids B (left) and C (right). (van den Ende, 1976)

The wall of the zygote becomes characteristically thick and spiny, and the resulting structure is called a zygospore. Meiosis usually coincides with germination of the zygospore, and a sporangium containing haploid spores is produced. Asexual reproduction occurs by the production of additional sporangia on the mycelia. A similar type of life cycle occurs in homothallic species (van den Ende, 1976).

Investigations of several mucoraceous species by Burgeff (1924) in which cultures were separated by a collodion membrane indicated that a diffusable, volatile chemical produced by each strain stimulated the development of zygophores in the other strain and also caused the zygophores to grow toward each other. Even members of different species are attracted toward each other, which suggests that one or more hormones is common to this order of fungi.

In 1967, the zygophore-inducing hormones were identified and named trisporic acid B and trisporic acid C (Figure 9.19). A third molecule, trisporic acid A, which presumably lacks the functional group on the side chain has also been isolated. However, it has very low biological activity and is thought to be merely a transformation product of trisporic acid C. Trisporic acids from *Blakeslea trispora* and *Mucor mucedo* are indistinguishable. Table 9.3 not only shows that both the cis and trans forms are effective but also that each trisporic acid can induce zygophore formation in both the plus and minus strains. Thus, these hormones are not sex-specific. They are also effective even on homothallic species (van den Ende, 1976).

The biosynthesis of trisporic acids provides a good example of reproduction biochemistry. Both the plus and minus strains must be present in order for trisporic acids to be produced; no hormone synthesis occurs in strains grown separately. Labeling studies using ^{14}C indicate that trisporic acid synthesis occurs in both the

Table 9.3 Comparative Bioassays on Plus and Minus Strains of *Mucor mucedo*

Compounds	Zygophores/μg	
	Plus	Minus
9-*cis*-Trisporic acid C	150	550
9-*trans*-Trisporic acid C	150	420
9-*cis*-Trisporic acid B	260	910
9-*trans*-Trisporic acid B	160	450
Trisporic acid A	50	120

Source: Bu'Lock et al. (1972).

plus and minus strains and that each mating type contributes about equally to trisporic acid production. Sutter and colleagues discovered that each strain secretes small amounts of a metabolite, now referred to as a prohormone, which diffuses to the opposite strain and is there converted to trisporic acid (Sutter, 1970; Sutter et al., 1973, 1974).

The proposed biosynthetic pathway for trisporic acid production in M. *mucedo* and B. *trispora* is illustrated in Figure 9.20. Products of the mevalonic acid pathway

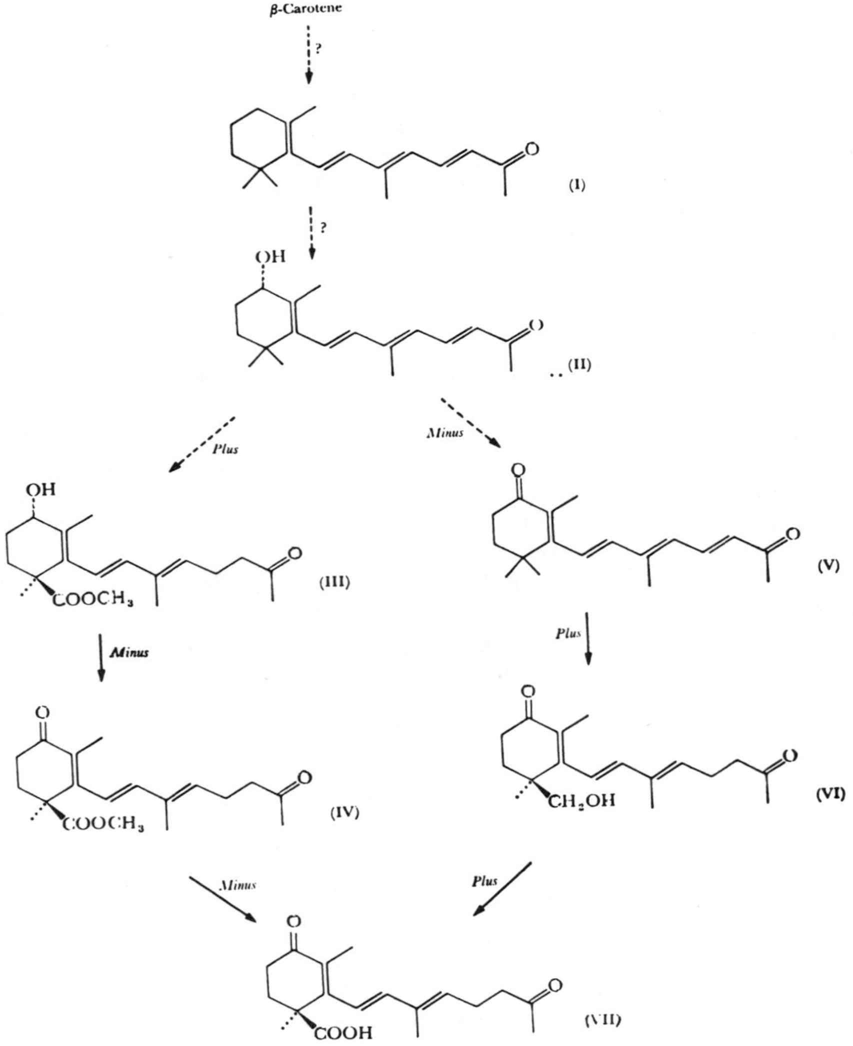

Figure 9.20 Proposed biosynthetic pathway of trisporic acid B in mated cultures of Mucor *mucedo* and Blakeslea *trispora*. (van den Ende, 1976)

such as β-carotene have been postulated as possible precursors for trisporic acids. In the plus strain, the prohormone compound III is produced and diffuses to the minus strain where compound IV and subsequently trisporic acid B (compound VII) is synthesized. On the other hand, the minus strain lacks the enzyme to synthesize compound III and instead produces compound V. This prohormone diffuses to the plus strain, which completes the synthesis of trisporic acid via compound VI. Thus, each strain synthesizes and secretes an "imperfect" hormone that can be converted to trisporic acid only by the opposite mating type (Ross, 1979). Once synthesized, trisporic acids have several effects. First, they derepress the genes responsible for zygophore production and for the synthesis of recognition factors on the zygophore surface. Trisporic acids also derepress the synthesis of prohormones. Not only does this result in increased trisporic acid synthesis, but the gradient of prohormones is responsible for the directional growth of the zygophores (Bu'Lock, 1976; van den Ende, 1976).

A similar hormone system is present in homothallic species as well. Here, attraction and fusion occurs between a small, budlike gametangium which has the same mating type as the vegetative mycelium and a larger, out-growing gametangium in which the opposite mating type is induced. There is much interest in understanding how this differential gene activation within a single coenocytic thallus is achieved (Bu'Lock, 1976).

Yeasts

Certain yeasts such as *Hansenula wingei*, *Saccharomyces cerevisiae*, and *Schizosaccharomyces pombe* reproduce sexually by conjugation. In this process, cells of opposite mating type adhere and form protuberances at the point of contact. The end walls of these protuberances dissolve, and a conjugation tube is formed. The nuclei fuse, usually in the center of this tube.

Mating of *Hansenula wingei* is triggered by nitrogen starvation and involves cellular recognition factors on the surface of the two mating strains. The recognition factors are glycoproteins that are constituitively present on the surface of haploid, but not diploid, cells. Their role is to bring the two strains into close contact in order to initiate cell fusion. Complementary agglutination results in wall deformation at the points of contact. The walls then extend, forming protuberances that develop into the conjugation tube. The two nuclei fuse within this tube.

The mating of *S. cerevisiae* occurs in a nutrient-rich medium and is controlled not only by agglutination factors but by diffusable hormonelike substances released by the cells. In this species the two mating types are designated "a" and "α." Strain α produces a peptide called α-factor which causes type a cells to become arrested in the G_1 phase of the cell cycle and to form protuberances. Strain a produces one or more a-factors which are proteinaceous but which produce the same effects on α cells. The α-factor is the more thoroughly studied of the two. One of its effects is to inhibit the initiation of DNA synthesis, thus causing a synchronous population of unbudded cells in the G_1 phase. Once mating has been triggered in this manner, cells of opposite mating type clump due to the presence of a- and α-agglutination factors on the surface of a and α cells, respectively. Both factors are glycoproteins with binding sites for the surface of the complementary cell. In some a strains the a-agglutination factor is inducible, probably by the α-factor. It has been suggested that these agglutinins are located on surface filaments that extend from the cell wall and that they are the first portions of the walls to make contact.

Little is known about the biochemical reactions accompanying these processes (Crandall et al., 1977).

SUMMARY

Fungi may reproduce asexually by fragmentation, by budding, or by the production of spores. Sexual reproduction involves the fusion of gametes and meiosis; often spores are produced as well. A multitude of environmental factors have been shown to affect reproduction. In general, conditions that favor growth do not usually favor reproduction as well. In many fungi, reproduction is triggered by the depletion of nutrients or the onset of other harsh environmental conditions. Specific nutrients involved in reproduction have been mentioned in previous chapters.

The asexual cycle of *Blastocladiella emersonii* illustrates the process by which sporangiospores are produced. The multinucleate sporangial cytoplasm is cleaved into numerous uninucleate spores characterized by a distinct ultrastructure. As the zoospore becomes motile and then encysts and as the cyst germinates, striking ultrastructural and biochemical changes occur.

Conidia, on the other hand, are produced singly by one of two general mechanisms. In blastic development, the conidium balloons out from the hyphal or conidiophore wall by a process similar to tip growth. In thallic development, an existing hyphal segment becomes modified into a conidium. Mycelia of *Aspergillus* species must become genetically competent before conidiogenesis can occur. In addition, different stages of conidiogenesis have different nutritional requirements.

Three examples of sexual reproduction in filamentous fungi have been discussed, each involving a hormone produced by one mating type which causes the attraction of the other mating type and triggers sexual differentiation. In *Achlya*, the steroid hormones antheridiol and oogoniol cause the directed growth response (chemotropism) and differentiation of hyphal strands. In *Allomyces*, the hormone sirenin causes the attraction (chemotaxis) of motile gametes. Members of the Mucorales also exhibit chemotropism, but in this order each mating type produces an "incomplete hormone" which is converted to active trisporic acids by the opposite mating type.

In yeasts such as *Hansenula wingei*, conjugation of compatible mating types requires the presence of recognition factors on the cell surface. In *Saccharomyces cerevisiae* diffusible sexual hormones are also involved.

REFERENCES

Anderson, J. G., and J. E. Smith. 1971. Synchronous initiation and maturation of *Aspergillus niger* conidiophores in culture. *Trans. Brit. Mycol. Soc.* **56**: 9–29.

Arsenault, G. P., K. Biemann, A. W. Barksdale, and T. C. McMorris 1968. The structure of antheridiol, a sex hormone in *Achyla bisexualis. J. Amer. Chem. Soc.* **90**: 5635–5636.

Axelrod, D. E. 1972. Kinetics of differentiation of conidiophores and conidia by colonies of *Aspergillus nidulans. J. Gen. Microbiol.* **73**: 181–184.

Axelrod, D. E., M. Gealt, and M. Pastushok. 1973. Gene control of developmental competence in *Aspergillus nidulans. Dev. Biol.* **34**: 9–15.

Barksdale, A. W. 1963a. The role of hormone A during sexual conjugation in *Achlya ambisexualis. Mycologia* **55**: 627–632.

Barksdale, A. W. 1963b. The uptake of exogenous hormone A by certain strains of *Achlya. Mycologia* **55**: 164–171.

Barksdale, A. W. 1966. Segregation of sex in the progeny of selfed heterozygote of *Achlya bisexualis. Mycologia* **58**: 802–804.

Barksdale, A. W. 1969. Sexual hormones of *Achlya* and other fungi. *Science* **166**: 831–837.

Barksdale, A. W., and L. L. Lasure. 1973. Induction of gametangial phenotypes in *Achlya. Bull. Torrey Bot. Club* **100**: 199–202.

Barstow, W. E., and J. S. Lovett. 1978. Ultrastructure of a reduced developmental cycle (minicycle) in *Blastocladiella emersonii. Exp. Mycol.* **2**: 145–155.

Bu'Lock, J. D. 1976. Hormones in fungi. In J. E. Smith and D. R. Berry (Eds.), *The Filamentous Fungi,* Vol. 2, pp. 345–368. New York: Wiley.

Bu'Lock, J. D., D. Drake, and D. J. Winstanley. 1972. Specificity and transformations of the trisporic acid series of fungal sex hormones. *Phytochemistry* **11**: 2011–2018.

Burgeff, H. 1924. Untersuchungen uber Sexualitat und Parasitismus bei Mucorineen. *Bot. Abhandlungen* **4**: 1–135.

Cantino, E. C., and G. L. Mills. 1976. Form and function in chytridiomycete spores. In D. J. Weber and W. M. Hess (Eds.), *The Fungal Spore: Form and Function,* pp. 501–556. New York: Wiley.

Cantino, E. C., and G. L. Mills. 1978. *Blastocladiella emersonii.* In M. S. Fuller (Ed.), *Lower Fungi in the Laboratory,* pp. 39–40. Athens: Department of Botany, University of Georgia.

Cantino, E. C., and L. C. Truesdell. 1970. Organization and fine structure of the side body and its lipid sac in the zoospore of *Blastocladiella emersonii. Mycologia* **62**: 548–567.

Carlile, M. J., and L. Machlis. 1965. The response of male gametes of *Allomyces* to the sexual hormone sirenin. *Am. J. Bot.* **52**: 478–483.

Cole, G. T., and R. A. Samson. 1979. *Patterns of Development in Conidial Fungi.* London: Pitman.

Crandall, M., R. Egel, and V. L. Mackay. 1977. Physiology of mating in three yeasts. *Adv. Microbial Physiol.* **15**: 308–392.

Edwards, J. A., J. S. Mills, J. Sundeen, and J. H. Fried. 1969. The synthesis of the fungal sex hormone antheridiol. *J. Am. Chem. Soc.* **91**: 1248–1249.

Emerson, R. 1941. An experimental study of the life cycles and taxonomy of *Allomyces. Lloydia* **4**: 77–144.

Galbraith, J. C., and J. E. Smith. 1969. Sporulation of *Aspergillus niger* in submerged liquid culture. *J. Gen. Microbiol.* **59**: 31–45.

Gooday, G. W. 1974. Fungal sex hormones. *Ann. Rev. Biochem.* **43**: 35–49.

Gwynne D. I., and B. P. Brandhorst. 1980. Antheridiol-induced differentiation of *Achlya* in the absence of detectable synthesis of new proteins. *Exp. Mycol.* **4**: 251–259.

Hawker, L. E. 1966. Environmental influences on reproduction. In G. C. Ainsworth and A. S. Sussman (Eds.), *The Fungi,* Vol. 2, pp. 435–469. New York: Acadmic.

Horgen, P. A. 1977. Steroid induction of differentiation: *Achlya* as a model system. In D. H. O'Day and P. A. Horgen (Eds.), *Eukaryotic Microbes as Model Developmental Systems,* pp. 272–293. New York: Dekker.

Kane, B. E., J. B. Reiskind, and J. T. Mullins. 1973. Hormonal control of morphogenesis in *Achlya:* Dependence on protein and ribonucleic acid syntheses. *Science* **180**: 1192–1193.

Klapper, B. F., and M. H. Klapper. 1977. A natural inhibitor of sexual attraction in the water mold *Allomyces. Exp. Mycol.* **1**: 352–355.

Lovett, J. S. 1975. Growth and differentiation of the water mold *Blastocladiella emersonii:* Cytodifferentiation and the rate of ribonucleic acid and protein synthesis. *Bacteriol. Rev.* **39**: 345–404.

Machlis, L. 1958. Evidence for a sexual hormone in *Allomyces. Physiol. Plant.* **11**: 181–192.

Machlis, L., W. H. Nutting, and H. Rapoport. 1968. The structure of sirenin. *J. Am. Chem. Soc.* **90:** 1674–1676.

Machlis, L., W. H. Nutting, M. W. Williams, and H. Rapoport. 1966. Production, isolation and characterization of sirenin. *Biochemistry* **5:** 2147–2152.

Machlis, L., and E. Rawitscher-Kunkel. 1967. Mechanisms of gametic approach in plants. In C. B. Metz and A. Monroy (Eds.), *Fertilization*, Vol. 1, pp. 117–161. New York: Academic.

McMorris, T. C., R. Seshadi, G. R. Weihe, G. P. Arsenault, and A. W. Barksdale. 1975. Structures of oogoniol-1, -2 and -3, steroid sex hormones of the watermold *Achlya*. *J. Am. Chem. Soc.* **97:** 2544–2545.

Mills, G. L., and E. C. Cantino. 1978. The lipid composition of the *Blastocladiella emersonii* gamma and the function of gamma-particle lipid in chitin formation. *Exp. Mycol.* **2:** 99–109.

Mills, G. L., and E. C. Cantino. 1979. Trimodal formation of microbodies and associated biochemical and cytochemical changes during development in *Blastocladiella emersonii*. *Exp. Mycol.* **3:** 53–69.

Mullins, J. T., and E. A. Ellis. 1974. Ultrastructural basis for the hormonal induction of antheridial hyphae. *Proc. Nat. Acad. Sci. USA* **71:** 1347–1350.

Murphy, M. N., and J. S. Lovett. 1966. RNA and protein synthesis during zoospore differentiation in synchronized cultures of *Blastocladiella*. *Dev. Biol.* **14:** 68–95.

Ng, A., J. E. Smith, and A. F. McIntosh. 1972. Conidiation of *Aspergillus niger* in continuous culture. *Arch. Mikrobiol.* **88:** 119–126.

Nolan, R. A., and A. K. Bal. 1974. Cellulase localization in hyphae of *Achlya ambisexualis*. *J. Bacteriol.* **117:** 840–843.

Nutting, W. H., H. Rapoport, and L. Machlis. 1968. The structure of sirenin. *J. Am. Chem. Soc.* **90:** 6434–6438.

Ojha, M., J. Haemmerli, and G. Turian. 1980. Induction of premature gametangial differentiation on surface colonies of *Allomyces arbuscula*. *Exp. Mycol.* **4:** 171–174.

Pommerville, J. C. 1977. Chemotaxis of *Allomyces* gametes. *Exp. Cell Res.* **109:** 43–51.

Raper, J. R. 1940. Sexuality of *Achlya ambisexualis*. *Mycologia* **32:** 710–727.

Raper, J. R. 1951. Sexual hormones in *Achlya*. *Am. Sci.* **39:** 110–120.

Raper, J. R. 1952. Chemical regulation of sexual processes in the Thallophytes. *Bot. Rev.* **18:** 447–545.

Ross, I. K. 1979. *Biology of the Fungi*. New York: McGraw-Hill.

Rozek, C. E., and W. E. Timberlake. 1980. Absence of evidence for changes in messenger RNA populations during steroid hormone-induced cell differentiation in *Achlya*. *Exp. Mycol.* **4:** 33–47.

Smith, J. E. 1978. Asexual sporulation in filamentous fungi. In J. E. Smith and D. R. Berry (Eds.), *The Filamentous Fungi*, Vol. 3, pp. 214–239. New York: Wiley.

Smith, J. E., and J. G. Anderson. 1973. Differentiation in the Aspergilli. *Symp. Soc. Gen. Microbiol.* **23:** 295–337.

Smith, J. E., and D. R. Berry. 1974. *An Introduction to the Biochemistry of Fungal Development*. New York: Academic.

Sutter, R. P. 1970. Trisporic acid synthesis in Blakeslea trispora. *Science* **168:** 1590–1592.

Sutter, R. P., D. A. Capage, T. L. Harrison, and W. A. Keen. 1973. Trisporic acid biosynthesis in separate plus and minus cultures of *Blakeslea trispora*: Identification by Mucor assay of two mating-type-specific components. *J. Bacteriol.* **114:** 1074–1082.

Sutter, R. P., T. L. Harrison, and G. Galasko. 1974. Trisporic acid biosynthesis in *Blakeslea trispora* via mating type-specific precursors. *J. Biol. Chem.* **249:** 2282–2284.

Thomas, D. des S., and J. T. Mullins. 1967. Role of enzymatic wall softening in plant morphogenesis: Hormonal induction in *Achlya*. *Science* **156:** 84–85.

Timberlake, W. E. 1976. Alterations in RNA and protein synthesis associated with steroid hormone-induced sexual morphogenesis in the water mold *Achlya*. *Dev. Biol.* **51:** 202–214.

Truesdell, L. C., and E. C. Cantino. 1970. Decay of gamma particles in germinating zoospores of *Blastocladiella emersonii*. *Arch. Mikrobiol.* **70:** 378–392.

Truesdell, L. C., and E. C. Cantino. 1971. The induction and early events of germination in the zoospores of *Blastocladiella emersonii*. In A. Monroy and A. A. Moscona (Eds.), *Current Topics in Developmental Biology*, Vol. 6, pp. 1–41. New York: Academic.

van den Ende, H. 1976. *Sexual Interactions in Plants: The Role of Specific Substances in Sexual Reproduction*. New York: Academic.

10

METABOLISM

The heterotrophic mode of nutrition in fungi dictates that carbon skeletons obtained from the environment must be metabolized in ways that provide both building blocks for cellular structures and sources of energy for maintenance, growth, and development. In earlier chapters we alluded to this flow of carbon through the network of metabolic pathways. In this chapter we shall provide an overview of the major pathways of degradation and synthesis in fungi with the purpose of showing how carbon-containing nutrients can be utilized for survival.

We shall first discuss glycolysis and respiration as a framework around which the remainder of metabolism is built. We shall then focus on the synthesis of polysaccharides and the metabolism of lipids. Discussion of protein and nucleic acid synthesis is postponed to Chapter 11, and the synthesis of aromatic and steroid compounds is discussed in Chapter 12.

GLYCOLYSIS AND FERMENTATION

Glycolysis is the physiological conversion of glucose to pyruvate without regard to mechanism. The glycolytic pathways in fungi, as in all cells, serve to furnish energy, precursors for the synthesis of a variety of compounds, and reducing power in the form of NADH and NADPH necessary for converting these precursors to appropriate intermediates or end products. The details of the enzymology of these pathways are well documented, and the available evidence indicates that in fungi they are essentially the same as in other organisms. Reviews of glycolysis in fungi have been published by Blumenthal (1965) and Cochrane (1976).

Two major glycolytic pathways operate in almost all fungal cells: the Embden–Meyerhof (EM) pathway and the pentose-phosphate (PP) pathway (also called the hexose-monophosphate shunt). These pathways are summarized in Figure 10.1.

Embden-Meyerhof Pathway

The EM pathway is composed of 10 enzymatically catalyzed steps that take place in the cytoplasm of the cell. Lehninger (1975) points out that there are two major

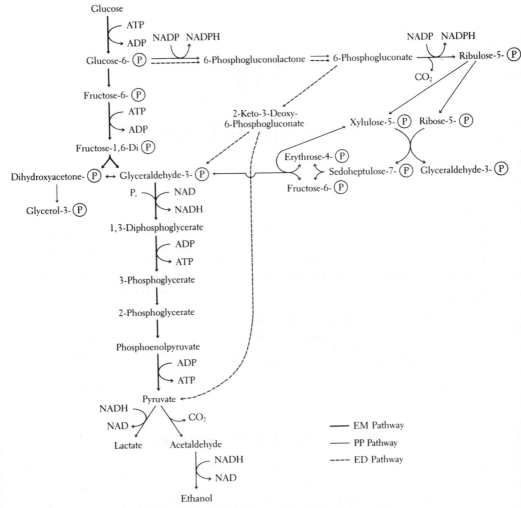

Figure 10.1   Major pathways of glycolysis and fermentation.

stages in this pathway. In the first stage glucose is primed for degradation by being phosphorylated and cleaved to form the 3-carbon sugar glyceraldehyde-3-phosphate. This stage also serves as a collection phase in which a number of other hexoses enter the glycolytic sequence. For example, galactose is phosphorylated and converted to glucose-6-phosphate, mannose is phosphorylated and converted to fructose-6-phosphate, and fructose is phosphorylated and converted to glyceraldehyde-3-phosphate. Polysaccharides such as glycogen and starch are hydrolyzed, phosphorylated, and converted to glucose-6-phosphate.

The second stage is common to all sugars and involves the conversion of glyceraldehyde-3-phosphate to pyruvate with the concomitant reduction of NAD and release of energy that is stored in molecules of ATP.

Notice that all the intermediates of the EM pathway are phosphorylated. Not only does phosphorylation provide the necessary energy to make the sugars reactive but the negative charge on the phosphate group prevents these molecules from readily diffusing through the plasma membrane. When one molecule of glucose has been metabolized by the EM pathway there has been an input of two molecules of ATP and an output of four ATPs, giving the cell a net gain of two ATPs. In addition, two molecules of NADH have been generated.

Pentose-Phosphate Pathway

The PP pathway can be thought of as a side branch of the EM pathway, which begins with glucose-6-phosphate and funnels back in at glyceraldehyde-3-phosphate and fructose-6-phosphate (Figure 10.1). This pathway begins with two oxidation-reduction reactions yielding NADPH, ribulose-5-phosphate, and CO_2. Ribulose-5-phosphate can then be fed into a series of reactions in which a pool of phosphorylated, interconvertible sugars are formed. Although theoretically the PP pathway could be used for the complete oxidation of glucose to CO_2 by six cycles through the pathway, in practice this occurs rarely, if at all. Instead, intermediates of the PP pathway are usually siphoned off for the synthesis of other compounds. For example, ribose-5-phosphate is a precursor for the synthesis of nucleotides, and erythrose-4-phosphate is one of the precursors for the synthesis of aromatic compounds via the shikimic acid pathway. Of even more importance is the generation of NADPH. Most cells have developed a system for using reducing power such that NADH is used in degradation reactions and NADPH for biosynthesis. The PP pathway thus functions as a major source of this biosynthetic reducing power. Lastly, pentoses that are encountered by the fungus in the environment, such as xylose and arabinose, may be metabolized via this pathway (Cochrane, 1976).

Evidence for the importance of the PP pathway in generating NADPH has been provided by Brody and Tatum (1966) using mutants of *Neurospora crassa*. The col-2 mutant has both a much reduced growth habit and an altered form of glucose-6-phosphate dehydrogenase, the first enzyme in the PP pathway. As a consequence, the level of NADPH in this mutant is much reduced (Brody, 1970). These results suggest that a properly functioning PP pathway is necessary for normal morphology and that low levels of NADPH may prohibit adequate rates of synthesis of cell wall polysaccharides or membrane lipids (Cochrane, 1976).

Other Glycolytic Pathways

The Entner–Doudoroff (ED) pathway is a common glycolytic pathway in bacteria, and there is some evidence that it is present in a few fungal species. As indicated in Figure 10.1, this pathway proceeds to 6-phosphogluconic acid as in the PP pathway, but this compound is then converted to 2-keto-3-deoxy-6-phosphogluconate and finally to glyceraldehyde-3-phosphate and pyruvate.

In *Aspergillus niger*, glucose can be oxidized to gluconate, which is then converted to pyruvate and glyceraldehyde via the intermediate 2-keto-3-deoxygluconate (Elzainy et al., 1973). This system is analogous to the ED pathway except that no phosphorylated intermediates are involved (Cochrane, 1976).

The Balance of Glycolytic Pathways

By the use of specifically labelled glucose it is possible to determine the extent to which these glycolytic pathways are operating in fungal species. This is done by measuring the yield of $^{14}CO_2$ and/or measuring the radioactivity in a key product of glycolysis. Blumenthal (1968) reviewed work in which these two approaches were used in fungi, and a summary appears in Table 10.1. Notice that in most fungi glucose is degraded mainly via the EM pathway. Although the PP pathway is present in most species studied, it is seldom the major pathway. In A. *niger*, activity of the PP pathway increases just prior to sporulation (Ng et al., 1972).

Another method of assessing the degree of participation of glycolytic pathways is by assaying for key enzymes. Although most of the enzymes of the EM pathway are common to the PP and ED pathways, certain enzymes are unique to each. For

Table 10.1   Quantitative Estimations of Pathways of Glucose Catabolism

Microorganism	Pathway Participation			Method
	EM (%)	HMP (%)	ED (%)	
Aspergillus niger	78	—	—	Product ^{14}C distribution
Caldariomyces fumago	—	35	65	Product ^{14}C distribution and specific activity
Candida utilis	50–96	4–50	—	Product specific activity
Claviceps purpurea	90–96	4–10	—	$^{14}CO_2$ yield
Fusarium lini	83	17	—	$^{14}CO_2$ specific activity
Neurospora crassa, mycelium	88–99	1–12	—	$^{14}CO_2$ yield
N. *crassa*, conidia	90	10	—	$^{14}CO_2$ yield
Penicillium chrysogenum	56–70	30–44	—	Product specific activity
P. *chrysogenum*	42	58	—	$^{14}CO_2$ specific activity
P. *chrysogenum*	77	23	—	$^{14}CO_2$ yield
P. *digitatum*	77–83	17–23	—	$^{14}CO_2$ yield
P. *urticae*	40–80	20–60	—	$^{14}CO_2$ yield
Rhizopus oryzae	100	—	—	Product C^{14} distribution
R. MX	100	—	—	Product C^{14} distribution and specific activity
Tilletia caries, mycelium	66	34	—	$^{14}CO_2$ yield
T. *caries*, spores	—	—	100	$^{14}CO_2$ yield
T. *contraversa*, spores	33	67	—	$^{14}CO_2$ yield
Verticillium alboatrum	48	52	—	$^{14}CO_2$ yield

Source: Blumenthal (1968).

example, phosphofructokinase is unique to the EM pathway, phosphogluconate dehydrogenase, transketolase, and transaldolase are unique to the PP pathway, and 6-phosphogluconate dehydratase and 2-keto-3-deoxy-6-phosphogluconate aldolase are unique to the ED pathway. However, claims for the lack of a pathway on the basis of the inability to demonstrate key enzymes may be misleading due to inactivation of the enzyme during isolation and purification.

Fermentation Pathways

All glycolytic pathways can take place under anaerobic as well as aerobic conditions. However, in the absence of oxygen, pyruvate (the principal end product), instead of being completely degraded to CO_2 in respiration, is shunted into one or two fermentation pathways. Certain fungi, particularly members of the Chytridiomycetes and Oomycetes, exhibit lactic acid fermentation. In this pathway pyruvate is reduced to lactic acid by NADH in the presence of lactic acid dehydrogenase (Figure 10.1). On the other hand, yeast and other higher fungi exhibit alcoholic fermentation. In this pathway pyruvate is first decarboxylated to acetaldehyde via pyruvate decarboxylase, and the acetaldehyde is then reduced to ethanol in the presence of alcohol dehydrogenase. In certain *Rhizopus* species, both ethanol and lactic acid are produced, indicating that both fermentation pathways are operational. Notice that fermentation pathways provide a means of utilizing the NADH formed in the EM pathway and generating the NAD necessary for continuation of the EM pathway's second stage (Cochrane, 1976).

Various nutritional factors have been shown to affect the rates of glycolysis and fermentation reactions. Becker and Betz (1972) used ethanol yield to compare rates of glycolysis in *Saccharomyces carlsbergensis* grown on glucose or fructose. They inferred that the rates of glycolysis are comparable with either sugar. However, more ethanol is produced with glucose because it is phosphorylated at a more rapid rate. The quantity of ethanol produced by *Helminthosporium maydis* incubated under anaerobic conditions varies with the type of nitrogen source in the growth medium. Ethanol production is highest with ammonium chloride and lowest with glutamic acid (Evans and Garraway, 1976).

Fungi incubated under anaerobic conditions often have different nutritional requirements than under aerobicity. *Saccharomyces cerevisiae* requires ergosterol and an unsaturated fatty acid (Andreasen and Stier, 1954), and *Mucor rouxii* must be supplied with thiamine and nicotinic acid (Bartnicki-Garcia and Nickerson, 1961). *Fusarium oxysporium* will grow without oxygen only if the medium contains one of several nutrients such as MnO_2, nitrate, selenite, ferric ion, or yeast extract. It is not known how these stimulate growth of *F. oxysporium* under anaerobicity but they may function as alternatives to O_2 as electron acceptors and/or may make the redox potential of the medium more positive (Gunner and Alexander, 1964).

Although a few fungi such as the aquatic species *Blastocladia ramosa* and *Aqualinderella fermentans* grow equally well under anaerobic and aerobic conditions, most fungi grow poorly if at all without O_2. This may be due to various reasons: An essential nutrient may not be able to be utilized, toxic products may accumulate, O_2

may be directly required for certain enzymes such as oxygenases, there may be inadequate redox potential of the medium, or the organism may be unable to function properly with the reduced ATP supply (Gunner and Alexander, 1964). As we shall see, aerobicity opens up a new series of metabolic pathways that are extremely advantageous to the organism in terms of energy yield and carbon skeletons. These are the pathways of respiration.

RESPIRATION

The metabolism of glucose to ethanol or lactic acid under anaerobic conditions results in a relatively small amount of energy for the cell. In the EM pathway there is a net gain of only two moles of ATP per mole of glucose, and there is still much energy remaining in the end products of fermentation. Under aerobic conditions, the energy in these end products can be released in the reactions of respiration.

Respiration is the complete oxidation of carbon molecules to CO_2 and H_2O in the presence of oxygen. If we consider glucose to be the principal starting compound, respiration results in a net gain of 36–38 moles of ATP per mole. Respiration can conveniently be divided into four parts: glycolysis, the formation of acetyl coenzyme A (acetyl-CoA), the tricarboxylic acid (TCA) or Krebs cycle, and the electron transport chain.

Glycolysis and the Formation of Acetyl-CoA

Regardless of the particular glycolytic pathway or pathways in operation, the pyruvate that is formed is transported from the cytoplasm into the mitochondrion and is there converted to acetyl-CoA. This is a multistep reaction catalyzed by a pyruvate dehydrogenase multi-enzyme complex. Pyruvate is first decarboxylated, and the resulting acetyl group is linked to coenzyme A by a reaction involving the conversion of NAD to NADH (Figure 10.2). The acetyl-CoA then enters the TCA cycle.

TCA Cycle

In the TCA cycle, the acetyl group of acetyl-CoA is transferred to oxaloacetate forming citrate (Figure 10.2). Citrate is then metabolized via a series of reactions resulting in the loss of two molecules of carbon dioxide and the regeneration of oxaloacetate. Oxaloacetate can then react with another acetylCoA for another round of the cycle. During each turn of the TCA cycle, one high-energy molecule (usually GTP) is formed in addition to three molecules of NADH and one molecule of FADH. Note that the cycle must turn twice for each molecule of glucose that enters glycolysis. With the exception of succinic dehydrogenase, which is bound to the inner mitochondrial membrane, the enzymes of the TCA cycle are found in the matrix of the mitochondria.

The TCA cycle has been shown to be present in representative fungi belonging to all the major taxonomic classes (Niederpruem, 1965). It is probably safe to assume that this pathway occurs in all fungi. Tests for the presence of the TCA cycle include

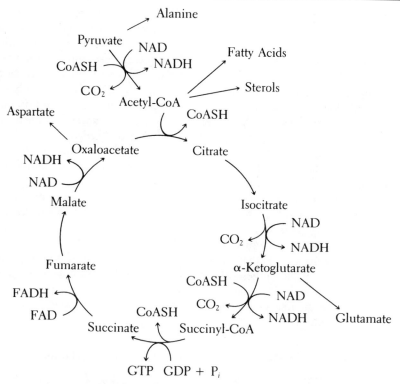

Figure 10.2 The tricarboxylic acid (Krebs) cycle.

the ability of isolated mitochondria to utilize intermediates of the cycle (Kawakita, 1970), the appearance of label in these intermediates after incubating the fungus with ^{14}C-acetate (DeMoss and Swim, 1957), and the presence of key TCA cycle enzymes in mitochondrial extracts (Kawakita, 1970).

Metabolites involved in feedback control of enzymes associated with the TCA cycle have been extensively studied and reviewed by Atkinson (1966a, 1966b, 1969). Key metabolites involved are adenylates (ATP, ADP, and AMP) and long-chain acyl-CoA derivatives. These metabolites can alter the rate and direction of reactions in the TCA cycle not only in fungi but in bacteria, plants, and animals as well (Atkinson, 1969; Gulyi, 1977). The changes brought about by these metabolites may affect in part other metabolic processes, leading to a change in the metabolism of the entire cell. This could lead, in turn, to changes in the rates and patterns of growth and development.

Atkinson (1969) has defined a concept called the energy charge (EC) as the ratio of cellular adenylates:

$$EC = \frac{[ATP] + \frac{1}{2}[ADP]}{[ATP] + [ADP] + [AMP]}$$

A cellular energy charge of 1.0 indicates that all the adenylates are in the ATP form; when the energy charge is 0 no ATP or ADP is present. Usually the energy charge of the cell is somewhere between these extremes. Atkinson (1969) points out that as the energy charge of a cell rises, the rates of certain reactions drop and the rates of others rise. With regard to the TCA cycle, citrate synthase from yeast is strongly inhibited by ATP (Hathaway and Atkinson, 1965), indicating that it is subject to feedback inhibition in cells with a high energy charge. On the other hand, isocitrate dehydrogenase is stimulated by AMP (Atkinson, 1966a), indicating that it is likely to be most active when the energy charge is low (the AMP or ADP levels are relatively high). The differential sensitivity of these two TCA cycle enzymes to adenylates represents an important regulatory mechanism with implications for control of the metabolic machinery of the entire cell.

In addition, Atkinson (1966b) notes that several TCA enzymes from mammalian heart and liver cells are subject to feedback inhibition by free fatty acids. Palmityl-CoA and other long-chain CoA esters are inhibitory to citrate synthase. Similar feedback mechanisms are presumed to operate in fungi.

In addition to being a key pathway in glucose degradation, the TCA cycle provides carbon skeletons for the biosynthesis of other compounds. For example, many amino acids are formed using TCA cycle intermediates as precursors. The amino acids glutamate and aspartate may be formed directly by amination of α-ketoglutarate and oxaloacetate, respectively.

Electron Transport Chain and Oxidative Phosphorylation

The NADH and FADH generated in glycolysis, the formation of acetylCoA, and the TCA cycle move to the inner mitochondrial membrane where they enter the electron transport chain (ETC). The components of this chain transfer electrons stepwise down an energy gradient. At three sites, the electrons release sufficient energy to power the phosphorylation of ADP forming ATP (Figure 10.3). This process is called oxidative phosphorylation. At the top of the ETC (energetically speaking) are the carriers flavoprotein and ubiquinone, which transfer both electrons and their associated hydrogen ions. The NADH from glycolysis and the TCA cycle transfer their electrons and hydrogens to the flavoprotein; but FADH, because of energy considerations, donates its electrons and hydrogens directly to ubiquinone. From ubiquinone the electrons are transferred to cytochrome b, then sequentially to other cytochromes, and finally to oxygen, which when reduced forms water. The cytochromes are proteins containing an iron-porphyrin prosthetic group. Iron-sulfur proteins serve as carriers also. Neither of these molecules is able to transport hydrogen ions. These hydrogen ions, ejected from the ETC following ubiquinone, generate across the mitochondrial membrane an electrochemical gradient that is thought to provide the driving force for ATP production. Notice that for each NADH that transfers its electrons to the ETC, three ATPs are formed; but because of its different point of entry, FADH bypasses the first oxidative phosphorylation site and results in the formation of only two ATPs.

A tabulation of the number of ATP molecules generated in respiration gives a

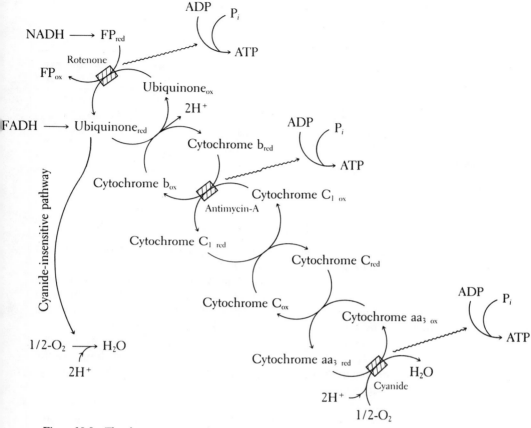

Figure 10.3 The electron transport chain and the cyanide-insensitive pathway. FP = Flavoprotein.

total of 38 per molecule of glucose. However, mitochondrial membranes are generally impermeable to NADH, and thus "shuttle systems" are necessary to transport reducing equivalents of the NADH generated in glycolysis into the mitochondrion. Depending on the type of shuttle present, either two or three ATPs may be formed from cytoplasmic NADH—hence the net result of either 36 or 38 ATPs, depending on the species studied (Lehninger, 1975).

In general, it appears that the mitochondrial reactions in fungi are similar to those in plants and animals (Watson, 1976), although the exact sites of the phosphorylation reactions are not precisely known. In addition, as indicated in Table 10.2, there are differences among fungal species in the concentrations of the various carriers. The elucidation of the components and phosphorylation sites in the ETC has been facilitated by the use of specific inhibitors such as rotenone, cyanide, and antimycin-A. The sites of inhibition of these compounds are shown in Figure 10.3.

Table 10.2   Respiratory Components of Mitochondria from Filamentous Fungi and Yeasts

	Concentrations Expressed As nmol/mg^{-1} Protein			
Component	Aspergillus niger	Aspergillus oryzae	Neurospora crassa	Saccharomyces carlsbergensis
Cytochrome aa$_3$	0.15	0.20	0.38	0.15
Cytochrome b	0.19	0.13	0.38	0.28
Cytochrome C + C$_1$	0.39	0.20	1.6	0.65
Ubiquinone	2.9		1.5	0.54
NAD	7.0		12.0	

Source: Watson (1976).

Cyanide-Insensitive Respiration

Recently, an alternative electron transport pathway has been demonstrated in a number of plants and eukaryotic microorganisms. This pathway contains no cytochromes, is insensitive to inhibition by antimycin-A and cyanide, and results in little if any phosphorylation. This "cyanide-insensitive pathway" apparently branches from the normal pathway at ubiquinone (Figure 10.3). *Moniliella tomentosa* utilizes the cytochrome ETC under most conditions, but the vast majority of electrons are shunted down the cyanide-insensitive pathway when the cells are grown in the presence of drugs such as chloramphenicol or ethidium bromide, which specifically interfere with mitochondrial functioning. Under other conditions, both pathways may function simultaneously. Incubation of *M. tomentosa* in *n*-propanol or *n*-butanol induces the appearance of the cyanide-insensitive pathway without affecting the normal pathway. Methanol and ethanol have no effect (Vanderleyden et al., 1978).

In *Saccharomyces cerevisiae* a single copper-containing protein of molecular weight 55,000 was found to be responsible for the cyanide-insensitive pathway. This "alternative oxidase" appears to be localized in membranous particles that occur outside the mitochondria (Ainsworth et al., 1980a,b). In contrast, the alternative oxidase in *M. tomentosa* is localized in particles that are firmly bound to the inner mitochondrial membrane (Vanderleyden et al., 1979).

The function of the cyanide-insensitive pathway is not known. It has been hypothesized that this pathway provides an alternative means of oxidizing NADH and FADH under conditions when normal electron transport is inhibited (Ainsworth et al., 1980a). In *M. tomentosa*, the functioning of this pathway is thought to be associated with the inhibitory effects of propanol and butanol on oxidative phosphorylation (Vanderleyden et al., 1978).

Notice that the only role for molecular oxygen in respiration is to serve as the terminal electron acceptor. Although this may seem a relatively mundane function it is crucially important for the operation of the TCA cycle and the ETC. When cells are deprived of oxygen, the carriers in the ETC are unable to pass on their electrons and thus remain in the reduced form. Similarly, NADH generated in glycolysis and the TCA cycle remain reduced, and the supply of NAD (in the oxidized form), necessary for several reactions in these pathways, is depleted. The principal alternative route for oxidizing NADH then becomes the dehydrogenase reactions of fermentation. Thus, without oxygen the ETC and the TCA cycle cease to function, but glycolysis and fermentation can continue with the concomitant diminution in ATP production.

Regulation of Respiration

Certain mechanisms in the respiratory pathway modulate the rate of glucose utilization. One key regulatory enzyme is phosphofructokinase (PFK), which occurs in glycolysis at the point where fructose-6-phosphate is phosphorylated to fructose-1,6-di-phosphate. PFK is an allosteric enzyme affected by energy charge; its activity is stimulated by ADP and inhibited by ATP. Thus, when the cellular activity of ATP is high, PFK is inhibited and the flow of carbon through glycolysis and ultimately the TCA cycle slows. As a result, ATP production diminishes. Once the cellular ATP level reaches a certain low level (i.e., low energy charge), PFK is activated and the rate of respiration increases. Citrate is also an inhibitory modulator of PFK, and the overproduction of citrate resulting from excess activity of the TCA cycle also decreases the flow of carbon in respiration.

This modulation of respiration via PFK also explains the increase in glucose utilization observed when a cell is transferred from aerobic to anaerobic conditions. In this case, both the TCA cycle and the ETC cease completely, but glycolysis is stimulated by the depletion of ATP and the corresponding accumulation of ADP. The rate of glucose utilization thus increases under anaerobic conditions, a phenomenon known as the "Pasteur effect." This increase in the rate of glycolysis thus compensates the cell somewhat for the relatively small number of ATP molecules produced via this pathway. When aerobic conditions are restored, the ATP level increases and the rate of glycolysis decreases once again. Aerobic conditions also cause the induction of enzymes in the ETC in S. cerevisiae (Brown and Beattie, 1978) and an increase in the number of mitochondria per cell and in the number of cristae per mitochondrion (Table 10.3) (Howell et al., 1971).

The decrease in the rate of glycolysis caused by aerobicity may not occur if the

Table 10.3 Ultrastructural Dimensions of Anaerobic and Aerobic Cultured Cells of a Facultative Anaerobe of *Neurospora*

Culture	Average Diameter (μ)			Average No. Mitochondria per Nucleus	Average No. Cristae per Mitochondria
	Cell	Nucleus	Mitochondrion		
Aerobic	3.5	1.01	0.42×0.18	7.8	6.7
Anaerobic	2.3	1.97	0.82×0.45	2.0	1.4

Source: Adapted from Howell et al. (1971).

concentration of carbon source in the medium is sufficiently high. For example, the respiration rate of S. *cerevisiae* increases with increasing glucose concentrations up to 6×10^{-3} M; above this concentration the rate of respiration declines and the rate of fermentation increases (Slonimski, 1956). The phenomenon of aerobic fermentation and decreased respiration brought about by high concentrations of carbon source is called the "Crabtree effect" for Henry Crabtree, who discovered it in 1929. It has been described for a number of fungi, particularly yeasts (Table 10.4). The extent of the Crabtree effect in S. *cerevisiae* varies with the carbon source. Little aerobic fermentation is seen with high concentrations of galactose, but the rate is

Table 10.4   Presence of the Crabtree Effect in Various Yeast Strains[a]

Organism	μliter of Gas/10^7 Organisms/10 min.		Ratio, Fermented Glucose/ Respirated Glucose	Crabtree Effect
	Respiration (O_2 uptake)	Aerobic Fermentation (CO_2 evolved)		
Saccharomyces cerevisiae	4.8	78.0	49.0	+
S. chevalieri	1.2	90.8	250.0	+
S. fragilis	24.5	1.9	0.23	−
S. italicus	0.0	94.5	∞	+
S. oviformis	0.0	61.2	∞	+
S. pasteurianus	1.9	58.7	93.0	+
S. turbidans	3.6	68.2	57.0	+
S. carlsbergensis	0.0	68.2	∞	+
Schizosaccharomyces pombe	0.0	40.6	∞	+
Candida utilis	30.0	0.0	0.0	−
C. tropicalis	27.7	0.9	0.1	−
C. monosa	21.0	0.0	0.0	−
Trichosporon fermentans	15.5	0.0	0.0	−
Hansenula anomala	24.1	0.0	0.0	−
Debaryomyces globosus	12.3	22.2	5.4	+
Pichia fermentans	24.3	1.3	0.16	−
Schwanniomyces occidentalis	9.1	0.0	0.0	−
Brettanomyces lambicus	1.2	9.3	23.0	+
Torulopsis dattila	0.0	52.0	∞	+
T. sphaerica	25.7	3.5	0.4	−
T. glabrata	—	—	—	+
T. colliculosa	10.7	39.2	11.0	+
T. sake	13.3	0.0	0.0	−
Nematospora coryli	21.1	29.2	4.1	+
Nadsonia fulvescens	—	—	38.0	+

Source: DeDeken (1966).

[a]*Note.* The absence of a respiration value means that it was too low to be measured in the presence of the high rate of aerobic fermentation.

high with glucose and fructose; mannose gives an intermediate rate (DeDeken, 1966). It appears that the Crabtree effect is due to the repression of enzymes of the TCA cycle and those in the pathways of cytochrome biosynthesis; structural changes in mitochondria have been observed as well (Barford and Hall, 1979). In addition, a competition between glycolysis and the ETC for inorganic phosphate has been implicated (Koobs, 1972). Recent studies using batch culture have indicated that the Crabtree effect may be due merely to a very slow adaptation to a fully respiratory condition, although specific concentration effects may also be involved (Barford and Hall, 1979).

LIPID METABOLISM

Lipids are excellent sources of stored energy in many fungal species, as well as being components of cellular membranes. In this section we focus on the synthesis and degradation of fatty acids; metabolism of terpenoids and sterols is discussed in Chapter 12.

Biosynthesis of Fatty Acids

Fatty acids are synthesized in the cytoplasm in a series of reactions catalyzed by a multienzyme fatty acid synthetase complex. Six of the enzymes in this complex are arranged around a central acyl carrier protein (ACP). The ACP has no catalytic function but instead serves as a point of attachment and rotates as a "swinging arm" to bring the substrate around sequentially to the six outer enzymes.

All the carbons in the fatty acid molecule come from acetylCoA, a molecule that we have discussed in relation to its transition role between glycolysis and the TCA cycle. The first step in fatty acid synthesis is the attachment of the acetyl group of acetyl-CoA to the first enzyme in the complex, β-ketoacyl-ACP synthase (Figure 10.4). These acetyl carbons will be the terminal carbons in the fatty acid molecule. All other acetyl groups to be added are incorporated not as acetyl-CoA but as malonyl-CoA, formed by the carboxylation of acetyl-CoA in the presence of ATP (Figure 10.4). This reaction is catalyzed by the biotin-requiring enzyme acetyl-CoA carboxylase.

MalonylCoA is picked up by the ACP, and CoA is liberated. The ACP-bound malonyl group then condenses with the acetyl group that was previously "parked" on the synthase enzyme to yield ACP-bound acetoacetate and liberates CO_2. With this reaction the "swinging arm" begins to rotate, and the acetoacetate is reduced twice and dehydrated to yield a four-carbon chain. The ACP then "parks" this chain on the synthase enzyme and picks up another malonyl-CoA molecule to take through the cycle. At each revolution a two-carbon unit is added to the chain until the chain reaches a length of 16–18 carbons. The fatty acid, as its CoA derivative, is then released from the ACP.

Notice that fatty acids synthesized via this pathway are 16–18 carbons long and unsaturated. Fatty acids longer than these may be formed by the addition of two-

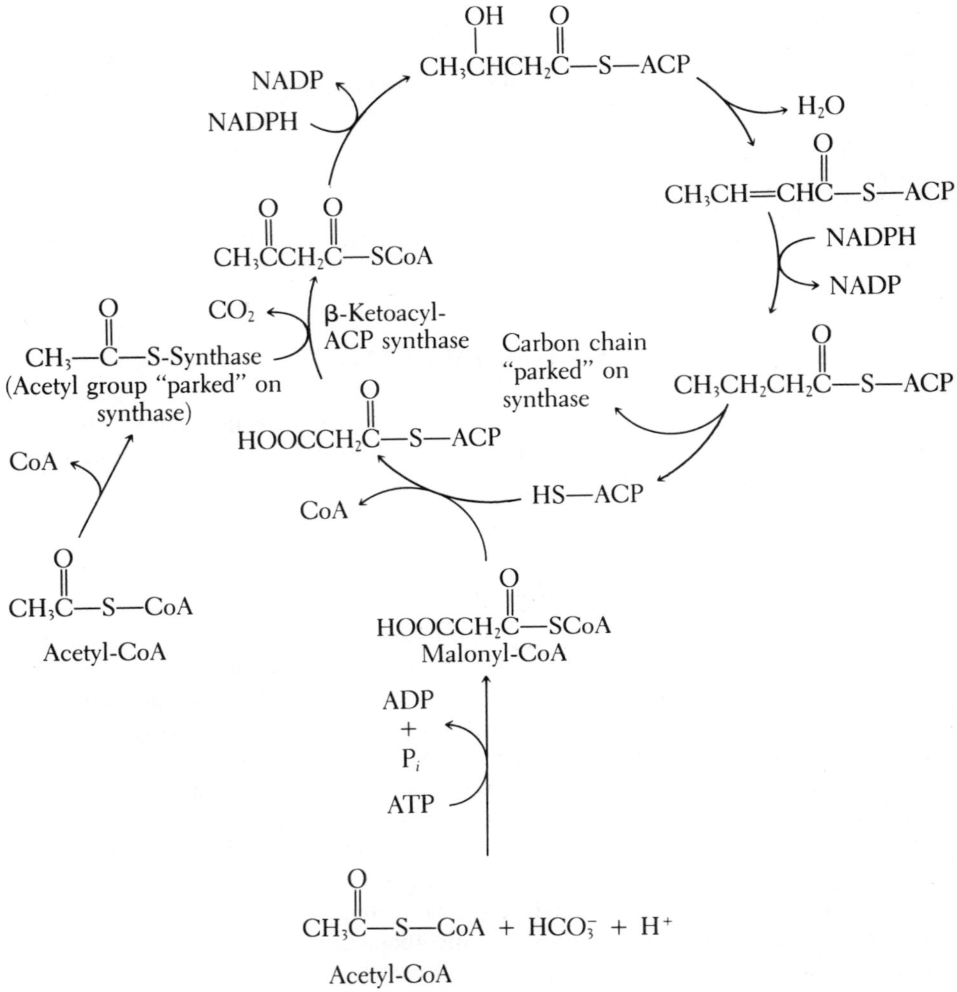

$$\underset{\text{OH}}{\overset{}{\text{CH}_3\text{CHCH}_2}}\overset{\overset{\text{O}}{\|}}{\text{C}}-\text{S}-\text{ACP}$$

NADP

NADPH

H$_2$O

$$\text{CH}_3\text{CH}=\text{CH}\overset{\overset{\text{O}}{\|}}{\text{C}}-\text{S}-\text{ACP}$$

NADPH

NADP

$$\text{CH}_3\overset{\overset{\text{O}}{\|}}{\text{C}}\text{CH}_2\overset{\overset{\text{O}}{\|}}{\text{C}}-\text{SCoA}$$

CO$_2$ β-Ketoacyl-
ACP synthase

Carbon chain
"parked" on
synthase

$$\text{CH}_3\text{CH}_2\text{CH}_2\overset{\overset{\text{O}}{\|}}{\text{C}}-\text{S}-\text{ACP}$$

$$\text{CH}_3-\overset{\overset{\text{O}}{\|}}{\text{C}}-\text{S-Synthase}$$
(Acetyl group "parked" on
synthase)

$$\text{HOOCCH}_2\overset{\overset{\text{O}}{\|}}{\text{C}}-\text{S}-\text{ACP}$$

CoA

HS—ACP

CoA

$$\text{CH}_3\overset{\overset{\text{O}}{\|}}{\text{C}}-\text{S}-\text{CoA}$$
Acetyl-CoA

$$\text{HOOCCH}_2\overset{\overset{\text{O}}{\|}}{\text{C}}-\text{SCoA}$$
Malonyl-CoA

ADP
+
P$_i$

ATP

$$\text{CH}_3\overset{\overset{\text{O}}{\|}}{\text{C}}-\text{S}-\text{CoA} + \text{HCO}_3^- + \text{H}^+$$
Acetyl-CoA

Figure 10.4   The reactions of fatty acid biosynthesis.

carbon units to the original fatty acid after it is released from the complex. However, these reactions have not been studied in detail in fungi. Monounsaturated fatty acids can be formed by the direct oxidation of saturated fatty acids. Polyunsaturated fatty acids are formed by a series of alternate desaturations and elongations (Walker and Woodbine, 1976).

Biosynthesis of Triglycerides (Triacylglycerols)

Triglycerides are often found as storage products in fungi. They are composed of two basic subunits: glycerol and fatty acids. Glycerol, as glycerol-3-phosphate, is normally formed from the glycolytic intermediate dihydroxyacetone phosphate:

$$\text{NADH + dihydroxyacetone phosphate} \longrightarrow \text{NAD + glycerol-3-phosphate}$$

CoA derivatives of three fatty acids then condense with glycerol-3-phosphate to yield the triglyceride.

Degradation of Fatty Acids

Fatty acids are oxidized to carbon dioxide and water in the matrix of the mitochondrion. Before degradation is initiated these molecules must be activated to their CoA derivatives in a reaction powered by ATP:

$$\text{RCOOH + ATP + CoASH} \rightleftharpoons \text{RCO-S-CoA + AMP + PP}_i$$

The activated fatty acid is sequentially dehydrogenated, hydrated, and hydrogenated again (Figure 10.5). The resulting β-ketoacyl-CoA then reacts with a second mole-

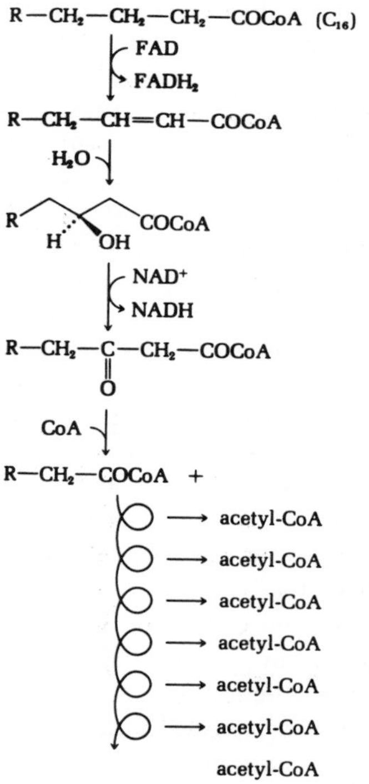

Figure 10.5 The fatty acid oxidation spiral. One acetyl-CoA is removed during each pass through the sequence. One molecule of palmitic acid (C$_{16}$) after activation to palmitoyl-CoA yields eight molecules of acetyl-CoA. (Lehninger, 1975)

cule of CoA in such a way as to cleave off acetyl-CoA, leaving the remaining fatty acylCoA with two carbons less than it started with. The cycle then repeats, each time cleaving off two-carbon units until degradation is completed. Because cleavage occurs at the β-carbon of the fatty acid, this scheme is called the β-oxidation pathway. Hydrogens and electrons carried by NAD and FAD enter the ETC, and the released acetyl-CoA units enter the TCA cycle.

Oxidation of fatty acids releases considerable energy to the cell. For example, for each molecule of palmitic acid (16-C) degraded by this pathway there is a net gain of 129 ATP molecules. Lipids are thus effective molecules for energy storage in cells.

BIOSYNTHESIS OF CARBOHYDRATES: GLUCONEOGENESIS AND THE GLYOXYLATE CYCLE

In order to construct cell walls and other cellular components, fungi must be able to synthesize glucose and other carbohydrates from small molecules other than carbon dioxide. This biosynthesis process is called gluconeogenesis, and its occurrence in fungi has been reviewed by Cochrane (1976).

One of the key pathways in gluconeogenesis is essentially a reversal of the EM pathway, with three notable exceptions. Consider the pathway involved in the conversion of pyruvate to glucose. In glycolysis, the conversion of PEP to pyruvate is essentially irreversible. Thus, there is an alternate route for converting pyruvate to PEP whereby pyruvate is first carboxylated to oxaloacetate (OAA) via pyruvate carboxylase using the energy of ATP (Figure 10.6). OAA is then phosphorylated and decarboxylated by PEP carboxykinase to yield PEP. Since pyruvate carboxylase is a mitochondrial enzyme, these steps are complicated by the problem of transporting pyruvate into the mitochondria and transporting OAA out (Lehninger, 1975).

Once formed, PEP can be converted to fructose-1,6-diphosphate by a reversal of the reactions of the EM pathway (Figure 10.6). However, an alternative route from fructose-1,6-diphosphate to fructose-6-phosphate is required because of the irreversibility of PFK. This reaction is carried out by fructose-diphosphatase, an enzyme present in animal cells and studied in only a few fungi (Cochrane, 1976). In animals, this enzyme is allosterically stimulated by citrate and 3-phosphoglyceraldehyde but inhibited by AMP. Thus, gluconeogenesis is stimulated when the levels of both energy and carbohydrate precursors are high.

Gancedo and co-workers (1965) noted that PFK is repressed when *Saccharomyces cerevisiae* is grown on a medium containing either 2% ethanol or 1% lactate plus 1% glycerol as the carbon source. In contrast, this enzyme is derepressed when glucose is the carbon source. The reverse was observed with fructose-diphosphatase (FDPase). That is, its level is derepressed with ethanol or lactate plus glycerol and repressed with glucose. The induction of FDPase is dependent on the presence of an energy source and is prevented by cycloheximide (Gancedo, 1971). The nature of the response of PFK and FDPase to glucose and nonsugar carbon sources facilitates their effective utilization. If both enzymes were active in the presence of glucose the FDPase would tend to antagonize PFK and would decrease the net flow of

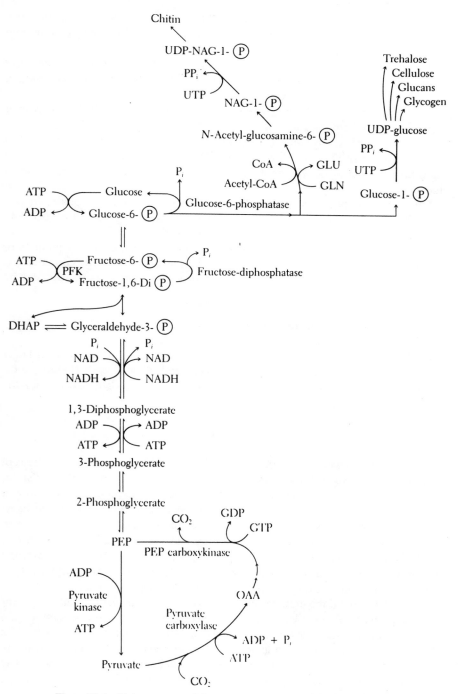

Figure 10.6 Major reactions of gluconeogenesis and polysaccharide synthesis.

carbon along the EM pathway. Similarly, if both enzymes were active in the presence of nonsugar sources, PFK would antagonize FDPase and the flow of carbon through gluconeogenesis would be impeded (Gancedo et al., 1965).

The studies of Foy and Bhattacharjee (1977) give a somewhat different perspective on the interacting roles of PFK and FDPase in regulating the flow of carbon through the EM pathway and gluconeogenesis. They noted that wild-type yeast cells grown in a nutrient medium containing glucose had over 100 times the PFK activity after 20 hours than did comparable cells on an ethanol medium. However, FDPase was present in cells grown in either carbon source, and the activity was approximately the same in each. Thus, the flow of carbon through the EM pathway appears to determine whether or not the net flow of carbon is in the direction of glycolysis or gluconeogenesis.

Fructose-6-phosphate can readily be converted to glucose-6-phosphate. In animal cells, this compound can then be dephosphorylated to glucose by glucose-6-phosphatase, but there are no unambiguous reports of this enzyme in fungi.

In any case, it is glucose-6-phosphate that is at the major branch point from gluconeogenesis to the pathways for polysaccharide synthesis. This compound is converted to glucose-1-phosphate, which then reacts with a nucleoside triphosphate such as UTP:

$$\text{UTP + glucose-1-phosphate} \rightleftharpoons \text{UDP-glucose + PP}_i$$

The UDP-glucose functions as an energized sugar donor for the synthesis of larger molecules such as trehalose, glycogen, cellulose and other glucans (Figure 10.6). Glucose-6-phosphate can also be converted to UDP-N-acetylglucosamine for the formation of chitin.

The Glyoxylate Cycle

Fungi can utilize a wide variety of compounds as carbon sources for the synthesis of carbohydrates so long as these sources can first be converted to pyruvate or PEP for entry into gluconeogenesis. For example, the amino acid alanine can function as a precursor by being deaminated to pyruvate. Glutamate can be deaminated to α-ketoglutarate which, along with other TCA cycle intermediates, can be converted to OAA and thence to PEP. Lipids can be degraded via β-oxidation to acetylCoA, which then enters the TCA cycle and forms OAA. However, in this last case, the two carbons entering from acetyl-CoA are lost as CO_2 as they pass around the TCA cycle. Can there thus be no net synthesis of carbohydrates from lipids?

In plants and fungi the presence of the glyoxylate cycle enables the cell to circumvent this problem. Acetyl-CoA is converted via TCA enzymes to isocitrate (Figure 10.7a). Isocitrate is then cleaved by isocitrate lyase to succinate and glyoxylate. The latter reacts with another molecule of acetyl-CoA in the presence of malate synthase to yield malate. Notice that this conversion of isocitrate to succinate and malate results in a bypass of those reactions in the TCA cycle where CO_2 is evolved. Thus,

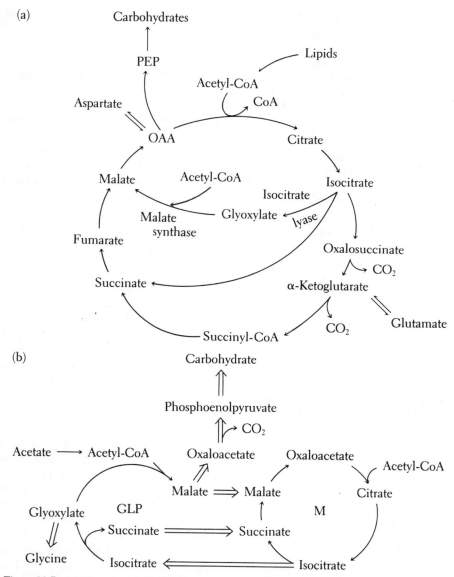

Figure 10.7 (a) The effect of the glyoxylate "shunt" in bypassing the points of CO_2 evolution in the tricarboxylic acid cycle. (b) Diagram to illustrate the possible metabolism of acetate in a fungus containing mitochondria (M) and glyoxysomelike particles (GLP) that lack any duplication of TCA cycle enzymes. Both malate and succinate must be transferred to the mitochondria to provide a net synthesis of TCA cycle intermediates (cytosol reactions and interparticular transfers shown by double lines). (Casselton, 1976)

lipids such as the triglycerides stored in spores can be used for the net synthesis of carbohydrates by means of the glyoxylate cycle.

As with the TCA cycle, glyoxylate cycle enzymes in fungi are subject to regulation by the carbon source supplied in the culture medium. Tabuchi and Igoshi (1978) reported that glucose represses the synthesis of these enzymes in *Candida lipolytica*. Derepression occurs when succinate or *n*-alkane is the carbon source.

There is considerable uncertainty, however, as to the intracellular location of the glyoxylate cycle enzymes in fungi. In plants, these enzymes are found in a micro-body named the glyoxysome. However, although glyoxysomelike particles are present in various fungi (Maxwell et al., 1977), in some species glyoxylate cycle enzymes are found in the soluble fraction as well. In addition, the glyoxysomelike particles in fungi appear to lack duplicates of the mitochondrial enzymes. Thus, it has been hypothesized (Figure 10.7b) that the reactions constituting the glyoxylate cycle actually involve the passage of intermediates back and forth between the glyoxysomelike particle and the mitochondrion before the cycle can be completed (Casselton, 1976).

SUMMARY: AN OVERVIEW OF METABOLISM

Knowledge of the pathways by which fungi metabolize carbon sources is central to an understanding of fungal nutrition. Although the major metabolic routes have been identified in many species, an overwhelming number of details are unknown.

Polysaccharides encountered by a fungus in the field are hydrolyzed extracellularly to subunits of a size that can be transported into the cell. In the cytoplasm, these subunits then feed into one or more glycolytic pathways—usually the Embden–Myerhof, pentose-phosphate, or Entner–Doudoroff pathways—and are converted to pyruvate. Under anaerobic conditions pyruvate is converted to ethanol and/or lactic acid by fermentation pathways. Pyruvate then enters the mitochondria, is converted to acetyl-CoA, and is fed into the TCA cycle where it is degraded to carbon dioxide plus electrons and hydrogens. These latter components are carried by NAD and FAD to the electron transport chain where oxidative phosphorylation occurs. Both a cyanide-sensitive and a cyanide-insensitive pathway for electron transport are present in many fungal species.

In addition to their roles in respiration, acetyl-CoA and the TCA cycle are at the hub of the metabolism of other compounds. Lipids such as triglycerides are also converted to acetyl-CoA via β-oxidation and enter the TCA cycle in this form. The TCA cycle also serves as the point of entry for most of the amino acids. In biosynthetic reactions, fatty acids are synthesized from acetyl-CoA subunits via a fatty acid synthetase complex. Many amino acids are synthesized from TCA intermediates. Carbohydrates are synthesized in gluconeogenesis via a reversal of the EM pathway with alternative routes to bypass irreversible reactions. The glyoxylate cycle operates in many species to permit the net synthesis of carbohydrate from lipid.

These reactions are under tight regulatory control. The energy charge in particular modulates the direction of carbon flow by affecting key regulatory enzymes such as PFK, FDPase, and citrate synthase. Other substrates such as fatty acyl-CoA and citrate also have a regulatory role. It is by such modulation that the patterns of growth and development are controlled.

REFERENCES

Ainsworth, P. J., A. J. S. Ball, and E. R. Tustanoff. 1980a. Cyanide-resistant respiration in yeast (*Saccharomyces cerevisiae*): 1. Isolation of a cyanide-insensitive NAD(P)H oxidoreductase. *Arch. Biochem. Biophys.* 202: 172–186.

Ainsworth, P. J., A. J. S. Ball, and E. R. Tustanoff. 1980b. Cyanide-resistant respiration in yeast (*Saccharomyces cerevisiae*): 2. Characterization of a cyanide-insensitive NAD(P)H oxidoreductase. *Arch. Biochem. Biophys.* 202: 187–200.

Andreasen, A. A., and T. J. B. Stier. 1954. Anaerobic nutrition of *Saccharomyces cerevisiae*: II. Unsaturated fatty acid requirement for growth in a defined medium. *K. Cell. Comp. Physiol.* 48: 317–328.

Atkinson, D. E. 1966a. Biological feedback control at the molecular level. *Science* 150: 851–857.

Atkinson, D. E. 1966b. Regulation of enzyme activity. *Ann. Rev. Biochem.* 35: 85–124.

Atkinson, D. E. 1969. Regulation of enzyme function. *Ann. Rev. Microbiol.* 23: 47–68.

Barford, J. P., and R. J. Hall. 1979. An examination of the Crabtree effect in *Saccharomyces cerevisiae*: The role of respiratory adaptation. *J. Gen. Microbiol.* 114: 267–275.

Bartnicki-Garcia, S., and W. J. Nickerson. 1961. Thiamine and nicotinic acid: Anaerobic growth factors for *Mucor rouxii*. *J. Bacteriol.* 82: 142–148.

Becker, J. U., and A. Betz. 1972. Membrane transport as controlling pacemaker of glycolysis in *Saccharomyces cerevisiae*. *Biochim. Biophys. Acta* 274: 584–597.

Blumenthal, H. J. 1965. Carbohydrate metabolism: 1. Glycolysis. In G. C. Ainsworth and A. S. Sussman (Eds.), *The Fungi*, Vol. 1, pp. 229–268. New York: Academic.

Blumenthal, H. J. 1968. Glucose catabolism in fungi. *Wallerstein Lab. Commun.* 31: 171–191.

Brody, S. 1970. Correlation between reduced nicotinamide adenine dinucleotide phosphate levels and morphological changes in *Neurospora crassa*. *J. Bacteriol.* 101: 802–807.

Brody, S., and E. L. Tatum. 1966. The primary biochemical effect of a morphological mutation in *Neurospora crassa*. *Proc. Natl. Acad. Sci. USA* 56: 1290–1297.

Brown, G. D., and D. S. Beattie. 1978. Formation of the yeast mitochondrial membrane: V. Difference in the assembly process of cytochrome oxidase and coenzyme QH2: cytochrome c reductase during respiratory adaptation. *Biochim. Biophys. Acta* 538: 173–187.

Casselton, P. J. 1976. Anaplerotic pathways. In J. E. Smith and D. R. Berry (Eds.), *The Filamentous Fungi*, Vol. 2, pp. 121–136. New York: Wiley.

Cochrane, V. W. 1976. Glycolysis. In J. E. Smith and D. R. Berry (Eds.), *The Filamentous Fungi*, Vol. 2, pp. 65–91. New York: Wiley.

DeDeken, R. H. 1966. The Crabtree effect: A regulatory system in yeast. *J. Gen. Microbiol.* 44: 149–156.

DeMoss, J. A., and H. E. Swim. 1957. Quantitative aspects of the tricarboxylic acid cycle in baker's yeast. *J. Bacteriol.* 74: 445–451.

Elzainy, T. A., M. M. Hassan, and A. M. Allam. 1973. New pathway for non-phosphorylated degradation of gluconate by *Aspergillus niger*. *J. Bacteriol.* 114: 457–459.

Evans, R. C., and M. O. Garraway. 1976. Effect of thiamine on ethanol and pyruvate production in *Helminthosporium maydis*. *Plant Physiol.* 57: 812–816.

Foy, J. J., and J. K. Bhattacharjee. 1977. Gluconeogenesis in *Saccharomyces cerevisiae*: Determination of fructose-1,6-biphosphate in cells grown in the presence of glycolytic carbon sources. *J. Bacteriol.* 129: 978–982.

Gancedo, C. 1971. Inactivation of fructose-1,6-diphosphatase by glucose in yeast. *J. Bacteriol.* 107: 401–405.

Gancedo, C., M. L. Salas, A. Giner, and A. Sols. 1965. Reciprocal effects of carbon sources on the levels of an AMP-sensitive fructose-1,6-diphosphate and phophosfructokinase in yeast. *Biochem. Biophys. Res. Commun.* **20**: 15–20.

Gulyi, M. F. 1977. Metabolic regulation of initial reactions of the tricarboxylic acid cycle. *UKR Biokhim. Zh.* **49**: 115–129.

Gunner, H. B., and M. Alexander. 1964. Anaerobic growth of *Fusarium oxysporium. J. Bacteriol.* **87**: 1309–1316.

Hathaway, J. A., and D. E. Atkinson. 1965. Kinetics of regulatory enzymes: Effect of adenosine triphosphate on yeast citrate synthase. *Biochem. Biophys. Res. Commun.* **20**: 661–665.

Howell, N., C. A. Zuiches, and K. D. Munkres. 1971. Mitochondial biogenesis in *Neurospora crassa*: I. An ultrastructural and biochemical investigation of the effects of anaerobiosis and chloramphenicol inhibition. *J. Cell. Biol.* **50**: 721–736.

Kawakita, M. 1970. Studies on the respiratory system of *Aspergillus oryzae*: II. Preparation and some properties of mitochondria from mycelium. *J. Biochem. (Tokyo)* **68**: 625–631.

Koobs, D. H. 1972. Phosphate mediation of the Crabtree and Pasteur effects. *Science* 178: 127–133.

Lehninger, A. L. 1975. Biochemistry. New York: Worth.

Maxwell, D. P., V. N. Armentrout, and L. B. Graves. 1977. Microbodies in plant pathogenic fungi. *Ann. Rev. Phytopathology* **15**: 115–134.

Ng, W. S., J. E. Smith, and J. G. Anderson. 1972. Changes in carbon catabolic pathways during synchronous development of *Aspergillus niger. J. Gen. Microbiol.* **71**: 495–504.

Niederpruem, D. J. 1965. Carbohydrate metabolism: 2. Tricarboxylic acid cycle. In G. C. Ainsworth and A. S. Sussman (Eds.), *The Fungi*, Vol. 1, pp. 269–300. New York: Academic.

Slonimski, P. P. 1956. Adaptation respiratoire: Developpment du systeme hemo-proteique induit par l'oxygene. Proceedings of the 3rd International Congress of Biochemistry, 1955, p. 242.

Tabuchi, T., and K. Igoshi. 1978. Regulation of enzyme synthesis of the glyoxylate, the citric, and the methylcitric acid cycles in *Candida lipolytica. Agric. Biol. Chem.* **42**: 2381–2386.

Vanderleyden, J., L. Hanssens, and H. Verachtert. 1978. Induction of cyanide-insensitive respiration in *Moniliella tomentosa* by the use of *n*-propanol. *J. Gen. Microbiol.* **105**: 63–68.

Vanderleyden, J., J. Kurth, and H. Verachtert. 1979. Characterization of cyanide-insensitive respiration in mitochondria and submitochondrial particles of *Moniliella tomentosa. Biochem. J.* **182**: 437–444.

Walker, P., and M. Woodbine. 1976. The biosynthesis of fatty acids. In J. E. Smith and D. R. Berry (Eds.), *The Filamentous Fungi*, Vol. 2, pp. 137–158. New York: Wiley.

Watson, K. 1976. The biochemistry and biogenesis of mitochondria. In J. E. Smith and D. R. Berry (Eds.), *The Filamentous Fungi*, Vol. 2, pp. 92–120. New York: Wiley.

11

THE SYNTHESIS OF NUCLEIC ACIDS AND PROTEINS

Genetic information encoded in molecules of DNA provides the basis for responses by fungi to nutrients and other environmental factors. Indeed, the existence of metabolic pathways and the manner in which their regulation results in different patterns of growth and development are due to the presence of discrete nucleotide sequences which are switched on at the proper time in the life cycle of the fungus.

Although there is considerable information concerning the synthesis of nucleic acids and proteins in bacteria such as *E. coli*, it is only in the past decade that many of the details have been investigated in fungi. There are still a great many gaps in our knowledge. In this chapter, we shall briefly discuss the major aspects of fungal DNA, RNA, and protein synthesis. A detailed review has recently been published by Berry and Berry (1976).

THE STRUCTURE OF CHROMATIN

In eukaryotic organisms, 80–99% of the genetic material is present in chromosomes as "chromatin," a complex of DNA and protein so named because of its reaction with certain biological stains. The remaining 1–20% of the DNA occurs in mito-chondria, although some strains of *Saccharomyces cerevisiae* contain up to 5% of their DNA in nuclear plasmids (Petes, 1980). In general, the DNA content of fungal cells is less than that of other eukaryotes (Table 11.1). *Saccharomyces cerevisiae*, with 1.4×10^7 nucleotide pairs per haploid genome, has the smallest DNA content of any known eukaryote, only three times larger than the genome of *E. coli* (Russell and Wilkerson, 1980). Note in Table 11.2 that the genomes of all fungi tested fall within the narrow range of 1.4×10^7 to 9.3×10^7 nucleotide pairs, several orders of magnitude lower than that of mammalian cells.

Table 11.1 The Approximate DNA Content of Some Cells and Viruses[a]

Species	DNA, pg per Cell (or Virion)	Number of Nucleotide Pairs (millions)
Mammals	6	5500
Amphibia	7	6500
Fishes	2	2000
Reptiles	5	4500
Birds	2	2000
Crustaceans	3	2800
Mollusks	1.2	1100
Sponges	0.1	100
Higher plants	2.5	2300
Fungi	0.02–0.17	20
Bacteria	0.002–0.06	2
Bacteriophage T4	0.00024	0.17
Bacteriophage λ	0.00008	0.05

Source: Lehninger (1975).

[a]The figures given for eukaryotic organisms are for somatic (diploid) cells.

Table 11.2 DNA Content per Haploid Genome of Selected Fungi

DNA Content per Haploid Genome		Nucleotide Pairs per Haploid Genome $\times 10^{-7}$
Species	Class	
Aspergillus nidulans	Ascomycete	4.1
Aspergillus sojae	Ascomycete	6.8
Neurospora crassa	Ascomycete	4.3
N. intermedia	Ascomycete	3.2
N. tetrasperma	Ascomycete	3.2
Ophiostoma multiannulatum	Ascomycete	4.5
Saccharomyces	Ascomycete	2.2
S. cerevisiae	Ascomycete	1.4
Talaromyces vermiculatus	Ascomycete	4.5
Coprinus lagopus	Basidiomycete	3.6
Ustilago maydis	Basidiomycete	5.3
Blastocladiella emersonii	Chytridiomycete	9.3
Mucor azygospora	Zygomycete	3.4
M. bacilliformis	Zygomycete	3.0
M. racemosus	Zygomycete	1.6
Phycomyces blakesleeanus	Zygomycete	2.9
Achlya ambisexualis	Oomycete	5.1

Source: Ullrich and Raper (1977).

Fungal DNA is similar to that of other eukaryotes in being composed of both unique (single-copy) and reiterated (multiple-copy or redundant) nucleotide sequences. The unique sequences code largely for mRNA; the reiterated sequences code for products such as rRNA, tRNA, and chromosomal proteins, which are required in large quantities. However, the function of a considerable amount of this reiterated DNA is unknown (Timberlake, 1978). Britten and Davidson (1969) proposed a model whereby these redundant genes would be involved in the regulation of gene expression. However, in N. crassa the reiterated genes are arranged in clusters, perhaps even on the same chromosome, instead of being interspersed among the unique genes. This clustering argues against their having a regulatory role, although reiterated sequences too short to be detected may be interspersed and serve a regulatory function (Krumlauf and Marzluf, 1979, 1980). On the other hand, reiterated sequences may merely be remnants of genetic experiments that failed (Ullrich and Raper, 1977). Fungi differ from plants and animals in having a much lower percentage of reiterated DNA (Van Etten, 1981; Long and Dawid, 1980). In some eukaryotes, as much as 80% of the total DNA is redundant; these genes may be repeated a hundred to a million times (Britten and Davidson, 1969). However, Dutta (1974) found only 10–20% of the DNA from a variety of fungal species to be reiterated with approximately 100–110 copies per gene (Table 11.3). In *Aspergillus nidulans* only 2–3% of the DNA is redundant with approximately 60 copies per haploid genome. Most of this DNA codes for rRNA (Timberlake, 1978).

Although the structure of DNA has been known since 1953 when Watson and Crick published their landmark paper, much less is known about the protein component and particularly the manner in which the DNA and proteins are associated to form a chromosome. There are two major kinds of proteins in chromatin: histone

Table 11.3 Summary of Occurrence of Repeated DNA Sequences in Fungi

Name	Percentage of Repeated DNA Sequence Based on		½ $C_o t$ Values from DNA:DNA Reassociation Kinetics	Calculated Molecular Weights in Daltons	No. of Copies
	^{32}P-labeled DNA	Optical Absorbency at 260 nm			
Basidiomycetes					
Coprinus lagopus H$_2$A$_6$B$_5$	12–15	15–18	0.060	3.6×10^7	112
Ascomycetes					
Neurospora crassa 74A	10–12	12–15	0.055	3.2×10^7	100
N. intermedia A	10–12	12–15	0.055	3.2×10^7	100
N. tetrasperma 85A	10–12	12–15	0.058	3.2×10^7	100
Phycomycetes					
Mucor azygospora	—	15–18	0.058	3.4×10^7	103
Rhizopus stolonifer	—	15–18	—	—	—
Allomyces arbuscula	15–20	—	—	—	—

Source: Dutta (1974).

proteins, which have a high content of basic amino acids such as arginine and lysine, and a wide assortment of acidic proteins lumped together under the rubric of nonhistone chromosomal proteins.

In fungi there are five major histone proteins, which have been given various names: H1 (also called F1 or I), H2A (F2A2 or IIb1), H2B (F2b or IIb2), H3 (F3 or III), and H4 (F2a1 or IV). These have been reported in all the major groups of fungi and in most cases appear similar to the histones of other eukaryotic organisms. Some fungi were thought not to contain any histones, but their lack may be due to the type of extraction procedure (Horgen and Silver, 1978).

The chromatin of fungi, like that of other eukaryotes, is arranged in a repeating pattern of nucleoprotein particles. These particles are arranged like beads on a string and have been termed nucleosomes (Figure 11.1). The nucleosome consists of DNA approximately 140 nucleotide pairs long, wrapped around the outside of a core of histones (two each of H2A, H2B, H3, and H4). Between the particles is a spacer region consisting of approximately 20 base pairs of DNA complexed with molecules of H1. The nucleosomes of most fungi are similar to those found in other organisms but generally have shorter spacer regions. The function of the nucleosome-spacer unit is not known. The length of the spacer is shorter in fungal cells growing exponentially (when more genes are presumably being transcribed) than in other

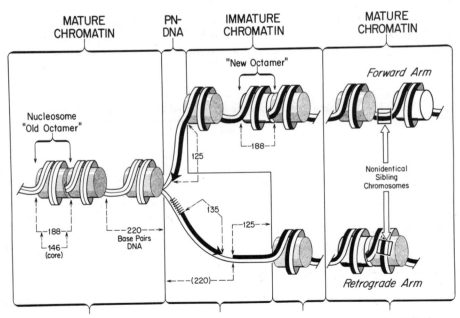

Figure 11.1 Replicating fork of a eukaryotic chromosome. Only those components for which there is substantial experimental evidence are represented. Mature (heavy shading) and immature (light shading) nucleosomes, an RNA-primed Okazaki fragment (||| →) on the retrograde arm, helix destabilizing (*HD*) protein, DNA polymerase (α-pol), and DNA ligase (lig-I) The sizes of the components are drawn to reflect their molecular weight. (DePamphilis and Wassarman, 1980)

phases of growth. This suggests that a relationship exists between length of the repeating units and transcriptional activity (Horgen and Silver, 1978; Petes, 1980).

Most histones are synthesized at the time the DNA is replicated. In *S. cerevisiae*, histone synthesis depends on simultaneous DNA replication, but DNA can be synthesized without concurrent histone synthesis (Moll and Wintersberger, 1976). Histones can also be modified by acetylation, methylation, and phosphorylation following translation. In *Physarum polycephalum*, the phosphorylation of H1 may trigger the condensation of chromosomes during mitosis (Bradbury et al., 1974a,b). In *Achlya*, treatment with the sex hormone antheridiol results in the acetylation of two histonelike proteins. This may be important in the selective gene activation preceding the differention of antheridial hyphae (Horgen and Ball, 1974).

In contrast to histones, nonhistone chromosomal proteins are quite diverse. Gel electrophoresis of these proteins often shows as many as 115 peptide bands. Some of these are enzymes including nucleases, nucleic acid polymerases, and enzymes associated with chromosomal metabolism. In *P. polycephalum* the contractile proteins actin and myosin are major components of nonhistone proteins (LeStourgeon et al., 1975). It is not known how these proteins are integrated into the nucleosome complex, but it appears that they may be involved in altering the expression of genes. In *P. polycephalum*, changes occur in the types of nonhistone proteins present in the organism as it sporulates and encysts. Such protein changes are reversed when the encysted plasmodium is incubated on a nutrient-rich medium which causes it to dedifferentiate (LeStourgeon and Rusch, 1973; Horgen and Silver, 1978).

DNA REPLICATION

DNA and histone synthesis occurs during the S (or synthetic) phase of the cell cycle (Figure 11.2). In most eukaryotes, this phase is preceded by a presynthetic "gap" called G_1, during which time the reactions prerequisite to DNA replication occur. The S phase is followed by a postsynthetic G2 phase in which preparations for the actual separation of the chromosomes occur. The relative lengths of these phases vary widely among fungal species and in some may even be absent. For example, actively growing plasmodia of *Physarum* lack the G_1 phase (Horgen and Silver, 1978).

So far as is known, DNA replication in fungi closely resembles that in other eukaryotes. In *Saccharomyces cerevisiae* replication occurs at multiple initiation sites which are spaced an average of every 30 μm along the chromosome (Figure 11.3). These initiation sites appear to be distinct regions of the chromosome but are not activated simultaneously. At these sites the DNA helix unwinds and nucleotides are incorporated on each single-stranded template. This process is catalyzed by DNA polymerase enzymes. Baker's yeast contains two DNA polymerases in the nucleus and a third in the mitochondria (Plevani and Chang, 1977). Resting uredospores of the bean rust pathogen *Uromyces phaseoli* appear to contain a single nuclear DNA polymerase that depends on the simultaneous presence of all four deoxyribonucleoside triphosphates (dATP, dCTP, dTTP, dGTP), a DNA template, and magnesium

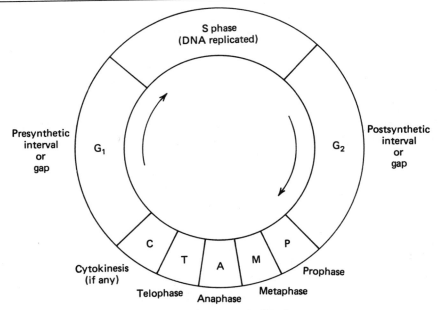

Figure 11.2 A typical cell cycle.

ions (Yaniv and Staples, 1978). In the mitochondria, several isozymes of this enzyme are also present (Staples and Yaniv, 1978). In both prokaryotes and other eukaryotes the nucleotides are first incorporated into short chains called Okazaki fragments via an RNA primer. The RNA is then excised and the fragments joined together by a DNA ligase (DePamphilis and Wassarman, 1980). Although nucleotide chains resembling Okazaki fragments have occasionally been observed in *S. cerevisiae*, it is unclear whether they are actually involved in DNA replication in this organism (Petes, 1980). The general pattern of DNA replication is illustrated in Figure 11.1.

In the normal strain of *Neurospora crassa*, DNA can also be synthesized using RNA as a substrate. Dutta and co-workers (1977) have identified an RNA-dependent DNA polymerase to which is bound to a segment of RNA. This enzyme, which resembles the "reverse transcriptase" of vertebrate cells, may be important in the infection of this organism by RNA viruses.

RNA SYNTHESIS: TRANSCRIPTION

It is fairly well accepted that when a particular set of genes is to be transcribed, one or more RNA polymerase enzymes bind to the DNA at specific initiation sites. The DNA helix then unwinds, and one strand serves as template for the synthesis of classes of RNA such as rRNA, tRNA, and mRNA. In this process, the nitrogen bases of ribonucleoside triphosphates form hydrogen bonds with the complementary bases on the DNA template. The pyrophosphate group is cleaved, and the ribonu-

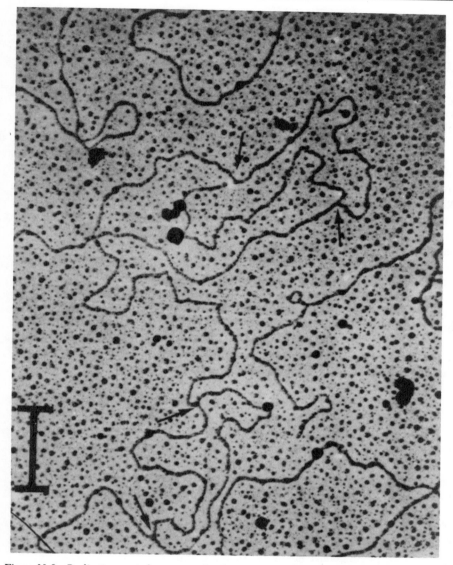

Figure 11.3 Replicating yeast chromosome showing a segment of yeast nuclear DNA with two active replication origins. The bar in the diagram is 1 μm. Arrows indicate the replication forks. (Petes, 1980)

cleoside monophosphates are linked together. Transcription continues until a termination code is encountered on the template; then the enzyme (or enzymes) and the new RNA strand detach, and the DNA helix re-forms. The unicellular and filamentous fungi are similar to other eukaryotes in having three principal RNA polymerases. In general, polymerase I preferentially transcribes rRNA, polymerase II transcribes mRNA, and polymerase III transcribes tRNA and some rRNA (Russell and

Wilkerson, 1980). Other RNA polymerases occur in the mitochondria and are responsible for the transcription of mitochondrial DNA (e.g., Horgen and Griffin, 1971).

The synthesis of RNA during spore germination has been studied in some detail. In species such as *Aspergillus oryzae* and *Peronspora tabacina*, there is a sequential synthesis of different RNA classes as germination proceeds (Ono et al., 1966; Hollomon, 1970). In others such as *Rhizopus stolonifer*, all the major classes are synthesized during the first few minutes (Roheim et al., 1974).

Once synthesized, the RNA strands can be modified in various ways, such as by methylation, by the addition of nucleotide "caps" at the 5' end, and by the addition of polyadenylate (poly A) segments to the 3' (or "tail") end. This modified RNA can then be processed further, either before or during its association with ribosomes. In *Rhizopus*, poly A segments approximately 100 nucleotides long are linked to the newly transcribed "precursor RNA" to form a poly A–RNA chain with a total length of 1200 bases. Before translation begins, 50 nucleotides are cleaved from the poly A segment and 300 nucleotides are cleaved from the opposite end. The rate of poly A synthesis is higher during the first hour of germination and decreases during later stages. In addition, the length of the poly A tail decreases to about 35 nucleotides as germination proceeds, suggesting that continued translation may result in a cleavage of these nucleotides (Freer et al., 1977). A similar system involving poly A addition has been reported for *Achlya ambisexualis*. In this species, there are three classes of poly A–RNA in which the poly A segments are 15, 30, and 50 nucleotides long. These also result from post-transcriptional modification and may serve to influence the function or metabolism of the RNA to which they are attached (Law et al., 1978).

Dormant *Rhizopus* spores also contain a substantial amount of poly A–RNA that was formed during spore formation. This "preformed RNA" codes for the synthesis of discrete proteins that set the germination process in motion (Van Etten and Freer, 1978).

In animal cells, most mRNA is transcribed in very long strands called heterogeneous nuclear RNA (HnRNA). While still in the nucleus, these strands are cleaved, sections are degraded, and the resulting fragments are passed into the cytoplasm to be translated. In contrast, fungi appear to lack appreciable HnRNA (Van Etten, 1981). In *Achlya*, for example, the composition of nuclear and polysomal poly A–RNA are essentially identical, indicating that little or no cleaving or degrading of mRNA occurs (Timberlake et al., 1977).

PROTEIN SYNTHESIS: TRANSLATION

Once an mRNA strand has been transcribed it passes into the cytoplasm and attaches itself to a ribosome. The process of polypeptide synthesis on the ribosome-mRNA complex is referred to as translation. It is in this process that the linear sequence of bases present in the mRNA is translated into a linear sequence of amino acids. Translation is achieved by means of specific adaptor molecules, the transfer RNAs (tRNAs), which carry individual amino acids from the cytoplasm to the ribosome-mRNA complex and form complementary hydrogen bonds with the triplets of bases on the mRNA. Peptide bonds are formed between adjacent amino acids, and this process culminates in the formation of a protein. Let us now examine these steps in detail.

Fungal Ribosomes

Ribosomes are particulate complexes of RNA and protein that serve as sites of protein synthesis. In general, fungal ribosomes are similar to those found in other eukaryotic cells and are classified according to their sedimentation characteristics using Svedberg units (S). Cytoplasmic ribosomes are 80S particles composed of one 60S spherical subunit and one 40S ellipsoidal subunit. Ribosomes in the mitochondria are 73S–80S and are similar in some respects to those in bacteria. The molecular weight of fungal ribosomes is roughly intermediate between that of bacteria and animal cells (Figure 11.4). A detailed review of the structure and composition of fungal ribosomes has recently been published (Russell and Wilkerson, 1980); in this section we focus attention on cytoplasmic ribosomes.

Each ribosomal subunit is composed of a core of rRNA surrounded by a protein coat. Approximately equal amounts of RNA and protein are present (Figure 11.4). In addition, divalent cations such as Mg are required for subunit binding, presumably to reduce the repulsion among negatively charged phosphate groups and/or to link together proteins or proteins and RNA. Each ribosome contains four RNA molecules: in the large subunit there are 5S, 5.8S, and 25–28S (referred to as 25S) molecules; in the small subunit there is a 17–18S (referred to as 17S) molecule. The role of each RNA molecule is uncertain. In a variety of eukaryotes the 17S molecules have a 3' end with triplets complementary to the termination, or "stop," sequences UAA, UAG, and UGA of mRNA, which suggests that this rRNA may be involved in the termination of translation. Both the 5S and 5.8S molecules have

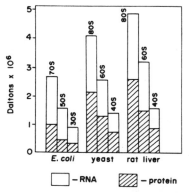

Figure 11.4 Relative RNA and protein content of cytoplasmic ribosomes from *E. coli*, yeast, and rat liver. (After Martini et al., 1973, and appearing in Russell and Wilkerson, 1980)

sequences complementary to the initiator, or "start," tRNA (in eukaryotes this is methionine–tRNA), and thus these molecules may be involved in the initiation of translation. Both the 17S and 25S molecules are methylated at several 2'-O-ribose locations, and nitrogen bases may be modified as well. These modifications may not only protect against degradation but may also induce a particular conformation in the rRNA for the binding of proteins. There are approximately 70 different proteins in each ribosome: 40 in the 60S subunit and 30 in the 40S subunit. Each protein appears to be present in only one copy, and thus it is likely that each has a unique role.

In all eukaryotes, the genes for rRNA production are present in multiple copies. In *Saccharomyces cerevisiae* and *S. carlsbergensis* there are 140 copies per haploid genome; in *Allomyces arbuscula* there are 270 copies. Such redundancy is not surprising when we consider that a single cell may contain many thousands of ribosomes. Most of these genes are linked on one or two chromosomes. In all fungi studied, the genes for the 17S, 5.8S, and 25S rRNA molecules are contiguous and are transcribed as a single pre-rRNA molecule. The 5S rRNA is transcribed separately. In some species such as *S. cerevisiae* the gene for the 5S molecule is linked to the other rRNA genes; in other species such as *Neurospora crassa*, this gene is located on a different chromosome (Figure 11.5).

Most of the ribosomal subunits are synthesized in the nucleus and then transported to the cytoplasm. In yeast, the adjacent 17S–5.8S–25S genes are transcribed into a pre-rRNA strand preferentially by RNA polymerase I. This strand is then modified while still in the nucleolus by 2'-O-ribose methylation and base modification. Concurrently, the pre-rRNA is cleaved into smaller segments in a manner which varies with the fungal species. The cleavage pattern in *S. carlsbergensis* is shown in Figure 11.6. Note that processing occurs by breaks in the intervening spacer regions, and then by excision of these short segments. The 5S rRNA is transcribed independently by RNA polymerase II and also modified. The ribosomal proteins are synthesized in the cytoplasm, modified by methylation and phosphorylation, and transported to the nucleus to be associated with rRNA immediately after rRNA synthesis (Figure 11.7).

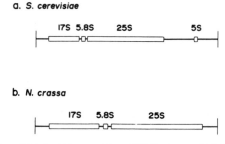

a. *S. cerevisiae*

17S 5.8S   25S   5S

b. *N. crassa*

17S   5.8S   25S

Figure 11.5 Organization of (a) the 5.9-megadalton (Mdal) ribosomal DNA repeat unit in *Saccharomyces cerevisiae* (Nath and Bollon, 1977), and (b) the 6.0-Mdal ribosomal DNA repeat unit in *Neurospora crassa* (Free et al., 1979). The boxed regions designate the mature RNA sequences and the solid lines represent spacer regions. (Russell and Wilkerson, 1980)

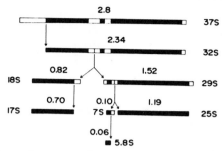

Figure 11.6   Detailed scheme of processing of the primary transcript of the ribosomal DNA of *Saccharomyces carlsbergensis* to produce mature 17S, 5.8S and 25S rRNAs (Planta et al., 1977). The number over each RNA species is the molecular weight of the RNA in megadaltons (Mdal). (Russell and Wilkerson, 1980)

The nutritional status of the organism has been shown to affect the rate of ribosome formation. For example, the rate of rRNA production in *N. crassa* incubated on nutrient broth is 10 times faster than in an ethanol medium (Alberghina et al., 1975). One can dramatically alter the rate of rRNA synthesis in this species simply by switching from nutrient-rich to nutrient-poor media (Sturani, et al., 1973).

In fungi, as in other eukaryotes, a significant number of ribosomes are membrane-bound. In *Dictyostelium discoideum*, for example, approximately 16% of the total ribosomes are bound to the endoplasmic reticulum or membrane vesicles. Although both free and bound ribosomes have a relatively rapid turnover rate in this species, each type turns over at a distinct point in the life cycle (Bourguignon and Katz, 1978).

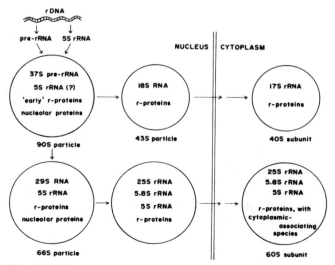

Figure 11.7   Scheme of ribosome maturation in *Saccharomyces carlsbergensis* (After Trapman et al., 1976, and presented in Russell and Wilkerson, 1980.)

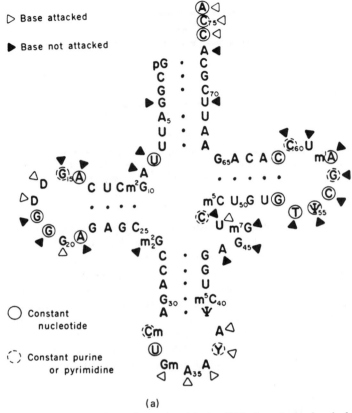

Figure 11.8 (a) Nucleotide sequence of yeast phenylalanine tRNA shown in the cloverleaf configuration. Circled bases are constant in all tRNAs, and dashed circles indicate which are occupied constantly by either purines or pyrimidines. The accessibility of various bases to chemical modification is indicated by open or closed triangles.

Charging of tRNA

In fungi as in other eukaryotes, tRNAs are relatively small molecules consisting of about 80 nucleotides folded into a precise, three-dimensional structure that resembles a cloverleaf (Figure 11.8a,b). All tRNAs have a guanosine residue at the 5′ end and a CCA sequence at the 3′ end to which the amino acid is attached. Several minor bases such as pseudouridylic acid and ribothymidilic acid are usually present; other bases may be methylated or otherwise modified. One of the arms of the cloverleaf is the anticodon loop, containing the three bases that will form complementary hydrogen bonds to the mRNA codon.

A specific amino acid is linked to a specific tRNA by a specific aminoacyl tRNA

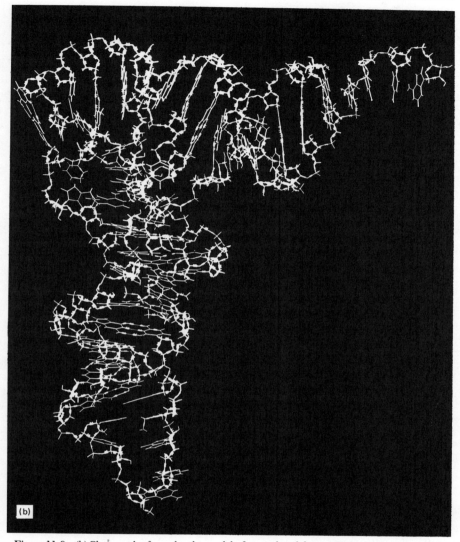

Figure 11.8 (b) Photograph of a molecular model of yeast phenylalanine tRNA. (Kim et al., 1974)

synthetase. This enzyme catalyzes the following two-step reaction, which "charges" the tRNA:

$$\text{ATP} + \text{amino acid} \longrightarrow \text{amino acid-AMP} + \text{PP};$$
$$\text{amino acid-AMP} + \text{tRNA} \longrightarrow \text{amino acid-tRNA} + \text{AMP}$$

The amino acid is first made reactive by binding to AMP. The aminoacyl group is then transferred to the 5′ (or CCA) end of the tRNA. In essence, the synthetase recognizes a specific amino acid and a specific tRNA and joins them together. In yeast, the rate-limiting step in this reaction is the release of the aminoacyl-tRNA complex from the enzyme (Fersht et al. 1978). There is usually more than one tRNA for each amino acid. In yeast, for example, there are five tRNAs for leucine, five for serine, 4 for glycine, and 4 for lysine (Berry and Berry, 1976).

Shearn and Horowitz (1969a,b) noted that the optimum conditions for charging amino acids to tRNA are not all the same. In *N. crassa*, the conditions for charging the respective tRNAs with alanine, glutamate, glutamine, glycine, and methionine are different from those of the other 14 amino acids studied. The most important factors appear to be magnesium ion concentration, levels of ATP and inorganic phosphate, and pH.

Other work suggests that aminoacyl-tRNA synthetases may vary in their specificity. Igloi et al. (1978) noted that in yeast valyl-tRNA synthetase will accept alanine, cysteine, isoleucine, serine, and threonine as substrates for both steps of the charging reaction. Similarly, phenylalanyl-tRNA synthetase will activate and transfer leucine, methionine and tyrosine. On the other hand, tyrosyl-tRNA synthetase is absolutely specific for tyrosine. These differences in specificities appear to be due to the ability of some of the synthetase enzymes to correct their mistakes by hydrolyzing the improper amino acid from the tRNA after charging. Thus, both valyl-tRNA synthetase and phenylalanyl-tRNA synthetase have a hydrolytic capacity but tyrosyl-tRNA synthetase does not. Enzymatic hydrolysis of a mis-acylated tRNA may play an important role in maintaining the specificity of the overall reaction.

Initiation, Elongation, and Termination of Protein Synthesis

Translation in both prokaryotes and eukaryotes is initiated by the mRNA codon AUG, the initiation codon. This codon binds the tRNA carrying methionine (Met-tRNA). In prokaryotes, the methionine is formylated resulting in N-formylmethionine-tRNA (fMet-tRNA). Although a short string of other nucleotides may precede the initiation codon on the mRNA, it is here that the first tRNA binds. Initiation is a complex process involving protein cofactors, enzymes, and energy in the form of GTP. In fungi, considerable progress has been made in elucidating the sequence of events in this process, and it appears that there are many similarities with other eukaryotic cells.

Gasior and co-workers (1979a,b) determined the principal steps in the initiation, elongation and termination of protein synthesis in *S. cerevisiae*. This sequence is diagrammed in Figure 11.9a–h. Once the Met-tRNA has been charged, it binds to the 40S ribosomal subunit via a protein initiation factor (IF) and the energy of GTP (a). (In both prokaryotic and eukaryotic cells it appears that the two ribosomal subunits are dissociated when the ribosome is not involved in translation.) The mRNA, possibly in the presence of another initiation factor, then also binds to the 40S subunit in a way such that the anticodon triplet forms complementary hydrogen bonds with the AUG initiation codon (b). Lastly, the 60S subunit joins the 40S-mRNA–Met-tRNA aggregation to form the 80S "initiation complex" (c). At this

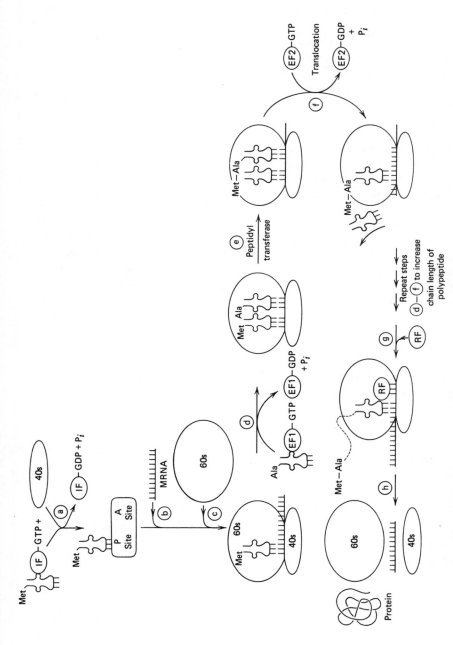

Figure 11.9 Scheme proposed for initiation, elongation, and termination of protein synthesis in *Saccharomyces cerevisiae*. See text for details. (Drawn from data in Gasior, 1979b.)

point the initiation Met-tRNA is bound to the peptidyl (or P) site on the 80S complex, and the initiation process is completed.

Elongation of the amino acid chain can now occur via the binding of an aminoacyl-tRNA to the next mRNA triplet at the acceptor, or A, site on the ribosome. This process requires a protein elongation factor, EF1, and the energy of GTP (d). A peptidyl transferase enzyme present in the 60S subunit then catalyzes the removal of methionine from the Met-tRNA and the formation of a peptide bond between the carbonyl group of the methionine and the amino group of the incoming aminoacyl-tRNA (e). In effect, the methionine is transferred from the first tRNA to the amino acid of the second. Elongation factor EF2 in the presence of GTP then causes the ribosome to move a distance of three nucleotides down the mRNA strand (f). This translocation step results in the expulsion of the discharged initiator tRNA and the movement of the dipeptide-tRNA from the A to the P site. A new aminoacyl-tRNA then attaches to the A site, the elongation process is repeated, and the peptide chain increases in length. As the ribosome moves down the mRNA strand, additional ribosomes may attach sequentially at the start codon and begin the translation process. This aggregation of several ribosomes on a single mRNA is called a polyribosome, or polysome, and provides a method for synthesizing several copies of a particular protein. In S. cerevisiae, polysomes with up to 10 or more ribosomes per mRNA have been reported. Polysomes are present even in ungerminated spores indicating that protein synthesis can occur even when transcription is not taking place. In conidia of N. crassa, for example, the percentage of polysomes increases from 3% in dry spores to 75% during the first 30 minutes of germination (Mirkes, 1974).

The elongation process ends when a termination codon is reached at the A site, causing a release factor (RF) to bind there (g). The ensuing reaction results in the cleaving of the polypeptide chain from the last tRNA, the release of the mRNA, and the dissociation of the ribosome into subunits (h) (Gasior et al., 1979a, b).

Post-Translational Changes in Proteins

Even after translation is completed the protein may be modified in various ways before it becomes functional in the cell. Some of these ways are mentioned briefly below.

1. **Changes in conformation.**   After it is released from the tRNA the protein is folded into a highly complex configuration with a distinctive three-dimensional structure. The extent of folding is a function of interactions among amino acids in the chain and interactions between amino acids and the environment such that the lowest free energy state is attained.

2. **Association with other proteins.**   Many enzymes are composed of more than one protein subunit. These are designated as dimers, trimers, tetramers, and so forth depending on the number of proteins involved. The association of these subunits is usually crucial for the proper functioning of the enzyme.

3. **Coenzyme-apoenzyme complexes.**   The combination of proteins with organic

or inorganic prosthetic groups is a widespread phenomenon and is often required for a protein to become functional. Many enzymes contain noncovalently bound prosthetic groups ranging from simple cations to vitamins and other growth factors. Numerous examples have been mentioned in Chapters 6 and 7. The prosthetic group may alter the folding of the polypeptide chain or may have a direct catalytic role.

4. Phosphorylation.   The catalytic functioning of many enzymes is modulated by phosphorylation and dephosphorylation. We have already mentioned the phosphorylation of ribosomal proteins. In addition, the activity of DNA-dependent RNA polymerase in yeast changes upon phosphorylation, presumably because the addition of a negatively charged phosphate group alters the association between the polymerase and the chromatin (Bell et al., 1976).

5. Glycosylation.   Many types of constituents covalently bound to proteins have been identified as carbohydrate groups. These may have structural or catalytic functions. An alpha-glucosidase from *Aspergillus niger* has been identified as a glycoprotein possessing 29 molecules of mannose, 6 of glucosamine and 14 of glucose (Rudick et al., 1979). Sometimes the amount of carbohydrate in a protein can be considerable. For example, a glycoprotein from *Histoplasma capsulatum* was found to have 60% carbohydrate and 40% protein (Sprouse, 1977).

6. Proteolysis.   Proteolytic enzymes may exert a significant influence on other proteins in the cell. This effect has been especially well documented with the conversion of zymogens into active enzymes by cleaving off a segment of the polypeptide chain. Yeast phosphofructokinase can be activated by treatment with proteinase (Afting, 1977). In *S. cerevisiae*, the vacuolar glycoprotein carboxypeptidase Y is synthesized from a large precursor that is activated by reacting with trypsin. This reaction results in the cleaving of four phenylalanine and at least 10 leucine residues (Hasilik and Tanner, 1978).

Synthesis of Nucleic Acids and Proteins in Mitochondria

Mitochondria from all eukaryotes studied contain DNA and the metabolic machinery necessary for DNA replication and protein synthesis. Mitochondrial DNA resembles that of bacteria in that it is circular, not complexed with histones, and lacks appreciable reiterated sequences. Unique RNA polymerases have been found in mitochondria, and mitochondria can actively incorporate labeled amino acids into protein in the absence of cytoplasmic RNA (Kuentzel and Blossey, 1974). Fungal mitochondria have ribosomes of the 73S size, slightly larger than those of bacteria but smaller than cytoplasmic ribosomes. It appears that at least some of the 73S particles are artifactual. Datema and co-workers (1974) isolated 80S ribosomes from mitochondria of *N. crassa* when magnesium and the ribonuclease inhibitor heparin were present in the isolation medium. When magnesium and heparin were absent, they obtained only 73S ribosomes.

Mitochondria have tRNAs and aminoacyl-tRNA synthetases that are distinct from those found in the cytoplasm (Emmet et al., 1972). Unlike cytoplasmic translation, the initial Met-tRNA to bind at the start codon is formylated, giving fMet-tRNA, and is thus similar to prokaryotic systems (Epler et al., 1970). Initiation and elongation factors from mitochondria of *N. crassa* are interchangeable with those from *Escherichia coli*, which suggests that mitochondrial and bacterial systems are similar in this respect (Sala and Kuntzel., 1970). Although differences exist between the protein synthesizing machinery of mitochondria and that of prokaryotic cells (see Borst and Grivell, 1978, for a review), the similarities are sufficient to support the hypothesis that mitochondria evolved from the parasitism of aerobic bacteria by eukaryotic cells.

Even though the machinery exists for decoding DNA to protein in mitochondria, 85–90% of the mitochondrial DNA is synthesized from nuclear DNA on cytoplasmic ribosomes. Apparently, mitochondrial DNA codes mostly for rRNA, tRNA, and translation factors (Berry and Berry, 1976: Watson, 1976). Why does the nucleus have most of the genes for mitochondrial activity? It has been hypothesized that numerous mitochondrial genes may have been transferred to the nucleus over long periods of evolutionary time. Evidence for such "jumping genes" has been reported for yeast by Farrelly and Butow (1983). They found four sections of nuclear DNA that were homologous with mitochondrial DNA. Presumably, portions of mitochondrial DNA were replicated and a copy integrated into the nuclear genome. Such gene transfers appear to be merely one aspect of the phenomenon of jumping genes already observed between one part of the nuclear genome and another.

SUMMARY

The mechanisms by which nucleic acids and proteins are synthesized in fungi appear generally similar to those in other eukaryotic organisms. However, various quantitative differences exist.

The quantity of DNA in fungal genomes is among the smallest of the eukaryotes. One explanation for the low DNA content may be the relatively low percentage of redundant, or reiterated, DNA sequences. The chromatin in fungal cells is organized into nucleosomes consisting of DNA strands wrapped around spherical aggregations of histone proteins. Nonhistone chromosomal proteins are also present.

One or more DNA polymerases exist that catalyze the formation of identical DNA strands. It is uncertain whether these strands are formed by the linkage of Okazaki fragments as in other eukaryotes.

Several different types of RNA polymerases catalyze the transcription of mRNA. Immediately after formation, certain RNAs may be modified by methylation, the addition of nucleotide caps, or the addition of polyadenylate sequences. These poly A "tails" may be shortened before or during translation. Although rRNA may be transcribed in long precursor strands that are later cleaved into segments, there is little such cleaving of mRNA strands. Thus, fungal cells lack appreciable heterogeneous nuclear mRNA.

Fungal cytoplasmic ribosomes are 80S in size and consist of a 60S subunit and a 40S subunit. These ribosomes are composed of approximately equal amounts of rRNA and protein. The genes for rRNA exist in multiple copies and most are closely linked. The genes for 17S, 5.8S, and 25S rRNAs are transcribed as a long precursor rRNA strand which subsequently is cleaved and has some of the bases modified. The 5S rRNA gene is transcribed separately.

When mRNA is to be translated, it binds to the 40S ribosomal subunit. Transfer RNA becomes charged with an amino acid in the presence of a specific aminoacyl tRNA synthetase and also binds to the 40S subunit. Finally the 60S subunit attaches to form the 80S initiation complex. In the presence of GTP and a number of protein factors, amino acids are transferred to the ribosome and are joined by peptide bonds. A release factor causes the termination of the translation process.

Once synthesized, the polypeptide can be modified by changing conformation, associating with other proteins, binding with coenzymes, or by phosphorylation, glycosylation, or proteolysis of certain segments.

Fungal mitochondria possess their own nucleic acid and protein synthesizing machinery somewhat resembling that of prokaryotic cells. However, most of the mitochondrial protein is synthesized from nuclear genes on cytoplasmic ribosomes.

REFERENCES

Afting, E. G., A. Lynen, H. Hinze, and H. Holzer. 1977. Effects of yeast proteinase A, proteinase B and carboxypeptidase Y on yeast phosphofructokinase. *Hoppe-Seyler's Z. Physiol. Chem.* **357:** 1771–1777.

Alberghina. F. A. M., E. Sturani, and J. R. Gohlke. 1975. Levels and rates of synthesis of ribosomal ribonucleic acid, transfer ribonucleic acid, and protein in *Neurospora crassa* in different steady states of growth. *J. Biol. Chem.* **250:** 4381–4388.

Bell, G. I., P. Valenzuela, and W. J. Rutter. 1976. Phosphorylation of yeast RNA polymerases. *Nature* **261:** 429–430.

Berry, D. R., and E. A. Berry. 1976. Nucleic acid and protein synthesis in filamentous fungi. In J. E. Smith and D. R. Berry (Eds.), *The Filamentous Fungi*, Vol. 2, pp. 238–291. New York: Wiley.

Borst, P., and L. A. Grivell. 1978. The mitochondrial genome of yeast. *Cell* **15:** 705–724.

Bourguignon, L. Y. W., and E. R. Katz. 1978. Isolation and characterization of the RNA of membrane-bound ribosomes in *Dictyostelium discoideum*. *J. Gen. Microbiol.* **106:** 93–102.

Bradbury, E. M., R. J. Inglis, and H. R. Matthews. 1974a. Control of cell division by very lysine rich histone (F1) phosphorylation. *Nature* **247:** 257–261.

Bradbury, E. M., R. J. Inglis, H. R. Matthews, and T. A. Langan. 1974b. Molecular basis of control of mitotic cell division in eukaryotes. *Nature* **249:** 553–556.

Britten, R. J., and E. H. Davidson. 1969. Gene regulation in higher cells: A theory. *Science* **165:** 349–357.

Datema, R., E. Agsteribbe, and A. M. Kroon. 1974. The mitochondrial ribosomes of *Neurospora crassa*: 1. On the occurrence of 80S ribosomes. *Biochem. Biophys. Acta* **335:** 386–395.

DePamphilis, M. L. and P. M. Wassarman. 1980. Replication of eukaryotic chromosomes: A close-up of the replication fork. *Ann. Rev. Biochem.* **49:** 627–666.

Dutta, S. K. 1974. Repeated DNA sequences in fungi. *Nucleic Acids Res.* **1:** 1411–1419.

Dutta, S. K., M. Beljanski, and P. Bourgarel. 1977. Endogenous RNA-bound RNA-dependent DNA polymerase activity in *Neurospora crassa*. *Exp. Mycol.* **1:** 173–182.

Emmet, N., C. M. Williams, L. Frederick, and L. S. Williams. 1972. Arginyl tRNA and synthetase of *Neurospora crassa*. *Mycologia* **64:** 499–506.

Epler, J. L., L. R. Shugart, and W. E. Barnett. 1970. N-formylmethionyl transfer ribonucleic acid in mitochondria from *Neurospora*. *Biochemistry* **9:** 3575–3579.

Farrelly, F., and R. A. Butow. 1983. Rearranged mitochondrial genes in the yeast nuclear genome. *Nature* **301**: 296–301.

Fersht, A. R., J. Gangloff, and G. Dirheimer. 1978. Reaction pathway and rate-determining step in aminoacylation of tRNA-ARG catalyzed by the arginyl-tRNA synthetase from yeast. *Biochemistry* **17**: 3740–3746.

Free, S. J., P. W. Rice, and R. Metzenberg. 1979. Arrangement of the genes for ribosomal ribonucleic acids in *Neurospora crassa*. *J. Bacteriol.* **137**: 1219–1226.

Freer, S. N., M. Mayama, and J. L. Van Etten. 1977. Synthesis of polyadenylate-containing RNA during the germination of *Rhizopus stolonifer* sporangiospores. *Exp. Mycol.* **1**: 116–127.

Gasior, E., F. Herrera, C. S. McLaughlin, and K. Moldave. 1979a. The analysis of intermediary reactions involved in protein synthesis in a cell-free extract of *Saccharomyces cerevisiae* that translates natural messenger ribonucleic acid. *J. Biol. Chem.* **254**: 3970–3976.

Gasior, E., F. Herrera, J. Sadnik, C. S. McLaughlin, and K. Moldave. 1979b. The preparation and characterization of a cell-free system from *Saccharomyces cerevisiae* that translates messenger ribonucleic acid. *J. Biol. Chem.* **254**: 3965–3969.

Hasilik, A., and W. Tanner. 1978. Biosynthesis of the vacuolar yeast glycoprotein carboxypeptidase Y: Conversion of precursors into the enzyme. *Eur. J. Biochem.* **85**: 599–608.

Hollomon, D. W. 1970. Ribonucleic acid synthesis during fungal spore germination. *J. Gen. Microbiol.* **62**: 75–87.

Horgen, P. A., and S. F. Ball. 1974. Nuclear protein acetylation during hormone-induced sexual differentiation in *Achlya ambisexualis*. *Cytobios* **10**: 181–185.

Horgen, P. A., and D. H. Griffin. 1971. RNA polymerase III of *Blastocladiella emersonii* is mitochondrial. *Nature, New Biol.* **234**: 17–18.

Horgen, P. A., and J. C. Silver. 1978. Chromatin in eukaryotic microbes. *Ann. Rev. Microbiol.* **32**: 249–284.

Igloi, G. L., F. von der Haar, and F. Cramer. 1978. Aminoacyl-tRNA synthetases from yeast: Generality of chemical proofreading in the prevention of misacylation of tRNA. *Biochemistry* **17**: 3459–3467.

Kim, S. H., F. L. Suddath, G. J. Quigley, A. McPherson, J. L. Sussman, A. H. J. Wang, N. C. Seeman, and A. Rich. 1974. Three-dimensional tertiary structure of yeast phenylalanine transfer RNA. *Science* **185**: 435–440.

Krumlauf, R., and G. A. Marzluf. 1979. Characterization of the sequence complexity and organization of the *Neurospora crassa* genome. *Biochemistry* **18**: 3705–3713.

Krumlauf, R., and G. A. Marzluf. 1980. Genome organization and characterization of the repetitive and inverted repeat DNA sequences in *Neurospora crassa*. *J. Biol. Chem.* **255**: 1138–1145.

Kuentzel, H., and H. C. Blossey. 1974. Translation products in vitro of mitochondrial messenger RNA from *Neurospora crassa*. *Eur. J. Biochem.* **47**: 165–171.

Lehninger, A. L. 1975. *Biochemistry*, 2nd ed. New York: Worth.

Law, D. J., C. E. Rozek, and W. E. Timberlake. 1978. Polyadenylate metabolism in *Achlya ambisexualis*. *Exp. Mycol.* **2**: 198–210.

LeStourgeon, W. M., A. Forer, Y. Z. Yang, J. S. Bertram, and H. P. Rusch. 1975. Contractile proteins: Major components of nuclear and chromosome non-histone proteins. *Biochim. Biophys. Acta* **379**: 529–552.

LeStourgeon, W. M., and H. P. Rusch. 1973. Localization of nucleolar and chromosomal residual acidic protein changes during differentiation in *Physarum polycephalum*. *Arch. Biochem. Biophys.* **155**: 144–158.

Long, E. O., and I. B. Dawid. 1980. Repeated genes in eukaryotes. *Ann. Rev. Biochem.* **49**: 727–764.

Martini, O. H. W., H. J. Gould, and H. W. S. King. 1973. 80S ribosomal proteins. *Biochem. Soc. Symp.* **37**: 51–68.

Mirkes, P. E. 1974. Polysomes, ribonucleic acid and protein synthesis during germination of *Neurospora crassa* conidia. *J. Bacteriol.* **117**: 196–202.

Moll, R., and E. Wintersberger. 1976. Synthesis of yeast histones in the cell cycle. *Proc. Natl. Acad. Sci. USA* **73**: 1863–1867.

Nath, K., and A. P. Bollon. 1977. Organization of yeast ribosomal RNA gene cluster via cloning and restriction analysis. *J. Biol. Chem.* **252**: 6562–6571.

Ono, T., K. Kimura, and T. Yanagita. 1966. Sequential syntheses of various molecular species of ribonucleic acid in the early phase of conidia germination in *Aspergillus oryzae*. *J. Gen. Microbiol.* 12: 13–26.

Petes, T. D. 1980. Molecular genetics of yeast. *Ann. Rev. Biochem.* 49: 845–876.

Planta, R. J., J. Retel, J. Klootwijk, J. H. Meyerink, P. DeJong, H. Van Kleuter, and R. C. Brand. 1977. Synthesis and processing of ribosomal nucleic acid in eukaryotes. *Biochem. Soc. Trans.* 5: 462–466.

Plevani, P., and L. M. S. Chang. 1977. Enzymatic initiation of DNA synthesis by yeast DNA polymerases. *Proc. Natl. Acad. Sci. USA* 74: 1937–1941.

Roheim, J. R., R. H. Knight, and J. L. Van Etten. 1974. Synthesis of ribonucleic acids during the germination of *Rhizopus stolonifer* sporangiospores. *Dev. Biol.* 41: 137–145.

Rudick, M. J., Z. E. Fitzgerald, and V. L. Rudick. 1979. Intra- and extracellular forms of α-glucosidase from *Aspergillus niger*. *Arch. Biochem. Biophys.* 193: 509–520.

Russell, P. J., and W. M. Wilkerson. 1980. The structure and biosynthesis of fungal cytoplasmic ribosomes. *Exp. Mycol.* 4: 281–337.

Sala, F., and H. Kuntzel. 1970. Peptide chain initiation in homologous and heterologous systems from mitochondria and bacteria. *Eur. J. Biochem.* 15: 280–286.

Shearn, A., and N. H. Horowitz. 1969a. A study of transfer ribonucleic acid in *Neurospora*: I. The attachment of amino acid and amino acid analogs. *Biochemistry* 8: 295–303.

Shearn, A., and N. H. Horowitz. 1969b. A study of transfer ribonucleic acid in *Neurospora*: II. Failure to detect transfer nucleic acid alterations in tyrosinase-depressed cultures. *Biochemistry* 8: 304–312.

Sprouse, R. F. 1977. Determination of molecular weight, isoelectric point, and glycoprotein moiety for the principal skin test-reactive component of histoplasmin. *Infect. Immunol.* 15: 263–271.

Staples, R. C., and A. Yaniv. 1978. Mitochondrial DNA polymerase from germinated bean rust uredospores. *Exp. Mycol.* 2: 290–294.

Sturani, E., F. Magnani, and F. A. M. Alberghina. 1973. Inhibition of ribosomal RNA synthesis during a shift-down transition of growth in *Neurospora crassa*. *Biochim. Biophys. Acta* 319: 153–164.

Timberlake, W. E. 1978. Low repetitive DNA content in *Aspergillus nidulans*. *Science* 202: 973–975.

Timberlake, W. E., D. S. Shumard, and R. B. Goldberg. 1977. Relationship between nuclear and polysomal RNA populations of *Achlya*: A simple eukaryotic system. *Cell* 10: 623–632.

Trapman, J., R. J. Planta, and H. A. Raue. 1976. Maturation of ribosomes in yeast: 2. Position of low molecular weight ribosomal RNA species in maturation process. *Biochim. Biophys. Acta* 442: 275–284.

Ullrich, R. C., and J. R. Raper. 1977. Evolutionary mechanisms in fungi. *Taxon* 26: 169–179.

Van Etten, J. L. 1981. Fungal nucleic acids. Symposium on fungal physiology and metabolism. Annual meeting of the Mycological Society of America, Bloomington, Indiana.

Van Etten, J. L., and S. N. Freer. 1978. Polyadenylate-containing RNA in dormant and germinating sporangiospores of *Rhizopus stolonifer*. *Exp. Mycol.* 2: 301–312.

Watson, K. 1976. The biochemistry and biogenesis of mitochondria. In J. M. Smith and D. R. Berry (Eds.), *The Filamentous Fungi*, Vol. 2, pp. 92–120. New York: Wiley.

Yaniv, Z., and R. C. Staples. 1978. DNA polymerase from nongerminated bean rust uredospores. *Exp. Mycol.* 2: 279–289.

12

SECONDARY METABOLISM

It is often useful to classify the biochemical pathways in fungi as being a part of either primary or secondary metabolism. The pathways of primary metabolism are involved in the degradation (catabolism) of molecules for energy production as well as in the synthesis (anabolism) of cellular building blocks such as lipids, carbohydrates, proteins, and nucleic acids. These reactions, most of which have been discussed in Chapter 10, appear to be basically the same for all organisms. The pathways of secondary metabolism, on the other hand, are usually anabolic rather than catabolic and result in the production of metabolites of unusual chemical structure with no obvious cellular function. In addition, the types of secondary metabolites produced are usually specific for a particular group of fungi (Turner, 1971; Martin and Demain, 1978).

The distinction between primary and secondary metabolites is not always clear. For example, in most fungi citric acid is classed as a primary metabolite because of its role as an intermediate in the TCA cycle; its steady-state concentrations are quite low. However, in *Aspergillus* species the metabolism is altered under certain conditions so that tremendous quantities of citrate are produced. The excess is excreted into the medium in such quantities that it can be used industrially. At these high concentrations citrate has no obvious cellular function and is thus classed as a secondary metabolite.

Fungi as a group produce a particularly diverse array of secondary metabolites. Many of these compounds, such as penicillin and griseofulvin, are immensely beneficial to humans. Others, such as aflatoxins and mushroom toxins, are detrimental. The pathways of secondary metabolism branch from those of primary metabolism at a relatively few locations. AcetylCoA is the most important precursor because of its role in the synthesis of terpenes, steroids, fatty acid metabolites, and polyketides. In addition, a variety of aromatic compounds are synthesized from the glycolytic intermediates phosphoenolpyruvate and erythrose-4-phosphate. Other secondary metabolites are derived from amides. Many of the enzymes in secondary metabolism are relatively nonspecific and thus form a branching network of alternative pathways. Because of this complexity, little is known about their regulation and enzymology.

In this chapter we briefly examine the pathways of secondary metabolism in fungi and then discuss the relationships between secondary metabolism, nutrition, and growth.

PATHWAYS OF SECONDARY METABOLISM IN FUNGI

Terpenes and Sterols

A wide variety of fungal metabolites including carotenoids, sterols, and certain toxins belong to the class of compounds called terpenes. Terpenes are derived from a head-to-tail condensation of five-carbon isoprene units.

$$\underset{\displaystyle C=C-C=C}{\overset{\displaystyle C}{|}}$$

The terminology for naming terpenes is based on the isoprene unit being considered a "hemiterpene." Thus, a condensation of two isoprene units yields a monoterpene, three units a sesquiterpene (i.e., "one and one-half terpenes"), four units a diterpene, eight units a tetraterpene, and so forth.

All terpenoids are synthesized via the mevalonic acid pathway. Although acetyl-CoA is the most common starting point for this pathway, leucine is also a possible precursor in some fungi (Figure 12.1). Two acetyl-CoA molecules condense to form acetoacetyl CoA, which then reacts with a third acetyl-CoA to form hydroxymethylglutaryl-CoA (HMG-CoA) and then mevalonic acid. In the alternative pathway, leucine is deaminated, carboxylated, and converted to HMG-CoA. Mevalonic acid is then phosphorylated, decarboxylated, and converted to isopentenylpyrophosphate (IPP). Note that IPP is the first molecule in the pathway containing the isoprene (hemiterpene) carbon skeleton.

An isomer of IPP, dimethylallylpyrophosphate, serves as the "chain-initiating unit" for the synthesis of larger terpenoids. Figure 12.2 gives an overview of these reactions. IPP functions as the "chain-building unit" by first reacting with dimethyl-allylpyrophosphate to form the 10-carbon, monoterpene derivative geranyl-pyrophosphate and then adding 5-carbon isoprene units sequentially to form terpenes as long as 50 carbons. The ubiquinone molecule found in the electron transport chain of fungi and other organisms is believed to be synthesized from a 50-carbon terpenoid-pyrophosphate.

The 15-carbon sesquiterpene derivative farnesyl-pyrophosphate is a particularly important intermediate because it serves as the precursor for other sesquiterpenes as well as triterpenes and sterols. Sesquiterpenes are the largest group of terpenoids. Figure 12.3 illustrates some of the diverse compounds of this class isolated from fungi. Condensation of two molecules of farnesyl-pyrophosphate yields the 30-carbon triterpene squalene. Squalene then reacts with molecular oxygen and cyclizes to form the steroid lanosterol (Figure 12.4a). Steroids are tetracyclic triterpenoids; the numbering and lettering system for these compounds is shown in Figure 12.4b.

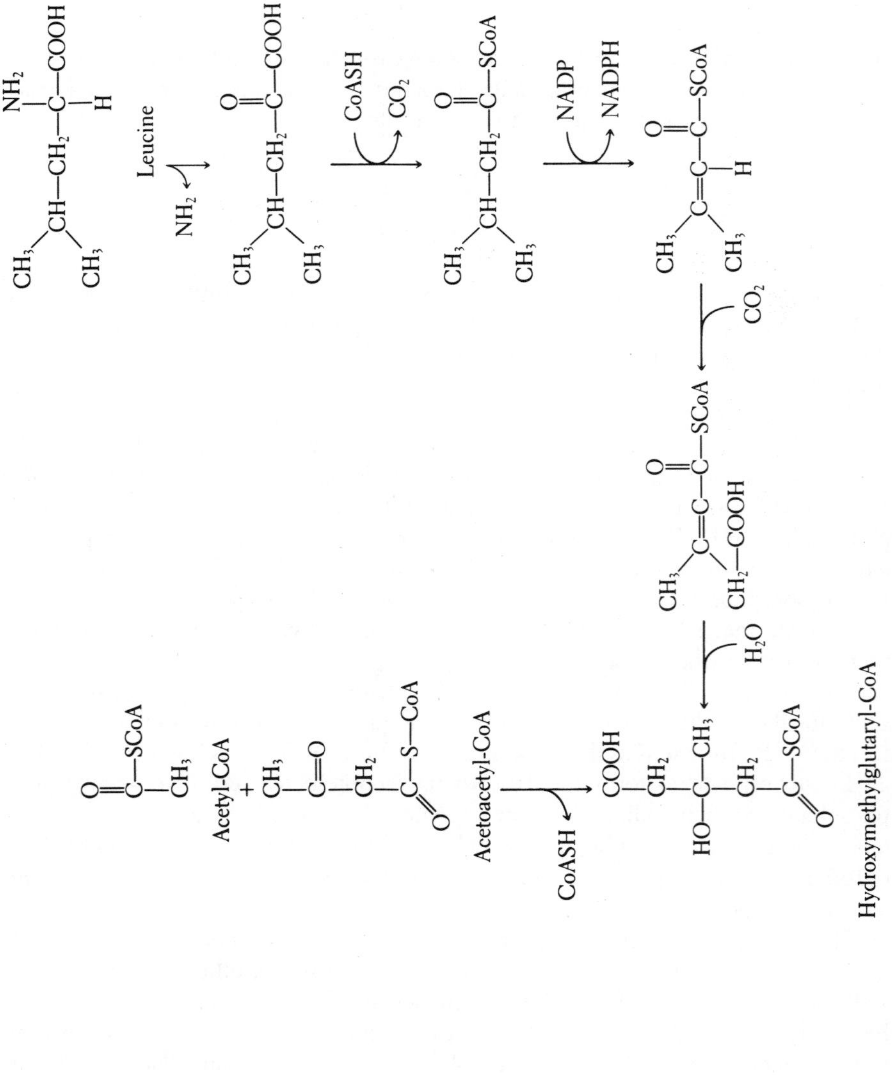

Leucine

CoASH → CO$_2$

NADP → NADPH

CO$_2$

Acetyl-CoA

+

Acetoacetyl-CoA

CoASH

H$_2$O

Hydroxymethylglutaryl-CoA

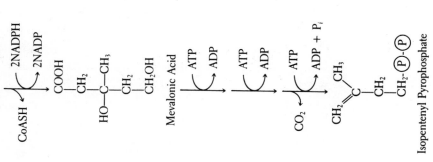

Figure 12.1 Formation of isopentenyl pyrophosphate in the mevalonic acid pathway.

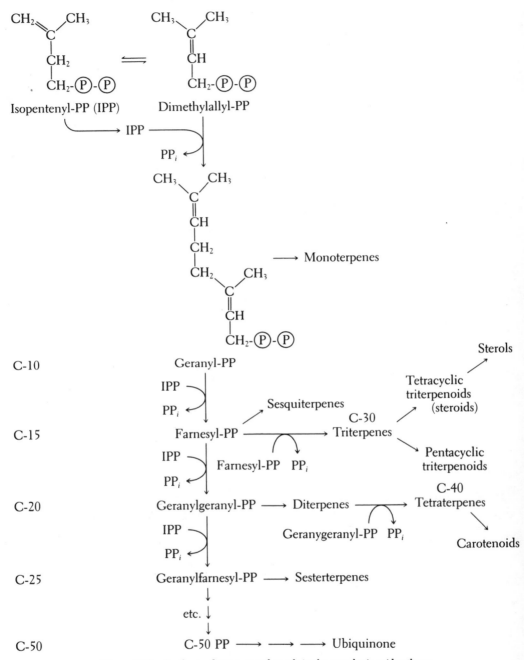

Figure 12.2 Synthesis of terpenes and sterols in the mevalonic acid pathway.

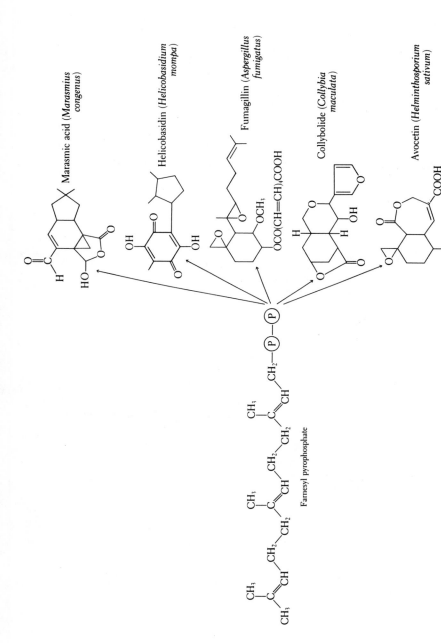

Figure 12.3 Some of the many sesquiterpenes produced by fungi. (From McCorkindale, 1976)

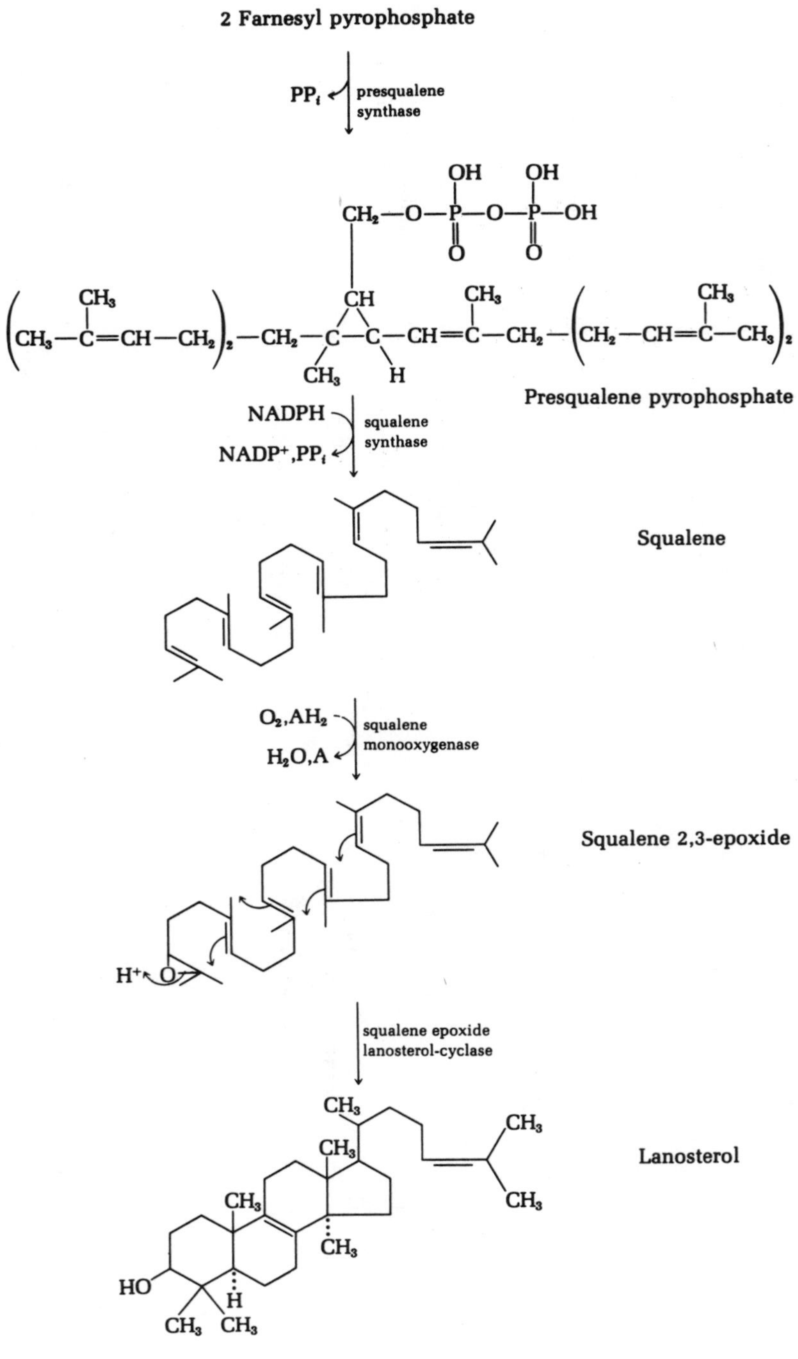
2 Farnesyl pyrophosphate
PP$_i$
presqualene synthase
Presqualene pyrophosphate
NADPH
NADP$^+$,PP$_i$
squalene synthase
Squalene
O$_2$,AH$_2$
H$_2$O,A
squalene monooxygenase
Squalene 2,3-epoxide
squalene epoxide lanosterol-cyclase
Lanosterol

(a)

Figure 12.4 (a) The synthesis of steroids.

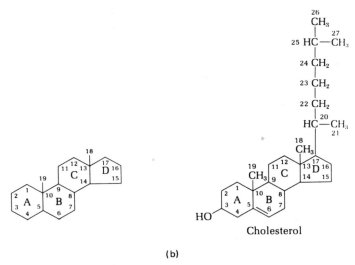

(b)

Figure 12.4   (b) The steroid ring system and cholesterol. (Lehninger, 1975)

Steroids with a hydroxyl group at position 3 on the A ring and a branched chain of eight or more carbons at position 17 are called sterols. Lanosterol is thus more properly classified as a sterol; this compound serves as the precursor for all fungal sterols, such as ergosterol, cholesterol, and fucosterol. These compounds may be modified by alkylation using S-adenosylmethionine, demethylation, dehydrogenation, and reduction. It appears that there is a network of interlinking biosynthetic pathways varying in activity according to the organism and the stage of the life cycle (McCorkindale, 1976).

The 20-carbon diterpene derivative geranylgeranyl-pyrophosphate serves as the precursor for a number of biologically important compounds. For example, the diterpene gibberellic acid produced by *Gibberella fujikuroi* functions as a hormone in green plants. In addition, two molecules of geranylgeranyl-pyrophosphate can condense in a tail-to-tail manner to form 40-carbon tetraterpenoids called carotenoids. The immediate product of such a condensation is the carotenoid phytoene (Figure 12.5). Phytoene is dehydrogenated in several steps to lycopene, which is then cyclized to β- or γ-carotene, both of which are common pigments in many, but not all, fungal cells. In addition, the oxygen-containing carotenoids called xanthophylls are occasionally present (Goodwin, 1972; 1976).

Various nutritional factors affect the synthesis of β-carotene. As is the case with many fungal metabolites, most of the β-carotene is usually synthesized after active growth ceases. However, as the culture ages further, carotene levels drop due to metabolic destruction. In *Phycomyces blakesleeanus*, maltose, glucose, fructose and xylose are equally effective in supporting growth; but the first two sugars stimulate carotene synthesis to a greater extent than the latter two (Garton et al., 1951). Intermediates of the TCA cycle and those nitrogen sources such as asparagine that can be deaminated to TCA intermediates also stimulate β-carotene synthesis, pre-

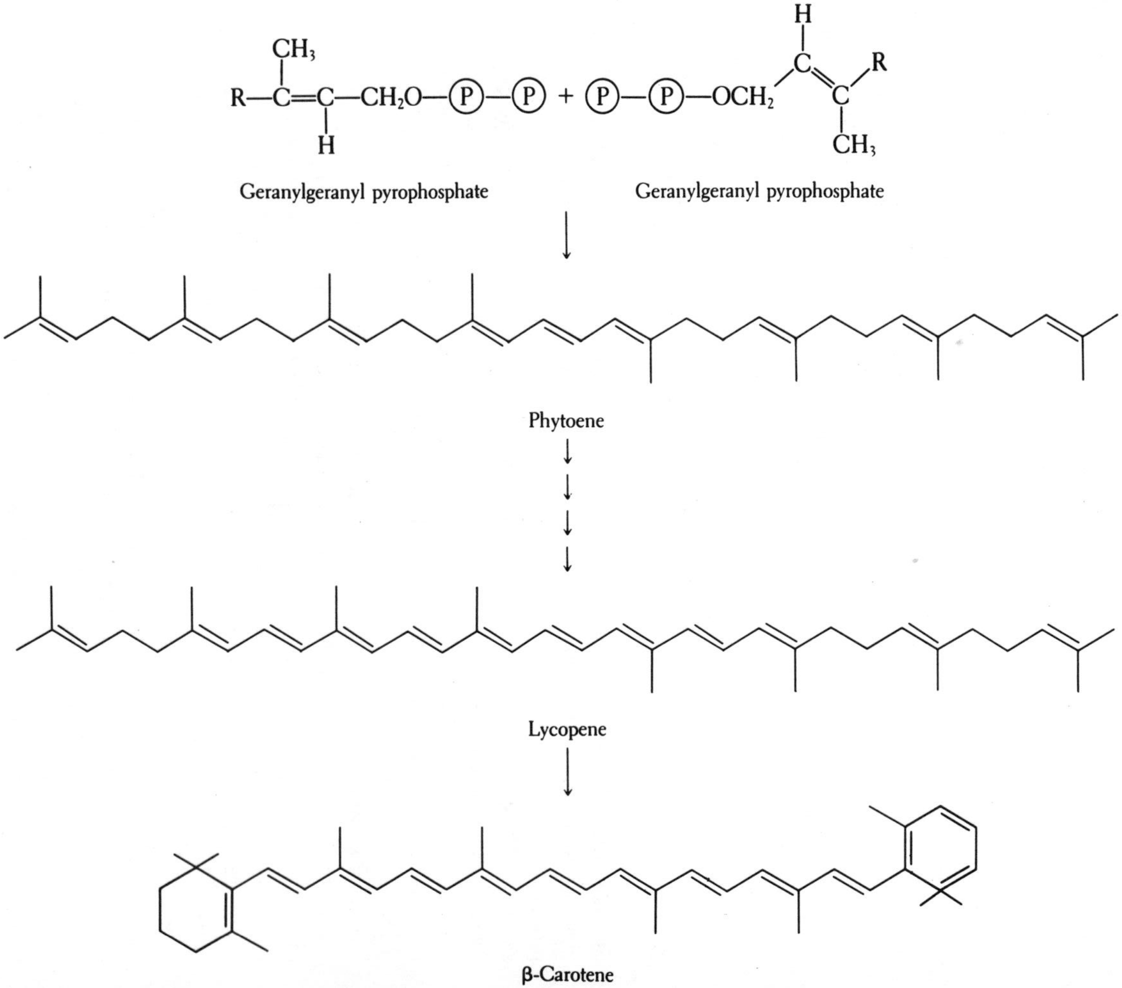

Figure 12.5 The synthesis of β-carotene. (Goodwin, 1976)

sumably because they can readily be converted to acetyl-CoA. Similarly, compounds that are readily converted to leucine stimulate synthesis (Goodwin, 1976).

The trisporic acid hormones present in the Mucoraceous fungi have been shown to be derived from β-carotene (Figure 9.20). As mentioned in Chapter 9, trisporic acids also stimulate β-carotene synthesis, apparently by enzyme derepression (van den Ende, 1976).

Metabolites Derived from the Polyketide Pathway

The pathways of polyketide biosynthesis are characteristic of fungi, particularly the Deuteromycetes. Indeed, more secondary metabolites in fungi are produced via this route than by any other pathway (Turner, 1976). Unfortunately, many of the details of their synthesis have yet to be worked out.

Polyketides are formed by the condensation of one molecule of acetyl-CoA with at least three molecules of malonyl-CoA (Figure 12.6). Intracellularly, it appears that the acetyl unit becomes bound to a protein in a manner similar to that found in fatty acid synthesis (Chapter 10). Three molecules of carbon dioxide are released, and the resulting tri-β-ketomethylene (triketide) chain cyclizes by a type of aldol condensation to form a variety of aromatic compounds such as orsellinic acid, dihydroxydimethylbenzoic acid, 6-methylsalicylic acid, and acetylphloroglucinol (Figure 12.6). With longer polyketides, a large number of aromatics are possible.

Once synthesized, these aromatics can be modified by reduction, hydroxylation, oxidation, decarboxylation, and methylation. Because of the different points at which these reactions can occur in a pathway, a tremendous variety of compounds are possible. In addition, polyketides can be linked with metabolites from other biosynthetic pathways. For example, a sesquiterpene side chain can be attached to a polyketide aromatic ring to form a triprenylphenol (Figure 12.7a). Cytochalasins are formed by the linkage of a polyketide with phenylalanine or tryptophan (Figure 12.7b) (Turner, 1976).

Metabolites Synthesized by the Shikimate-Chorismate Pathway

The shikimate-chorismate pathway is a common biosynthetic route in fungi as well as bacteria and plants and is responsible for the synthesis of a wide variety of aromatic compounds. The pathway begins with the condensation of phosphoenolpyruvate (PEP) and erythrose-4-phosphate, both glycolytic intermediates, to form the cyclized product dehydroquinic acid (Figure 12.8). This product is converted by a multienzyme complex to shikimic acid and then to chorismic acid. From chorismate are synthesized the aromatic amino acids phenylalanine, tyrosine, and tryptophan, the aromatic moiety of ubiquinone, and the p-aminobenzoic acid moiety of folic acid.

Of particular evolutionary significance are the five enzymatic steps between chorismate and tryptophan. Several of the enzymes in this pathway must associate with each other before a particular reaction can proceed. For example, the enzyme catalyzing the fourth step in the pathway must form an association with the enzyme catalyzing the first step in the pathway in order for that first step to proceed. Hutter

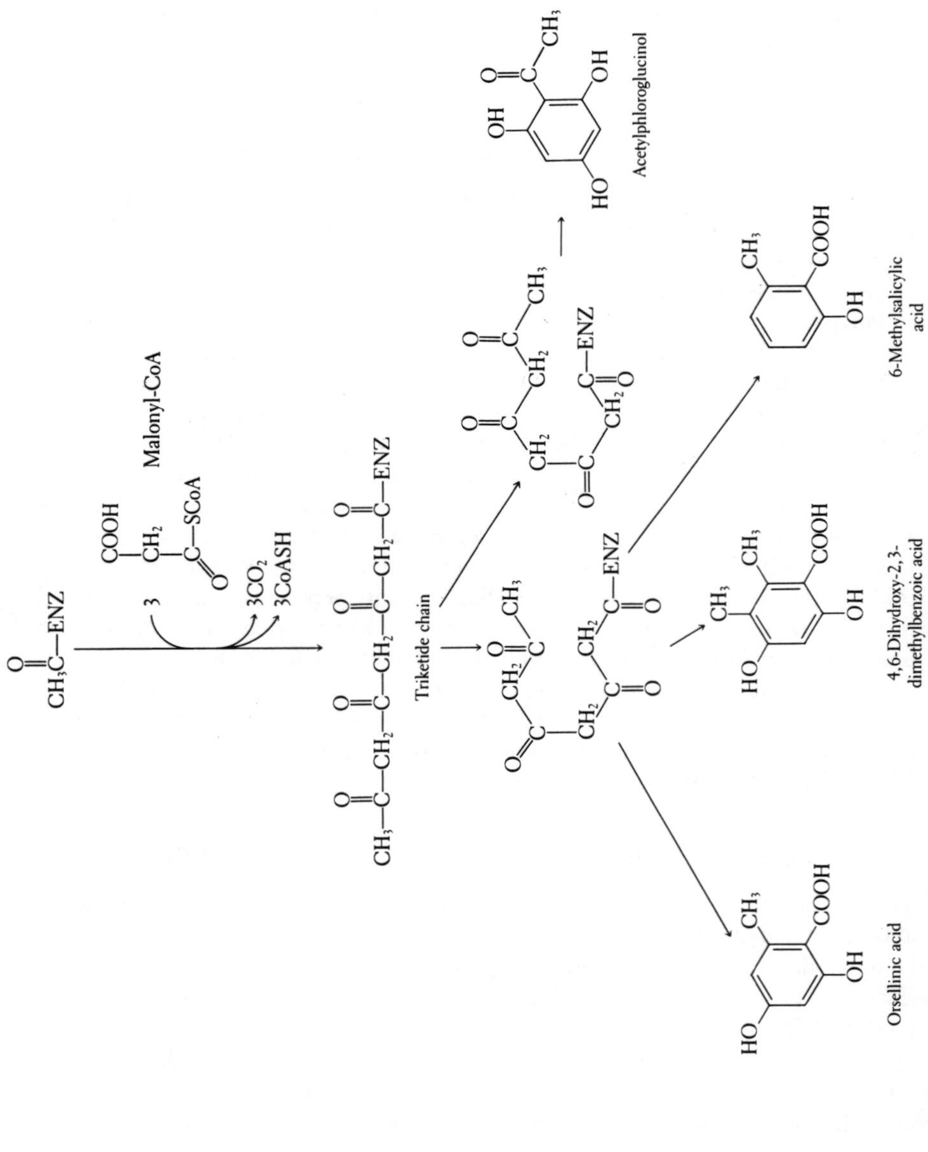

CH₃C—ENZ — rendered: $CH_3C{-}ENZ$ with O

Malonyl-CoA

$COOH$ / CH_2 / $C{-}SCoA$

$3CO_2$ / $3CoASH$

Triketide chain

Acetylphloroglucinol

6-Methylsalicylic acid

4,6-Dihydroxy-2,3-dimethylbenzoic acid

Orsellinic acid

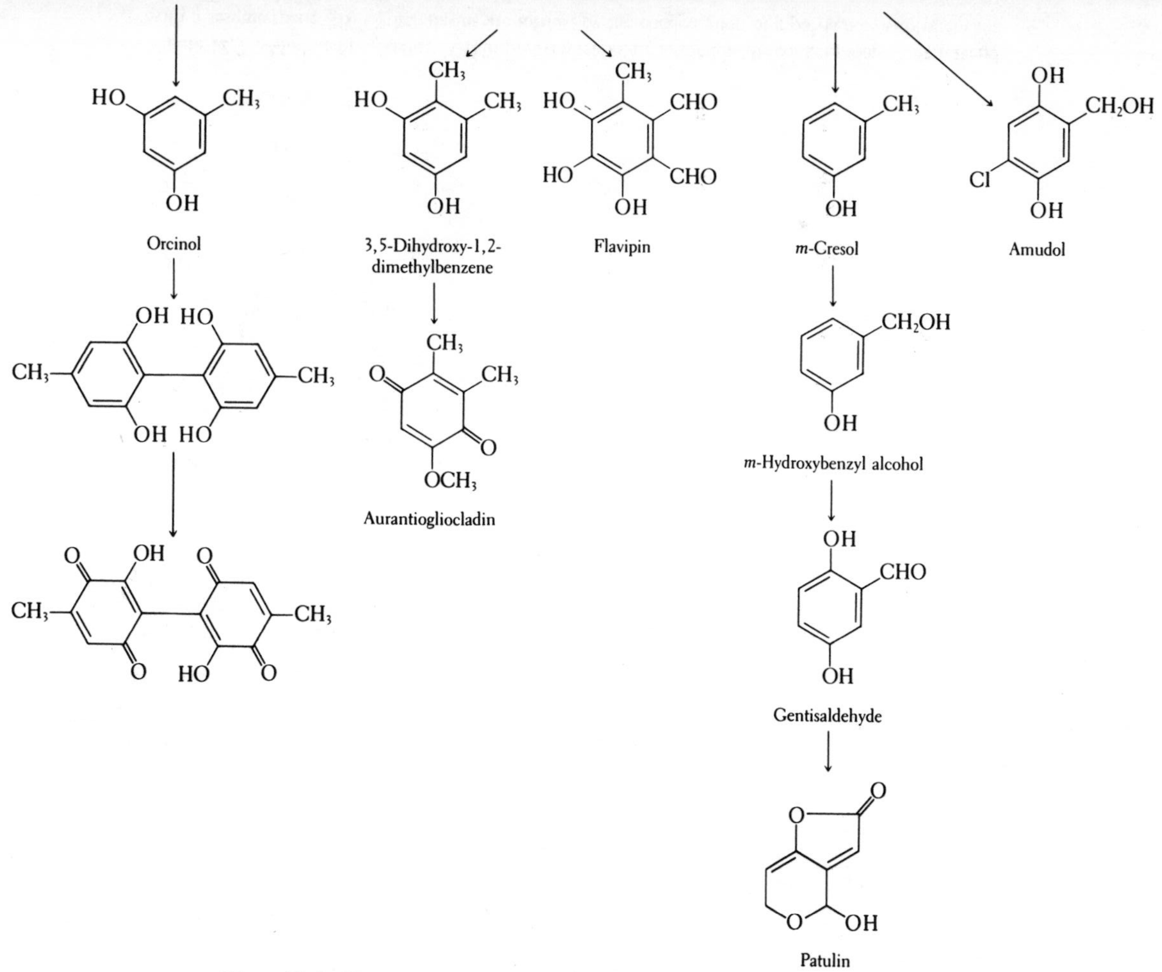

Figure 12.6 Representative reactions in the polyketide pathway. (From Turner, 1976.)

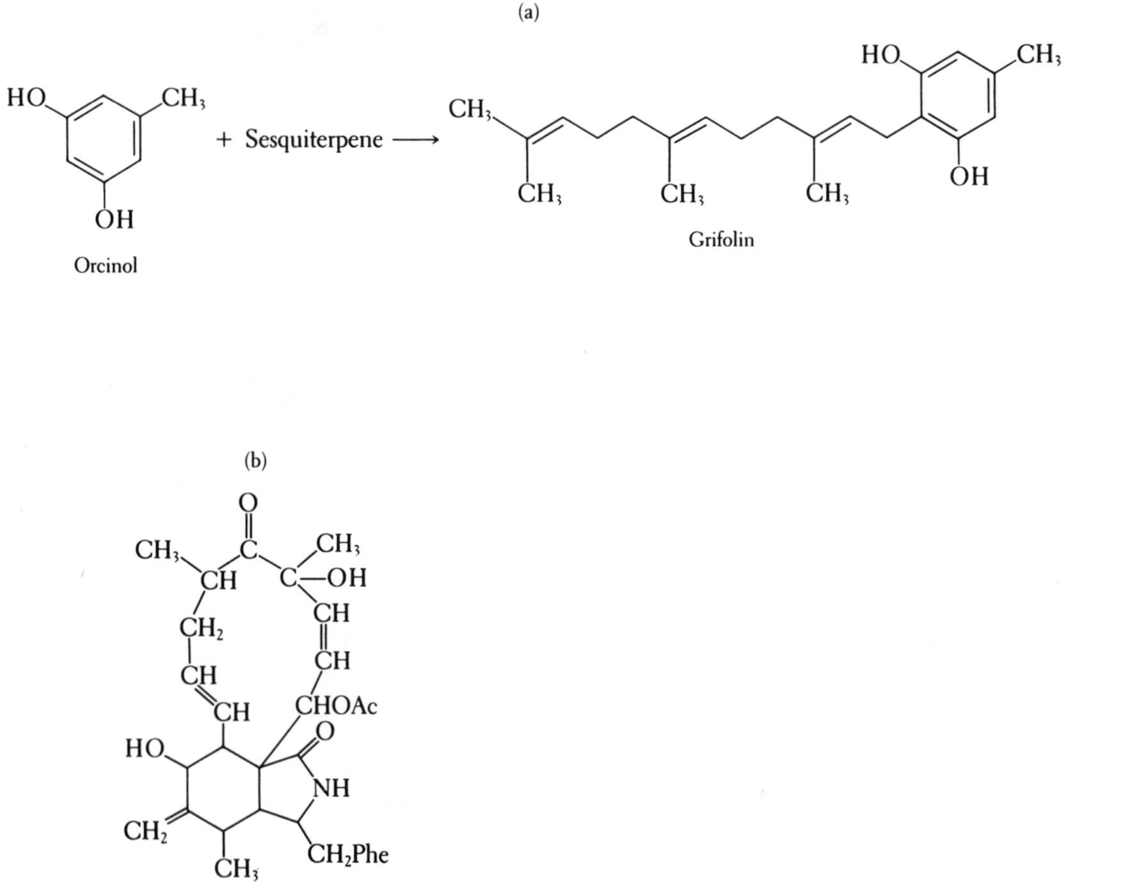

(a)

Orcinol

+ Sesquiterpene ⟶

Grifolin

(b)

Cytochalasin-D

Figure 12.7 Modified polyketides. (A) Triprenyl phenols are formed by the condensation of a tetraketide with a sesquiterpene. (B) Cytochalasins are formed by the condensation of a polyketide with phenylala-

348

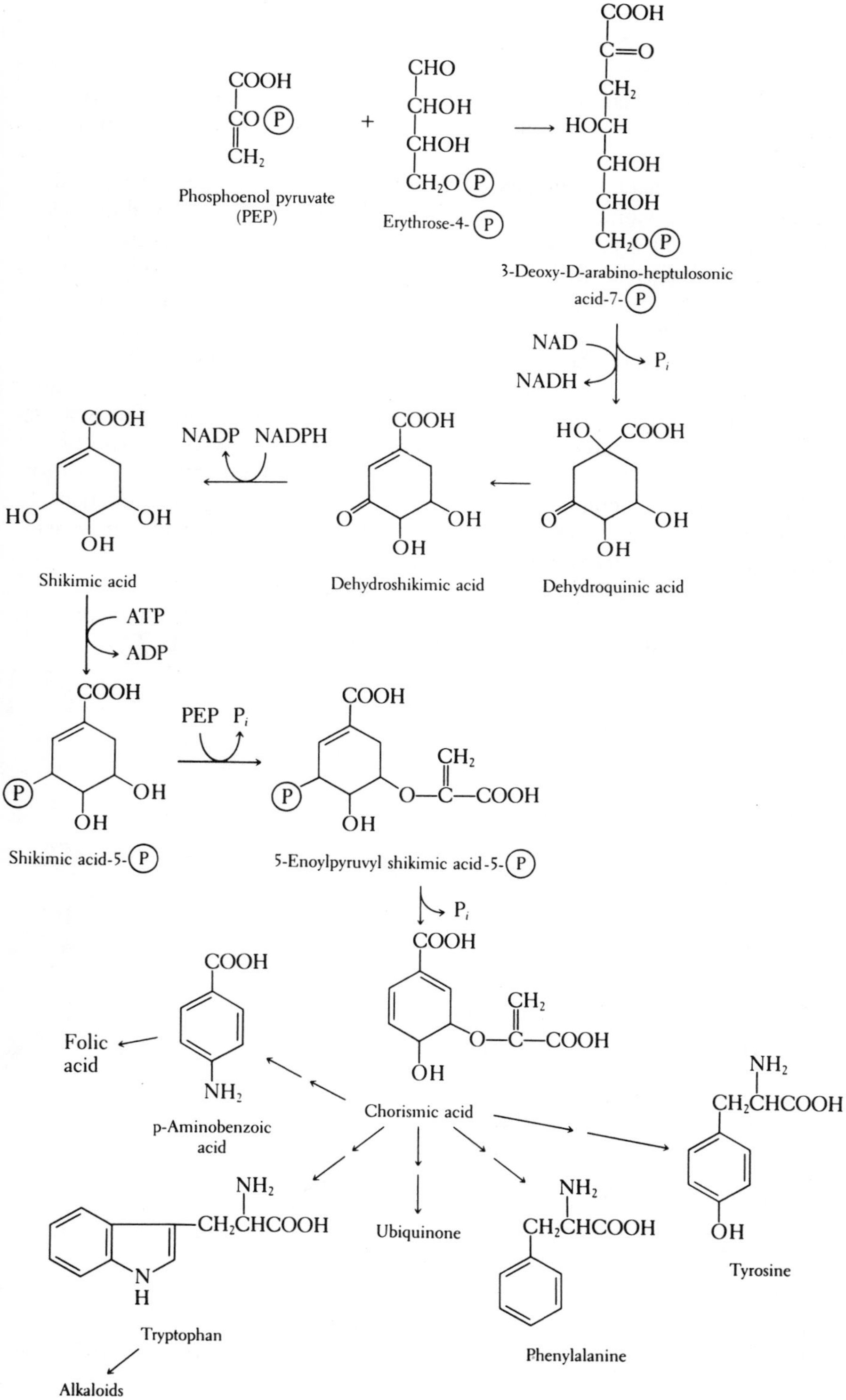

Figure 12.8 The Shikimate–Chorismate pathway.

and DeMoss (1967) examined 23 fungal species and found four different types of associations in this pathway that suggested evolutionary relationships among the taxomonic groups.

Each of the aromatic amino acids in turn serve as precursors for the synthesis of more complex compounds. For example, many alkaloids are formed by the condensation of tryptophan with a terpenoid chain. This chain then cyclizes in a manner that causes a nitrogen atom to be incorporated as part of a heterocyclic ring. Of particular interest are ergot alkaloids such as ergosterine and lysergic acid amide (Figure 12.9), which are produced in the sclerotia of *Claviceps purpurea*. Other alkaloids of current interest include the hallucinogens psilocin and psilocybin.

Phenylalanine serves as the precursor for cinnamic acid and its derivatives (Figure 12.10), some of which are inhibitory and others stimulatory to fungi. The first step in this conversion is catalyzed by phenylalanine-ammonia-lyase (PAL), an enzyme present in many Basidiomycetes and some Ascomycetes. Cinnamic acid, in turn, can be converted to coumarin, a stimulator of rust uredospore germination. In addition, cinnamic acid can be converted to *p*-coumaric acid and caffeic acid. Tyrosine provides an alternate route in the biosynthesis of these compounds, which serve as precursors in lignin biosynthesis. In addition, the derivatives methyl-*cis*-ferulate and methyl-3,4-dimethoxy-*cis*-cinnamate have attracted much attention as self-inhibitors of rust uredospore germination (Chapter 7) (Towers, 1976).

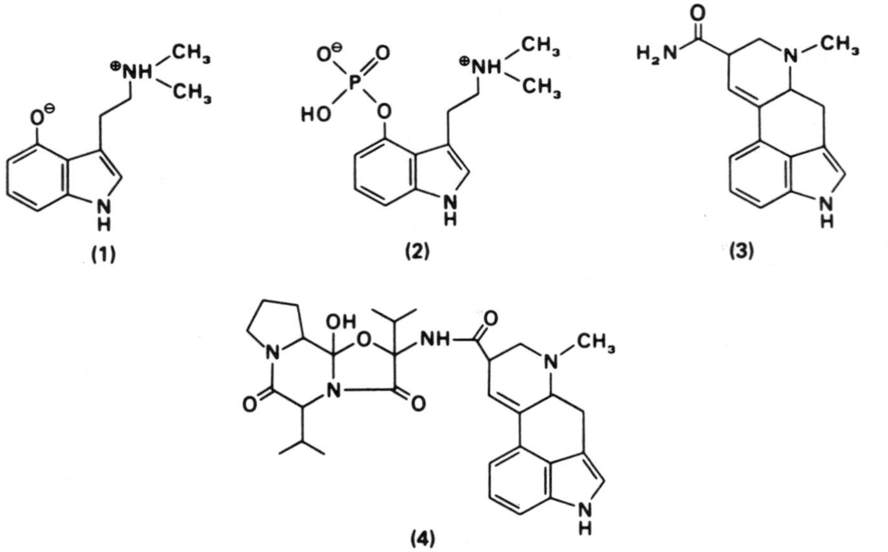

Figure 12.9 Some fungal alkaloids. (1) Psilocin. (2) Psilocybin. (3) Lysergic acid amide. (4) Ergocristine. (Turner, 1971)

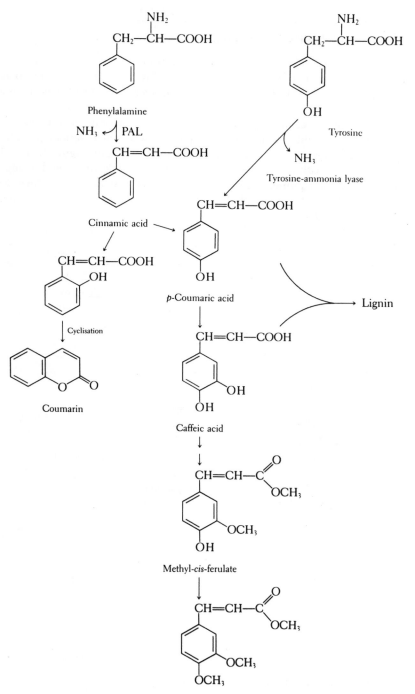

Methyl-3,4-dimethoxy-*cis*-cinnamate

Figure 12.10 The synthesis of cinnamic acid and its derivatives.

Metabolites Derived from Nonaromatic Amino Acids

In addition to the secondary metabolites derived from phenylalanine, tyrosine, and tryptophan, there are many compounds synthesized from nonaromatic amino acids (Wright and Vining, 1975). In this section we shall mention only a few of the major groups.

The penicillins and cephalosporins are a very important group of antibiotics. The basic penicillin molecule is a heterocyclic ring system to which various side chains are attached; cephalosporins have a slightly different ring system (Figure 12.11). The precise biosynthetic pathways for these compounds are not known. Penicillin appears to be synthesized from valine, α-aminoadipic acid, and cysteine/cystine; these form a tripeptide that later cyclizes. Cephalosporins may be synthesized from a penicillin precursor (Turner, 1971).

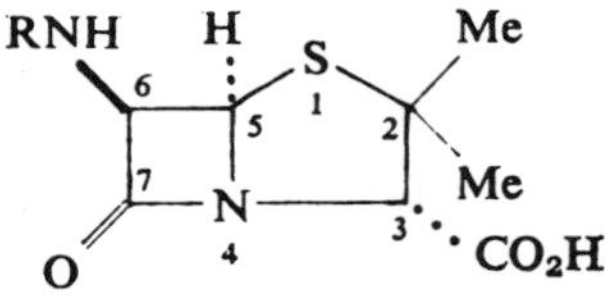

90

a R = H	6-aminopenicillanic acid (6-APA)
b R = Me CH$_2$CH=CHCH$_2$CO	penicillin F
c R = Me(CH$_2$)$_4$CO	dihydropenicillin F
d R = PhCH$_2$CO	penicillin G
e R = Me(CH$_2$)$_6$CO	penicillin K
f R = D-α-aminoadipoyl	penicillin N (cephalosporin N, synnematin B)
g R = L-α-aminoadipoyl	isopenicillin N
h R = p-HOC$_6$H$_4$CH$_2$CO	penicillin X

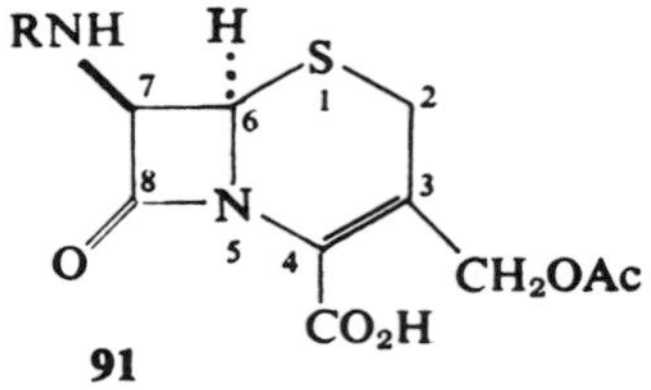

91

a R = H	7-aminocephalosporanic acid
b R = D-α-aminoadipoyl	cephalosporin C

Figure 12.11 Penicillins and cephalosporins. (Turner, 1971)

Certain basidiomycetes produce a variety of toxic and hallucinogenic compounds. We have already mentioned LSD and psilocybin, which are derived form the shikimate-chorismic pathway. In addition, muscarine, the toxin produced by *Amanita muscaria*, is derived directly from glucose. Another group of poisonous compounds, the amatoxins and phallotoxins produced by *A. phalloides* and others, are cyclic peptides. Phallotoxins such as phalloidin are composed of seven amino acids. Amatoxins such as α-, β-, and γ-amanitin contain eight (Figure 12.12). Both types of toxins affect liver cells; the amanitins act on the kidneys as well. Phalloidin disrupts cell membranes, and the amanitins inhibit RNA synthesis (Litten, 1975).

The siderochromes, which have already been discussed in relation to their importance in iron transport, are composed of three hydroxamic acid residues. Hydroxamic acids are derived from acetylated N-hydroxyornithine and are joined to glycine and serine residues to form a cyclic peptide (Figure 5.13).

Other Secondary Metabolites

In some fungi, secondary metabolites may be produced in pathways other than those described above. For example, the toxic metabolite kojic acid produced by several *Aspergillus* species is derived directly from glucose (Figure 12.13a). Polyacetylenes are straight-chain compounds containing conjugated triple-triple or double-triple bonds (Figure 12.13b). These compounds are formed by the desaturation of fatty acids and are produced almost exclusively by Basidiomycetes. Other compounds such as itatartaric acid (Figure 12.13c) are derived from intermediates of the TCA cycle (Turner, 1971).

Judging from the relatively few fungi that have been screened for secondary metabolites, it is likely that a great many additional types of compounds will be discovered in the future.

RELATIONSHIP OF SECONDARY METABOLISM TO NUTRITION AND GROWTH

The pathways of secondary metabolism do not usually function continuously throughout the fungal life cycle but often become active once the growth rate slows. Borrow and co-workers (1961) identified three general phases in the production of many secondary metabolites (Figure 12.14). In the balanced growth phase the rate of uptake and utilization of nutrients is maximal, and the fungus grows exponentially. Secondary metabolites are rarely produced in this phase. As soon as a nutrient is depleted to the point where it becomes limiting, the growth rate slows and the storage phase begins. In this phase, cell division ceases but the dry weight may still increase due to the accumulation of storage products such as lipids and polysaccharides. It is in this phase that the synthesis of secondary metabolites begins. Eventually the fungus enters the maintenance phase, when the dry weight becomes constant, the production of secondary metabolites slows, and cell autolysis sets in.

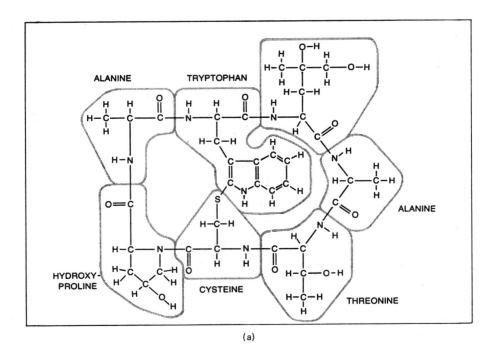

(a)

(b)

Figure 12.12 (a) Phalloidin, a phallotoxin. (b) α-amanitin, an amatoxin. (Litten, 1975)

354

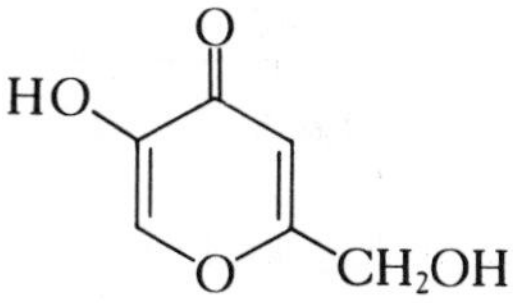

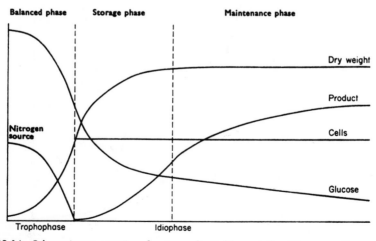

Figure 12.13 (a) Kojic acid isolated from *Aspergillus* species. (b) A polyacetylene from *Polyporus anthracophilus*. (c) Itatartaric acid from *Aspergillus terreus* and A. *itaconicus*. (Turner, 1971)

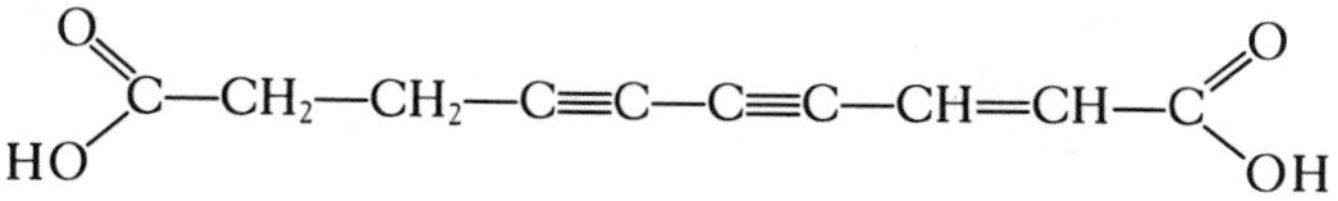

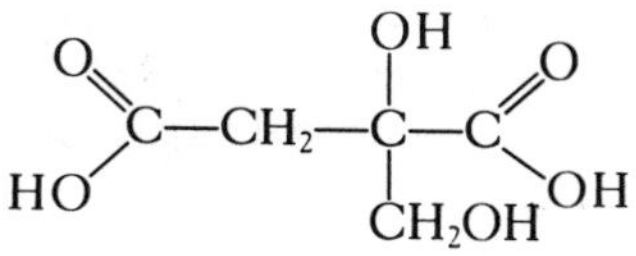

Balanced phase   Storage phase   Maintenance phase

Dry weight

Product

Nitrogen
source

Cells

Glucose

Trophophase   Idiophase

Figure 12.14 Schematic representation of a nitrogen-limited fermentation. The nomenclature of Borrow et al. is shown at the top of the figure and that of Bu'Lock at the bottom. (Turner, 1971)

In a modification of Borrow's scheme, Bu'Lock (1965) has proposed the term "tropophase" for the period of exponential growth and "idiophase" for the period of growth limitation and secondary metabolite production (Figure 12.14) (Turner, 1971).

There is considerable experimental evidence to generally support the proposed stages of secondary metabolite formation, but variation exists depending on the particular metabolite and the species. Figure 12.15 depicts the time course of gibberellin and bikaverin production by *Gibberella fujikuroi*. Notice that both the diterpenoid gibberellin and the polyketide bikaverin pigments are produced after the growth rate has begun to decline. The bikaverins are produced first, followed by the gibberellins some 20 hours later. Experiments have shown that gibberellin synthesis requires a lower nutrient level—that is, more severe growth limitation—than does bikaverin synthesis (Bu'Lock, 1975). This indicates that the patterns of secondary metabolism are determined not only by the existence of growth limitation but also by the intensity.

In some cases the relationship between the onset of limited growth and the production of secondary metabolites may be obscured by cultural conditions. Early studies dealing with the biosynthesis of the polyketide patulin by *Penicillium urticae* suggested that trophophase and idiophase were not distinct—that is, patulin synthesis appeared to begin during the exponential phase of growth instead of during

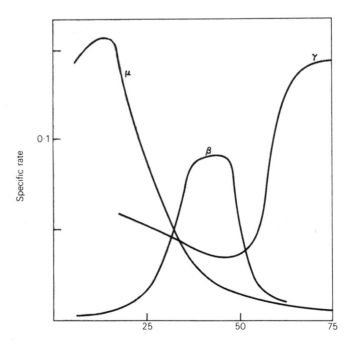

Figure 12.15 Specific rates (per mycelial dry weight) for a nitrogen-limited batch culture of *Gibberella fujikuroi*. μ (biomass) in h⁻¹, β (bikaverins) in mg/g-h, γ (gibberellins) in mg/g-h). (Bu'Lock, 1975)

growth limitation. However, Grootwassink and Gaucher (1980) showed that this overlap was due to a metabolically heterogeneous population of cells. At the time of assay, some of the hyphae were growing exponentially while others had reached the phase of stationary growth. When a technique was developed to obtain a homogeneous mycelial mass, patulin was found to be synthesized after exponential growth had ceased (Figure 12.16).

Patulin can be synthesized using either a glucose-nitrate or a glucose-yeast extract medium. The carbon source is required for patulin synthesis, but it is the depletion of the nitrogen source that actually triggers production. Other environmental factors such as oxygen supply and inoculum size also affect patulin synthesis, but neither phospate concentration nor the age of the cells has any effect. By carefully monitoring the time-course of patulin synthesis, Grootwassink and Gaucher (1980) demonstrated that two enzymes in the patulin biosynthetic pathway, 6-methylsalicylic acid (6-MSA) synthetase and m-hydroxybenzyl-alcohol dehydrogenase (HBAD) appear sequentially prior to the appearance of patulin itself. On either glucose-nitrate or glucose-yeast extract both enzymes appear in the mycelia after the deceleration phase of growth has begun. 6-MSA synthetase, the first enzyme in the pathway, appears 3–4 hours before HBAD, the fourth enzyme in the pathway (Figure 12.17). Both enzymes are synthesized *de novo*, and they may be induced sequentially by their own substrates (Grootwassink and Gaucher, 1980).

The triggering of secondary metabolism by the onset of limited growth suggests there may be competition between the two processes. Bu'Lock (1967) suggests that the exhaustion of a nutrient such as nitrogen or phosphorus may cause primary

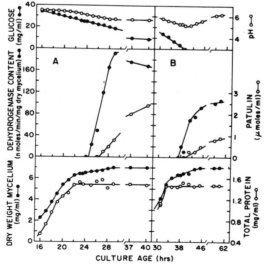

Figure 12.16 Time course of shake flask cultures of *Penicillium urticae* grown in yeast extract-glucose-buffer medium (A) and nitrate-glucose-buffer medium (B) from a spore inoculum. (Grootwassink and Gaucher, 1980)

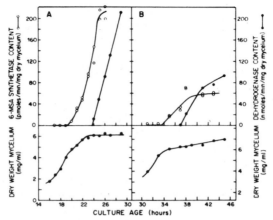

Figure 12.17 Sequential appearance of 6-MSA synthetase and *m*-hydroxylbenzyl alcohol dehydrogenase in shake flask cultures of *Penicillium urticae*. (A) Grown in yeast extract-glucose-buffer medium. (B) Grown in nitrate-glucose-buffer medium. (Grootwassink and Gaucher, 1980).

metabolic intermediates to accumulate. This buildup triggers the induction of enzymes of secondary metabolic pathways. Secondary metabolism may thus be thought of as a means of removing excess intermediates so that primary processes can remain operational in times of environmental stress (Bu'Lock, 1961).

However, synthesis of many types of metabolites is not triggered by nutrient limitation, and thus this hypothesis has been criticized as being overly simplistic. For example, the synthesis of the sesquiterpenoid abscisic acid (a well-known inhibitory plant hormone) by *Cercospora rosicola* increases in parallel with mycelial dry weight for the first 6 days (Figure 12.18). However, its rate of synthesis continues to increase even after the growth rate declines. Abscisic acid synthesis in this organism is affected by several nutritional factors. Thiamine stimulates biosynthesis, and individual amino acids have various effects. Abscisic acid synthesis is greatest when monosodium glutamate or asparagine is the nitrogen source and is least when ammonium nitrate or leucine are used (Norman et al., 1981).

The production of secondary metabolites can be thought of as a type of biochemical differentiation that may accompany morphological differentiation. The production of the farnesyl derivative mycophenolic acid by *Penicillium brevicompactum* is correlated with the emergence of aerial hyphae. When cultures are grown between pieces of dialysis tubing on an agar medium, no aerial hyphae are formed and no mycophenolic acid is produced. If the top layer of dialysis tubing is removed, aerial mycelia differentiate and mycophenolic acid is synthesized. Analysis of the aerial hyphae indicates that the majority of the mycophenolic acid is produced in these hyphae (Bartman et al., 1981). This correlation between secondary metabolism and morphology has been reported for a variety of fungi (Martin and Demain, 1978).

In *Cephalosporium acremonium*, the production of cephalosporin is regulated by the concentrations of both glucose and phosphate and is correlated with the forma-

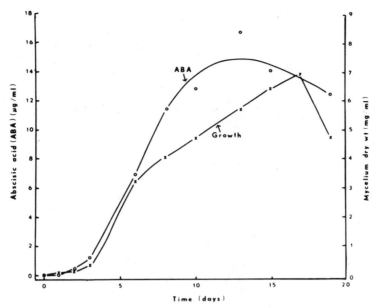

Figure 12.18 Effect of time on growth and ABA production of *Cercospora rosicola* in liquid shake culture containing monosodium glutamate. (Norman et al., 1981)

tion of arthrospores. As indicated in Figure 12.19, cephalosporin production increases approximately parallel with tissue dry weight. The onset of antibiotic production occurs at a time when phosphate is depleted from the medium; glucose is depleted considerably later. At high phosphate concentrations (380 mg phosphorus/liter) cephalosporin synthesis is approximately 50% lower than at lower phosphate concentrations (85 mg phosphorus/liter). However, the sensitivity of the fungus to phosphate depends on the presence of the carbon source. High phosphate concentrations stimulate glucose uptake, and the inhibitory effect of high phosphate concentration can be overcome by increasing the glucose concentration. In this case, therefore, the production of a secondary metabolite may be brought about either by nutrient limitation or by an oversupply of carbon. By accelerating the rate of glucose metabolism, high levels of phosphate may lead to an accumulation of products that can support antibiotic synthesis after the sugar is consumed (Kuenzi, 1980).

As is the case with mycophenolic acid, the synthesis of cephalosporin is correlated with morphological differentiation. As antibiotic synthesis begins, the hyphae swell and differentiate into arthrospores. There is a 1:1 correlation between the cephalosporin concentration and the percentage of arthrospores (Matsumura et al., 1980).

Nutrition and aerobicity are key factors affecting aflatoxin production by *Aspergillus parasiticus*. When this species is incubated in a glucose–mineral salts medium in stationary culture the synthesis of both aflatoxin and lipids increases roughly parallel with dry weight for the first 5 days (Figure 12.20). The greatest accumulation

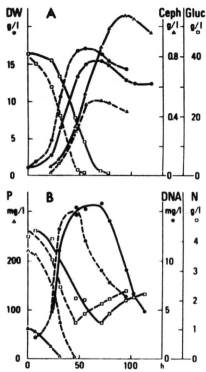

Figure 12.19 (A and B) Effect of the phosphate concentration on the fermentation kinetics of *Cephalosporium acremonium*. The cells were grown in two media containing 85 mg P/liter (solid lines) and 380 mg P/liter (broken lines), respectively. Growth was monitored by measuring dry weight (*DW*) and DNA content. The content of cephalosporins (*Ceph*), glucose, P, and N (as urea + ammonia) were determined in the supernatant after sedimentation of the cells by centrifugation. (Kuenzi, 1980)

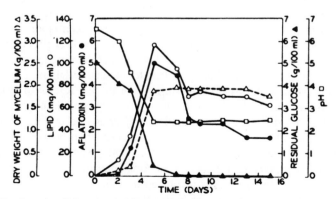

Figure 12.20 Growth of glucose uptake, changes in pH, and formation of aflatoxin and lipid by *Aspergillus parasiticus* in a glucose-salts medium at 28° C without agitation. (Shih and Marth, 1974a)

of these metabolites occurs shortly after most of the glucose is consumed and the exponential phase of growth ends. Indeed, the rate of glucose utilization increases sharply during aflatoxin synthesis. However, the levels of both toxin and lipid drop off as incubation continues past day 5. The same general trends are observed if the fungus is instead incubated in shake culture, although there is greater growth but less toxin and lipid production compared with stationary culture (Shih and Marth, 1974a).

Using [1-^{14}C]-glucose, Shih and Marth (1974a) monitored the flow of label into aflatoxin. Not only was there less label in the toxin from the fungus incubated under shake versus stationary conditions, but the efficiency of incorporation was also lower (Table 12.1). One explanation of these results is a shift in the relative participation of the Emden-Meyerhoff and pentose-phosphate glycolytic pathways. The PP may be favored by the relatively high aerobicity of the shake culture; in this pathway carbon from carbon 1 of glucose is rapidly lost as CO_2 and would thus not be available for aflatoxin synthesis. Under the less aerobic stationary culture the EMP may be favored, resulting in no selective loss of carbon from C-1 and thus a greater efficiency of incorporation of label from [1-^{14}C]-glucose into aflatoxin. Incubation of 3-day-old tissue in the presence of the respiratory inhibitor sodium azide (NaN_3) also stimulates aflatoxin and lipid biosynthesis (Table 12.2). This suggests that a block in the electron transport chain may lead to the accumulation of a carbon compound such as acetate, which can be siphoned off into the pathway for both aflatoxin and lipid. Thus, in static culture oxygen may quickly become limiting, triggering a decrease in the rate of respiration, a switch from the PP to the EMP as the major glycolytic pathway, and an accumulation of a primary intermediate such as acetate, which can be used for increased aflatoxin synthesis (Shih and Marth, 1974a).

Table 12.1   Effect of Aeration on Glucose Catabolism and Biosynthesis of Aflatoxin by *Aspergillus parasiticus*[a]

Type of Aeration	[1-^{14}C]Glucose Added (cpm)	Dry Mycelial Weight (g)	Aflatoxin Produced (μg/g of mycelium)	^{14}C Incorporated into Aflatoxin (Expressed as cpm/g of Mycelium)	^{14}C Incorporated into Aflatoxin (Expressed as cpm/μg of Toxin)
Shaking[b]					
Trial 1	2.68×10^6	0.28	2141	10,588	4.95
Trial 2	2.68×10^6	0.29	2695	14,333	5.30
Still					
Trial 1	2.68×10^6	0.32	6750	48,974	7.26
Trial 2	2.68×10^6	0.30	5827	40,919	7.02

Source: Shih and Marth (1974a).

[a]A. *parsiticus* was pregrown in a glucose–salts medium for 3 days. The washed mycelium was incubated with 20 ml of replacement medium containing [1-^{14}C]glucose at 28°C for 3 days.

[b]Rotary shaker (390 rev/min).

Table 12.2   Effect of Metabolic Inhibitors on Formation of Aflatoxin and Lipid by *A. parasiticus* Incubated at 28° C for 6 Days

| | Dry Wt (g) | pH | Aflatoxin (μg/100 ml) | | | | | Toxin (μg/g dry wt) | Lipid (mg/g dry wt) | Residual glucose (mg/ml medium)[a] |
			B₁	B₂	G₁	G₂	Total			
Control	1.96	2.35	1607	740	397	105	2849	1460	27.6	1.1
NaN₃[b]	1.084	2.70	3779	1328	565	141	5813	5360	54.7	1.9

Source: Shih and Marth, 1974a.

[a]Initial glucose concentration was 50 mg/ml.

[b]NaN₃ at a concentration of 1×10^{-4} M was added to the culture 3 days after it was inoculated with mold spores.

Other factors such as nutrient concentration and temperature also affect aflatoxin production but in ways that are distinct from their effects on growth. For example, the greatest amounts of aflatoxin and lipid are produced at a glucose concentration that is supraoptimal for growth but at a concentration of $(NH_4)_2SO_4$ that is suboptimal for growth. The temperature optimum for toxin and lipid synthesis is also less than that for growth (Shih and Marth, 1974b).

The precise role of the carbon source in aflatoxin biosynthesis has been studied by several groups of workers including Abdollahi and Buchanan (1981a,b). They found that although *A. parasiticus* grows well in a medium in which a protein hydrolysate such as peptone is the sole carbon source, no aflatoxins are produced. However, when the fungus was incubated on peptone for 48 hours and then trans-

Table 12.3   Effect of Peptone and Glucose on the *de novo* Synthesis of Aflatoxins by *Aspergillus parasiticus*[a]

| Carbon Source | | Aflatoxin (μg/25 ml) | | | | Aflatoxin/Mycelium (μg/mg) |
Basal Medium	Replacement Medium	B₁	B₂	G₁	Total	
Peptone	Peptone	ND[b]	ND	ND	—	—
Peptone	Glucose	300	266	ND	566 ± 77[c]	1.35
Peptone	Glucose + cycloheximide	ND	ND	ND	—	—
Glucose	Glucose + cycloheximide	687	24	14	725 ± 21	1.87

Source: Abdollahi and Buchanan (1981a).

[a]Cultures were grown in peptone or glucose basal medium for 48 hr, transferred to replacement medium containing peptone, glucose, or glucose + cycloheximide (150 μg/ml), and reincubated for 24 hr. Values are means of three replications with two determinations per replicate.

[b]No detectable aflatoxin.

[c]Mean ± standard deviation.

ferred to a replacement medium in which glucose is substituted for peptone, abundant aflatoxin is produced (Table 12.3). This initiation of toxin synthesis can be inhibited by both cycloheximide (Table 12.3) and actinomycin-D (Figure 12.21) indicating that glucose or a product of its metabolism induces the *de novo* synthesis of one or more of the enzymes responsible for aflatoxin synthesis. Thus, glucose not only serves as a source of carbon for toxin biosynthesis but also has a regulatory role (Abdollahi and Buchanan, 1981a).

In addition to glucose, aflatoxin synthesis is induced by ribose, xylose, fructose, sorbose, mannose, glactose, maltose, sucrose, raffinose, and glycerol (Table 12.4). In each case, induction is blocked by cycloheximide. No aflatoxin production is detected in replacement media containing lactose, lactate, acetate, pyruvate, citrate, oleic acid, or α-methyl-D-glucose. These findings suggest that to act as an inducer a molecule must be readily metabolizable via glycolytic pathways. The ability of glycerol to induce toxin synthesis adds support to the idea that the EMP is more important than the PP in aflatoxin synthesis. Even though glycerol can be converted to dihydroxyacetone-phosphate and then via the enzymes of gluconeogenesis to the start of the PP, a more direct glycolytic route is via the EMP (Abdollahi and Buchanan, 1981b).

Note in Table 12.4 that acetate, a presumed precursor of aflatoxins, is not an

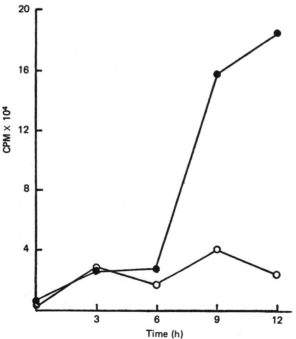

Figure 12.21 Incorporation of ³H-uridine in *Aspergillus parasiticus* initially incubated in peptone basal medium for 42 hours and then transferred to replacement medium containing (●) glucose and (O) glucose + actinomycin-D (150 μg/ml). (Abdollahi and Buchanan, 1981a)

Table 12.4 Effect of Various Carbon Sources on the Induction of Aflatoxin Production by *Aspergillus parasiticus*[a]

Replacement Medium		Aflatoxin (μg/25 ml)				
Carbon Source	Cycloheximide	B_1	B_2	G_1	G_2	Total
Peptone	−	ND[b]	ND	ND	ND	ND
	+	ND	ND	ND	ND	ND
Glucose	−	97	42	6	ND	145
	+	ND	ND	ND	ND	ND
Galactose	−	38	ND	1	ND	49
	+	ND	ND	ND	ND	ND
Fructose	−	71	48	ND	ND	119
	+	ND	ND	ND	ND	ND
Mannose	−	233	134	29	ND	396
	+	ND	ND	ND	ND	ND
Sorbose	−	187	39	226	ND	452
	+	ND	ND	ND	ND	ND
Ribose	−	123	59	66	ND	248
	+	ND	ND	ND	ND	ND
Xylose	−	21	7	ND	ND	28
	+	ND	ND	ND	ND	ND
Sucrose	−	177	72	32	1	282
	+	ND	ND	ND	ND	ND
Lactose	−	ND	ND	ND	ND	ND
	+	ND	ND	ND	ND	ND
Maltose	−	340	120	160	ND	620
	+	ND	ND	ND	ND	ND
Raffinose	−	448	92	344	5	889
	+	ND	ND	ND	ND	ND
Glycerol	−	123	35	31	ND	189
	+	ND	ND	ND	ND	ND
Lactic acid	−	ND	ND	ND	ND	ND
	+	ND	ND	ND	ND	ND
Acetate, Na[+]	−	ND	ND	ND	ND	ND
	+	ND	ND	ND	ND	ND
Pyruvate, Na[+]	−	ND	ND	ND	ND	ND
	+	ND	ND	ND	ND	ND
Citric acid	−	ND	ND	ND	ND	ND
	+	ND	ND	ND	ND	ND
Oleic acid	−	ND	ND	ND	ND	ND
	+	ND	ND	ND	ND	ND
α-Methyl-D-glucoside	−	ND	ND	ND	ND	ND
	+	ND	ND	ND	ND	ND

Source: Abdollahi and Buchanan (1981b).

[a]The cultures were first grown in peptone medium with agitation for 48 h and then transferred to replacement medium containing the carbon source (6%) + cycloheximide (150 μg/ml.) The cultures were then incubated first with agitation for 2 h and then without agitation for 72 h and analyzed for aflatoxins. The data are the means of two replicates with two determinations per replicate.

[b]No detectable aflatoxin.

inducer. Acetate, as well as pyruvate and citrate, may simply speed up the activity of the TCA cycle, thus siphoning carbon away from the pathways of aflatoxin synthesis (Buchanan and Ayres, 1976). Neither cAMP nor cGMP are inducers either. That a large number of carbon sources induce toxin synthesis suggests they do not function directly as inducers but rather indirectly through a common intermediate. For example, Gupta et al. (1976) have demonstrated that cultures producing aflatoxin have a high energy charge during the late exponential phase of growth. Perhaps the utilization of certain carbohydrates leads to an elevated energy charge, which in turn induces the pathways of aflatoxin synthesis (Abdollahi and Buchanan, 1981b).

SUMMARY

Fungi as a group produce a wide assortment of compounds, most of which have unusual chemical structures and unknown cellular functions. These are called secondary metabolites because they do not appear to be directly involved in the primary metabolic processes of energy production and the synthesis of cellular building blocks.

Secondary metabolites are produced via three principal pathways. The mevalonic acid pathway is responsible for the synthesis of terpenoids such as sterols, gibberellins, carotenes, and trisporic acids. In this pathway, molecules of increasing size are formed by the sequential additon of isopentenyl-pyrophosphate units or their derivatives.

The polyketide pathway is responsible for the synthesis of a wide variety of aromatic compounds. Polyketides are formed by the condensation of acetate units in a manner similar to fatty acid synthesis. Important polyketide metabolites include patulin, triprenylphenols, and cytochalasins.

Aromatics can also be synthesized via the shikimate-chorismate pathway in which phosphoenolpyruvate and erythrose-4-phosphate are precursors. Important products of this pathway include the amino acids phenylalanine, tyrosine and tryptophan; cinnamic acid derivatives such as coumarin and methyl-*cis*-ferulate; the antibiotics penicillin and cephalosporin; the mushroom toxins phalloidin and amanitin; the alkaloids LSD and psilocybin; and the siderochromes.

Many secondary metabolites are synthesized in response to nutrient depletion once the growth rate slows. Borrow et al. termed this phase the "storage" phase, which is followed by the "maintenance" phase when metabolite production slows. Bu'Lock calls the phase of secondary metabolite production the "idiophase." It has been suggested that secondary metabolism provides a way for the cell to remove excess intermediates during limited growth so that they do not become toxic. In many fungi the production of secondary metabolites is correlated with morphological differentiation.

In certain species, the production of secondary metabolites occurs in a manner parallel with growth instead of beginning once the growth rate slows. In *Cephalosporium acremonium* there is a complex interaction between phosphate, glucose, and cephalosporin production. In *Aspergillus parasiticus*, aeration and the carbon source are key factors affecting aflatoxin biosynthesis. In this species certain sugars induce the *de novo* synthesis involved in the aflatoxin synthetic pathway.

Because of the diverse array of secondary metabolites and the intricacy of their biosynthetic pathways, much work is still needed in this area. Very little is known about the regulation of these pathways, and the precise steps in the synthesis of most compounds are uncertain. Because of the beneficial or detrimental nature of many secondary metabolites, a tremendous potential exists for economically important discoveries in the field of secondary metabolism.

REFERENCES

Abdollahi, A., and R. L. Buchanan. 1981a. Regulation of aflatoxin biosynthesis: Characterization of glucose as an apparent inducer of aflatoxin production. *J. Food Sci.* **46**:143–146.

Abdollahi, A., and R. L. Buchanan. 1981b. Regulation of aflatoxin biosynthesis: Induction of aflatoxin production by various carbohydrates. *J. Food Sci.* **46**: 633–635.

Bartman, C. D., D. L. Doerfler, B. A. Bird, A. T. Remaley, J. N. Peace, and I. M. Campbell. 1981. Mycophenolic acid production by *Penicillium brevicompatum* on solid media. *Appl. Envion. Microbiol.* **41**: 729–736.

Borrow, A., E. G. Jefferys, R. H. J. Kessell, E. C. Lloyd, P. B. Lloyd, and I. S. Nixon. 1961. The metabolism of *Gibberella fujikuroi* in stirred culture. *Can. J. Microbiol.* **7**: 227–276.

Buchanan, R. L., and J. C. Ayres. 1976. Effect of sodium acetate on growth and aflatoxin production by *Aspergillus parasiticus* NRRL 2999. *J. Food Sci.* **41**: 128–132.

Bu'Lock, J. D. 1961. Intermediary metabolism and antibiotic synthesis. *Adv. Appl. Microbiol.* **3**: 293–342.

Bu'Lock, J. D. 1965. Aspects of secondary metabolism in fungi. In Z. Vanek and Z. Hostalek (Eds.), *Biogenesis of Antibiotic Substances*, pp. 61–71. New York: Academic.

Bu'Lock, J. D. 1967. *Essays in Biosynthesis and Microbial Development*, New York: Wiley.

Bu'Lock, J. D. 1975. Secondary metabolism in fungi and its relationships to growth and development. In J. E. Smith and D. R. Berry (Eds.), *The Filamentous Fungi*, Vol. 1, pp. 33–58. New York: Wiley.

Garton. G. A., T. W. Goodwin, and W. Lijinsky. 1951. General conditions governing β-carotene synthesis by the fungus *Phycomyces blakesleeanus*. *Biochem. J.* **48**: 154–163.

Goodwin, T. W. 1972. Carotenoids in fungi and nonphotosynthetic bacteria. *Prog. Ind. Microbiol.* **11**: 29–88.

Goodwin, T. W. 1976. Carotenoids. In J. E. Smith and D. R. Berry (Eds.), *The Filamentous Fungi*, Vol. 2, pp. 423–444. New York: Wiley.

Grootwassink, J. W. D., and G. M. Gaucher. 1980. De novo biosynthesis of secondary metabolism enzymes in homogeneous cultures of *Penicillium urticae*. *J. Bacteriol.* **141**: 443–455.

Gupta, S. K., K. K. Maggon, and T. A. Venkitasubramanian. 1976. Effect of zinc on adenine nucleotide pools in relation to aflatoxin biosynthesis in *Aspergillus parasiticus*. *Appl. Environ. Microbiol.* **32**: 753–756.

Hutter, R., and J. A. DeMoss. 1967. Organization of the tryptophan pathway: A phylogenetic study of the fungi. *J. Bacteriol.* **94**: 1896–1907.

Kuenzi, M. T. 1980. Regulation of cephalosporin synthesis in *Cephalosporium acremonium* by phosphate and glucose. *Arch. Microbiol.* **128**: 78–83.

Lehninger, A. L. 1975. *Biochemistry.* New York: Worth.

Litten, W. 1975. The most poisonous mushrooms. *Sci. Am.* **232**: 90–101.

Martin, J. F., and A. L. Demain. 1978. Fungal development and metabolite formation. In J. E. Smith and D. R. Berry (Eds.), *The Filamentous Fungi*, Vol. 3, pp. 426–259. New York: Wiley.

Matsumura, M., T. Imanaka, T. Yoshida, and H. Taguchi. 1980. Morphological differentiation in relation to cephalosporin C synthesis by *Cephalosporium acremonium*. *J. Ferment. Technol.* **58**: 197–204.

McCorkindale, N. J. 1976. The biosynthesis of terpenes and steroids. J. E. Smith and D. R. Berry (Eds.), *The Filamentous Fungi*, Vol. 2, pp. 369–422. New York: Wiley.

Norman, S. M., V. P. Maier, and L. C. Echols. 1981. Influence of nitrogen source, thiamine and light on biosynthesis of abscisic acid by *Cercospora rosicola* Passerini. *Appl. Environ. Microbiol.* **41**: 981–985.

Shih, C., and E. H. Marth. 1974a. Aflatoxin formation, lipid synthesis and glucose metabolism by *Aspergillus parasiticus* during incubation with and without aeration. *Biochim. Biophys. Acta* **338**: 286–296.

Shih, C., and E. H. Marth. 1974b. Some cultural conditions that control biosynthesis of lipid and aflatoxin by *Aspergillus parasiticus*. *Appl. Microbiol.* **27**: 452–456.

Towers, G. H. N. 1976. Secondary metabolites derived through the shikimate-chorismate pathway. In J. E. Smith and D. R., Berry (Eds.), *The Filamentous Fungi*, Vol. 2, pp. 460–474. New York: Wiley.

Turner, W. B. 1971. *Fungal Metabolites*. New York: Academic.

Turner, W. B. 1976. Polyketides and related metabolites. In J. E. Smith and D. R. Berry (Eds.), *The Filamentous Fungi*, Vol. 2, pp. 445–474. New York: Wiley.

van den Ende, H. 1976. *Sexual Interactions in Plants: The Role of Specific Substances in Sexual Reproduction*. New York: Academic.

Wright, J. L. C., and L. C. Vining. 1976. Secondary metabolites derived from non-aromatic amino acids. In J. E. Smith and D. R. Berry (Eds.), *The Filamentous Fungi*, Vol. 2, pp. 475–502. New York: Wiley.

AUTHOR INDEX

SUBJECT INDEX

DATE DUE